Odours in Wastewater Treatment

Odours in Wastewater Treatment

Measurement, Modelling and Control

Edited by

Richard Stuetz
School of Water Sciences, Cranfield University, UK

and

Franz-Bernd Frechen
Department of Sanitary and Environmental Engineering,
University of Kassel, Germany

IWA Publishing

Published by IWA Publishing, Alliance House, 12 Caxton Street, London SW1H 0QS, UK

Telephone: +44 (0) 20 7654 5500; Fax: +44 (0) 20 7654 5555; Email: publications@iwap.co.uk
Web: www.iwapublishing.com

First published 2001
Reprinted 2007

British Library Cataloguing in Publication Data
A CIP catalogue record for this book is available from the British Library

Library of Congress Cataloging- in-Publication Data
A catalog record for this book is available from the Library of Congress

ISBN: 1900222469
ISBN13: 9781900222464

Contents

Preface

The release of unpleasant odours from wastewater treatment works can have an impact on the local population. Public concerns over the release of odours from these facilities have increased in recent years. This is the direct result of the encroachment of housing on land surrounding sewage works, the raised awareness of public rights over environmental issues and the expectation of the public towards privatised water companies. Consequently, careful management is required to avoid the creation and release of annoyance odours during wastewater treatment.

Odorous compounds that are present or formed in sewer networks and during wastewater treatment can become an annoyance when they are released into the environment. To avoid the formation of odorous compounds requires an understanding of the processes involved. To control and prevent their release, the mechanisms by which odours are formed and then released and dispersed into the atmosphere must be understood.

In Part I of this book, the reader is introduced to how humans perceive odours, the biological mechanisms involved and their interpretation in relation to the number of complaints. An overview of the philosophy and basics that form the background for regulations and policies used to enforce environment protection is presented. Part II of the book describes the formation of odours and volatiles in sewer networks and sources of odours in wastewater treatment. Particular attention is focused on the role of microbial interactions and the physical factors that lead to odour release during treatment.

The accurate sampling and measurement of odours is essential for assessing the emission of odours as well as evaluating the efficiency of abatement technologies. Part III provides an account of the techniques used to sample odours from wastewater processes and presents the different analytical methods used to measure odours or odorants directly in the field or indirectly at a laboratory. Special attention is given to the recent draft European standard for olfactometry, the application of absorbents for concentrating odour mixtures and the use of novel sensor arrays for surrogate odour measurements.

Part IV of this book covers the practical aspects of assessing and predicting the release of nuisance odours from wastewater treatment in order to provide effective control. The techniques used to predict the emission of odours from different wastewater sources are discussed with a special focus on the use and benefits of the Odour Emission Capacity measurement. Methodologies for assessing the dispersion of odorous emissions from a wastewater source are presented. Practical examples of the use of H_2S contour maps, dispersions and odour models as well as experiences with monitoring nuisance are presented by the authors.

The chapters in Part V provide an overview of the technologies currently used to contain and treat odorous compounds. The suppression of odour formation by the addition of chemicals to sewer and wastewater and the containment of odorous atmospheres using process covers are discussed. The different mechanisms involved in the chemical, physical and biological treatment of odours are presented as well as the results of such different types of deodorization technologies..

The book has been written for engineers and scientists who are working, researching or generally interested in the fields of odour regulation, formation, measurement, modelling and treatment. The content of the individual chapters reflects the interdisciplinary nature of the subject matter. We believe that the problem of odour nuisance, odour formation and odour abatement is of increasing interest, and from this viewpoint this book may be the first, but surely not the last project dealing with this topic. We also do hope that experiences from different countries as well as expertise from different disciplines will work together even more in the future to help with establishing a nuisance-free environment, and that this book may be a step towards this aim.

We thank all the contributors of this book for their contributions and wish to acknowledge the assistance of Alan Click and Alan Peterson of IWA Publishing for their help, support and patience throughout the preparation of the book.

Richard Stuetz
Franz-Bernd Frechen
March 2001

List of Contributors

Teresa J. Bandosz
Department of Chemistry, The City College of City University of New York, New York, NY 10031, USA

Marc A. Boncz
Sub-Department of Environmental Technology, Agricultural University of Wageningen, P.O. Box 8129, 6700 EV, Wageningen, The Netherlands

Robert Bowker
Bowker and Associates Inc., 477 Congress Street, Portland, ME 04101, USA

Harry Bruning
Sub-Department of Environmental Technology, Agricultural University of Wageningen, P.O. Box 8129, 6700 EV, Wageningen, The Netherlands

Joanna E. Burgess
School of Water Sciences, Cranfield University, Cranfield, MK43 0AL, UK

Tom Card
Environmental Management Consulting, 100 292nd Avenue SE, Fall City, WA 98024, USA

Bart De heyder
Aquafin nv, Dijkstraat 8, 2630 Aartselaar, Belgium

Richard A. Fenner
Water Engineering Research Group, University of Hertfordshire, Hatfield, AL10 9AB, UK

Franz-Bernd Frechen
Dept of Sanitary and Environmental Engineering, University of Kassel, Kur-Wolters-Strasse 3, D-34125, Germany

Peter Gostelow
School of Water Sciences, Cranfield University, Cranfield, MK43 0AL, UK

Phil Hobbs
Institute of Grassland & Environmental Research, North Wyke, Okehampton, EX20 2SB, UK

John Hobson
WRc, Frankland Road, Blagrove, Swindon, SN5 8YF, UK

Thorkild Hvitved-Jacobsen
Environmental Engineering Laboratory, Aalborg University, Sohngaardsholmsvej 57, 9000 Aalborg, Denmark

John Jiang
Centre for Water and Waste Technology, School of Civil and Environmental Engineering, The University of New South Wales, Sydney, NSW, 2052, Australia

Ralph Kaye
Centre for Water and Waste Technology, School of Civil and Environmental Engineering, The University of New South Wales, Sydney, NSW, 2052, Australia

Lawrence Koe
Dept of Civil Engineering, National University of Singapore, 10 Kent Ridge Crescent, Singapore, 119260

Piet N.L. Lens
Sub-Department of Environmental Technology, Agricultural University of Wageningen, P.O. Box 8129, 6700 EV, Wageningen, The Netherlands

Philip Longhurst
School of Water Sciences, Cranfield University, Cranfield, MK43 0AL, UK

Alun McIntyre
Entec, Northumbria House, Regent Centre, Newcastle-upon-Tyne, NE3 3PE, UK

Simon A. Parsons
School of Water Sciences, Cranfield University, Cranfield, MK43 0AL, UK

Wim H. Rulkens
Sub-Department of Environmental Technology, Agricultural University of Wageningen, P.O. Box 8129, 6700 EV, Wageningen, The Netherlands

Jan Sipma
Sub-Department of Environmental Technology, Agricultural University of Wageningen, P.O. Box 8129, 6700 EV, Wageningen, The Netherlands

Robert W. Sneath
Bio-Engineering Division, Silsoe Research Institute, Wrest Park, Silsoe, MK45 4HS, UK

Richard M. Stuetz
School of Water Sciences, Cranfield University, Cranfield, MK43 0AL, UK

Amos Turk
Department of Chemistry, The City College of City University of New York, New York, NY 10031, USA

Herman Van Langenhove
Dept of Organic Chemistry, Ghent University, Coupure Links 653, B-9000, Ghent, Belgium

Alison J. Vincent
Hyder Consulting, Hyder Consulting, P.O. Box 4, Pentwyn Road, Nelson, CF46 6YA, UK

Jes Vollertsen
Environmental Engineering Laboratory, Aalborg University, Sohngaardsholmsvej 57, 9000 Aalborg, Denmark

Gong Yang
WRc, Frankland Road, Blagrove, Swindon, SN5 8YF, UK

Part I

INTRODUCTION

1

Odour perception

Richard M. Stuetz, Peter Gostelow and Joanna E. Burgess

1.1 INTRODUCTION

Smell (or olfaction) is perhaps the most interesting and the most routinely used sense to assess quality and yet is understood the least. Its practical applications can be severely limited by the fact that our sense of smell is subjective, tires easily and is expensive and difficult to utilise. Our sense of smell is linked to our emotions and aesthetics, which have a direct and perhaps a detrimental effect on our response to certain environmental odours (such as sewage odours). However, despite the importance of our perception of odours, we have significant problems in comparing one person's experience of a smell with that of another and even more difficulty in trying to quantify these effects.

This chapter will review our current understanding of olfaction and provide some details on the molecular interactions involved in transferring sensory responses in our nose to the brain for processing. It will also discuss the change

 Odours in Wastewater Treatment: Measurement, Modelling and Control edited by Richard Stuetz and Franz-Bernd Frechen. ISBN: 1 900222 46 9

in attitudes towards the emission of unpleasant odours from wastewater treatment works, which has resulted in water utilities having a poor public image in relation to odour pollution.

1.2 HUMAN PERCEPTION OF ODOURS

Humans (as well as animals) perceive odours through the interaction of molecules, given off by odorous materials, with sensory cells located in our nose. This interface between our sensory cells and volatile molecules generates a nerve impulse for that specific interaction, which is used for future interpretations. The interaction enables us to derive information about our environment, it allows us to detect and discriminate between different odours, but also allows us to indicate the intensity of an odour that can permit us to move away from hazardous or unpleasant environments or move toward favourable ones.

1.2.1 Perception of odours

Human responses to an odour are highly subjective; different people find different odours offensive at different concentrations. This results from the way different individuals perceive odours. A simple model to describe human odour perception is shown in Figure 1.1. The process is visualised in two stages, the physiological reception and psychological interpretation, which results in a mental impression of a specific odour. Another more complex model (to describe odour perception) is provided by Cheremisinoff (1988). This suggests that as an odour is perceived, the following stages occur: stimulation - the stimulus or odour is present; discrimination - determination of what the odour is; association - cue reduction; mediation - autonomic involvement with effect; inference - short memory involvement with odour; subception - unconscious involvement with odour and adaptation - determination of the odour as a relevant stimulus.

ODORANT → Reception (physiological) → Interpretation (psychological) → ODOUR IMPRESSION

Figure 1.1. Odour perception (Frechen 1994).

The sensitivity of the physiological reception of an odour differs from person to person (Gostelow *et al.* 2001). The perceived intensity of an odour is not linearly related to its concentration (Gardner and Bartlett 1999). We can identify two broad types of behaviour for odour intensity: (i) odours where the perceived intensity increases rapidly for a relative small concentration change, but the dynamic range (in terms of concentration) is small and (ii) odours where the perceived intensity rises slowly with increasing concentration, but the dynamic range is large.

Human olfactory thresholds for different compounds (Table 1.1) varies widely due the chemical nature of the compounds and between subjects depending on age, gender and state of health. Several studies have shown that odour sensitivity declines with age (Fortier *et al.* 1991; Patterson *et al.* 1993; Cain *et al.* 1995; Bliss *et al.* 1996) and is also worse for subjects who smoke or have poor health (Fortier *et al.* 1991; Griep *et al.* 1995, 1997). The effects of gender on odour perception have also been investigated, however the differences were not statistically significant (Fortier *et al.* 1991; Cain *et al.* 1995; Bliss *et al.* 1996).

Table 1.1. Olfactory thresholds for a range of odorants associated with wastewater treatment processes (Vincent and Hobson 1998).

Substances	Compound	Odour description	Odour threshold (ppb)
Sulfurous	Hydrogen sulphide	Rotten eggs	0.5
	Methyl mercaptan	Decayed cabbage, garlic	0.0014–18
	Ethyl mercaptan	Decayed cabbage	0.02
	Sulfur dioxide	Pungent, acidic	
	Dimethyl sulphide	Decayed vegetables	0.12–0.4
	Dimethyl disulphide	Putrefaction	0.3–11
	Thiocresol	Skunk, rancid	
Nitrogenous	Ammonia	Sharp, pungent	130–15300
	Methylamine	Fishy, rotten	0.9–53
	Ethylamine	Ammonical	2400
	Dimethylamine	Fish	23–80
	Pyridines	Disagreeable, irritating	
	Scatole	Faecal, repulsive	0.002–0.06
	Indole	Faecal, repulsive	1.4
Acids	Acetic	Vinegar	16
	Butyic	Rancid	0.09–20
	Valeric	Sweat	1.8–2630
Aldehydes and Ketones	Formaldehyde	Acrid, suffocating	370
	Acetaldehyde	Fruit, apple	0.005–2
	Butyraldehyde	Rancid, sweaty	4.6
	Isobutyaldehyde	Fruit	4.7–7
	Valeraldehyde	Fruit, apple	0.7–9
	Acetone	Fruit, sweet	4580
	Butanone	Green apple	270

An additional influence on odour sensitivity is prior exposure to an odorant. This can have two effects: (i) on extended exposure, the perceived odour intensity decreases, known as olfactory fatigue or *adaptation* (Dravnieks and Jarke 1980) whereas (ii) on repeated exposure odour intensity can increase (Cain 1980; Leonardos 1980; Laska and Hudson 1991). This is a result of the person becoming familiar with the specific odorant and subsequently their ability to identify the odour is increased. The precise time-scale for adaptation and recovery from an odorant depends on the concentration and the structure of the compound. Additional effects on odour perception include (i) a reduction in perception after the addition of a second odorant (*cross-adaptation*) and (ii) where one compound enhances the perceived intensity of another (*synergistic*), which appears to be restricted to low odour concentrations (Gardner and Bartlett 1999).

The psychological interpretation of odours leads to the judgement about how strong an odour is whether it is pleasant or unpleasant and also the impression of what the odour may or may not be associated with (Gostelow *et al.* 2001). Annoyance odours are usually associated with hazardous or unpleasant environments. The odours that emanate from a wastewater or sludge treatment works are generally associated with the biological decay of organic material. Although the odours themselves are not directly a problem, their association with decaying material indicates that it is something that would best be avoided, as decaying matter itself can represent a health risk.

The perception of odours can also be linked to emotional experiences. The memory of an event, whether the emotional experience was happy or sad, can be linked to factors associated with the experience such as the pleasantness or unpleasantness of a smell (Cheremisinoff 1988). Therefore, the association of odours with particular sources or events is a learning process, which enables individuals to derive information about their environment that can be used for future interpretations.

1.2.2 Classification of odours

Odours that humans perceive are not due to a single compound but are rather the results of a combined impact of a mixture of separate compounds. This impact can vary with time because the volatility and diffusivity of the different compounds also vary (Gardner and Bartlett 1999). Odours associated with wastewater emissions are made of a number of compounds (Table 1.1). Hydrogen sulphide (H_2S) is the most important of these compounds, however the interactions of H_2S with other compounds (particularly those derived from industrial discharges to the sewer) can lead to odour problems that produce even more unpleasant odours (Vincent and Hobson 1998).

Owing to the complexity of odour mixtures and the subjectivity of perceived intensity of odours, the development of techniques for magnitude matching (whereby the judgement of a sensory magnitude is made by reference to a known stimulus) has assisted in making comparison between groups of subjects (Gardner and Bartlett 1999). There are two types of thresholds that can be identified: (i) the threshold to detection - the minimum concentration at which the assessor can detect a difference between a sample and a blank and (ii) the threshold for recognition - the minimum concentration at which the assessor can correctly identify the odour qualities of the compound. These threshold values are dependent on the solvents used to present the samples and the methodology for measurement; consequently tabulated varies for odour thresholds vary widely (Gardner and Bartlett 1999). Some typical examples for threshold values for compounds associated with wastewater treatment are shown in Table 1.1.

Alternatively, odours can be classified by the use of descriptors. However, to date no unique or wholly satisfactory scheme has emerged (Gardner and Bartlett 1999). Amoore (1963a,b) initially proposed that there were seven primary odours (based of a study from 600 organic compounds): camphor, musk, floral, peppermint, ether, pungent and putrid. However, subsequent studies have shown that this number varies and is dependent on the product sector or application and on how familiar the odours are to the assessor or if the assessor has been given some training (Gardner and Bartlett 1999; Wright 1982). More detailed information on odour descriptors can be found in *The Atlas of Odour Character Profiles* (Dravanieks 1985) which includes a list of 146 odour descriptors. However, the most complete collection of 830 odour descriptors has been compiled by the American Society of Testing and Materials (ASTM) (Ohloff 1994). Some examples of odour descriptors for classifying compounds associated with wastewater treatment are shown in Table 1.1.

1.2.3 Mechanisms and processes involved in olfaction

Odours are detected by olfactory receptor cells in the olfactory epithelium, located in the upper reaches of the nasal cavity. Figure 1.2 shows an overview of the different anatomical components and their physical location with respect to the brain (Gardner and Bartlett 1999). During normal respiration, only 3% of the airflow enters this region (Gardner and Bartlett 1999). However, when an odour is detected, sniffing can significantly increase the airflow into the upper reaches of the nasal cavity and direct it over the olfactory epithelium. This interaction between odorous molecules and receptor cells generates an electrical signal, which propagates down the axon of the olfactory receptor cells and into the olfactory bulb for signal processing (Wright 1982).

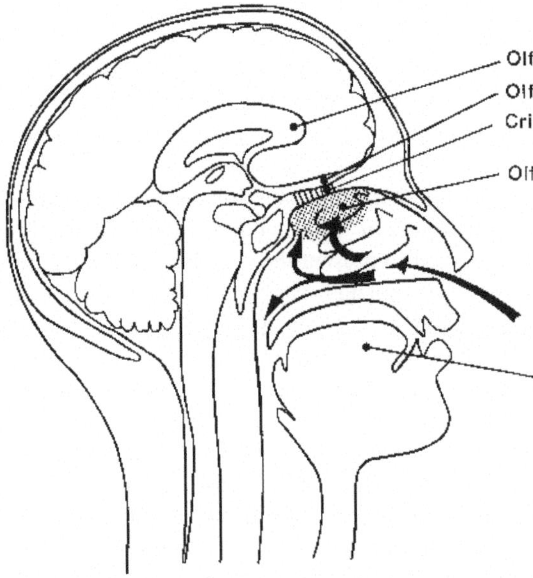

Figure 1.2. The anatomy of the human olfactory system (Gardner and Bartlett 1999).

1.2.3.1 The olfactory epithelium

Olfactory receptor cells are bipolar cells whose dendrites terminate in 10 or more olfactory cilia that interweave and form a network in the mucous layer of the epithelium (Davson 1968; Wright 1982). These cilia provide an increased surface area for odour sensing and are the sites where molecular reception with the odorants occurs and sensory transduction starts. Each receptor cell is connected by its own nerve fibre axon, which transmits a neural impulse to the olfactory bulb (Wright 1982).

The fundamental molecular mechanisms involved in the interactions between odorants and cilia are not fully understand. However, it is generally agreed that olfactory receptor proteins in the membranes of the cilia initiate an enzyme cascade across the membrane when stimulated by an odorant in the olfactory mucus. This molecular process involves the interaction of odorant binding proteins (OBPs) that facilitate the transfer of odorants across the mucous layer to the receptors and G-proteins that aid in binding the odorant to a receptor in the olfactory membrane.

Several theories have been postulated to account for the activation of this chemosensory process. However, any theory must be able to explain the threshold of smell, concentration/intensity relationships above the threshold, differences in odour quality and adaptation of odours (Koe and Brady 1986). The three most prominent theories of olfaction, the stereochemical theory, the vibrational theory and the electron tunnelling theory are summarised in Table 1.2.

Table 1.2. Theories relating odorant quality to molecular structure (Amoore 1964; Wright 1982).

Theory	Description
Stereochemical	This theory suggests that molecules are smelled when they fit into a compliementary receptor site within the olfactory epithelium. This 'lock and key' hypothesis was based on enzyme kinetic type mechanisms.
Vibrational	This theory suggests that the olfactory receptors are sensitive to the vibrational frequencies of the molecules. The hypothesis is analogous to infra-red spectrometry.
Electron tunnelling	This theory suggests that olfactory receptors respond to the vibration of the molecule and not their shape. The hypothesis is based on inelastic electron tunnelling, whereby when an odorant occupies a binding site, electrons can lose energy by exciting their vibrational mode.

1.2.3.2 The olfactory bulb, cortex and higher brain

The nerve impulse or *action potential* generated in the olfactory receptors is then transmitted along a single unbranched axon, the *olfactory neurone* that form up into bundles (of 10–100 axons) which penetrate the cribriform plate and terminate in the olfactory bulb (Figure 1.3). Although the olfactory neurones express a specific receptor and are randomly distributed within a particular region of the olfactory epithelium, they converge on synaptic glomeruli (Gardner and Bartlett 1999). The glomeruli are connected in groups that converge into mitral cells. The architecture of the olfactory bulb results in 1:1000 convergence of the olfactory receptor neurones to the mitral cells.

This convergence increases the sensitivity of the signal being passed on to the olfactory cortex. These signals are projected directly to the higher sensory centres in the cerebral cortex where the signal is decoded and olfactory interpretation and response occurs.

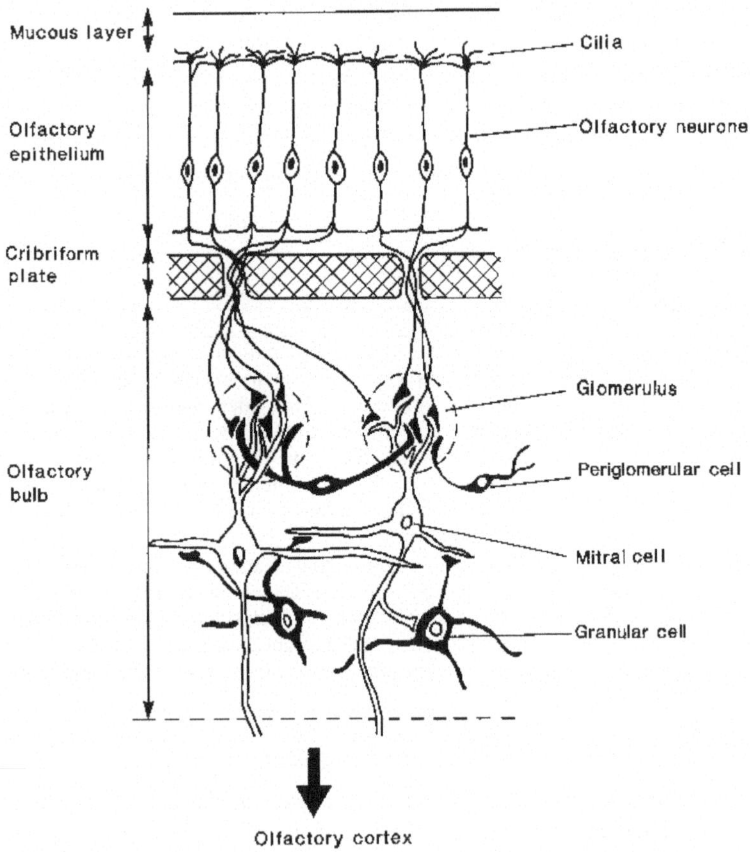

Figure 1.3. The connections between the different parts of the mammalian olfactory system (Gardner and Bartlett 1999).

1.3 ODOUR COMPLAINTS

Public concern over the release of odours from wastewater treatment works has been known for some time (Gostelow *et al.* 2001). However, gaseous emissions have traditionally received the least attention compared with the generation of liquid or solid wastes from sewage and sludge treatment works. This is mostly due to the fact that gaseous emissions pose fewer public health or environmental risks than liquid effluents and sewage sludge. Gaseous emissions and particularly odours can have the greatest impact on the population in the vicinity of a wastewater treatment works (Frechen 1988; Wilson *et al.* 1980). Although odour emissions may not lead to direct health-related problems, they can affect the quality of life (Brennan, 1993), which in turn can lead to indirect problems such as psychological stress (Wilson *et al.* 1980). As a result sewage and sludge treatment works have a poor public image in relation to odour pollution.

1.3.1 Changing trends in odours complaint data

Odour complaints from agricultural, landfill and wastewater treatment works have varied considerably over the last decade. Figure 1.4 shows the number of odour complaints (per million population) in England and Wales between 1989–2000[1]. The data show an increase in the number of complaints from agricultural practices and industrial processes (which includes sewage and sludge treatment works) for 1989/90 to 1995/96, followed by a decrease since 1995/96, which is greater for industrial processes.

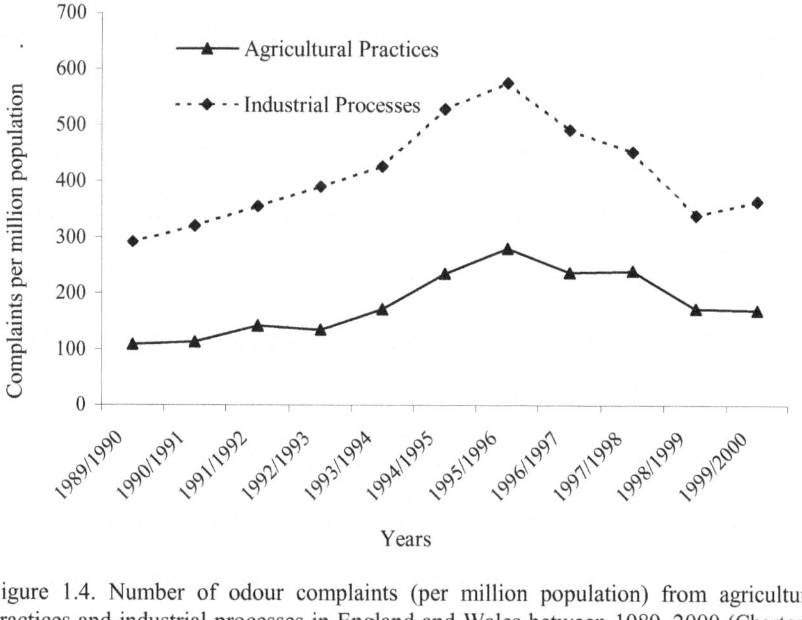

Figure 1.4. Number of odour complaints (per million population) from agricultural practices and industrial processes in England and Wales between 1989–2000 (Chartered Institute of Environmental Health 2000).

---

[1] Odour complaints to local authorities.

Several reasons for the increase in odour complaints (particularly towards sewage treatment works) have been proposed. Foremost among these is the encroachment of housing on lands surrounding sewage treatment works (Balling and Reynolds 1980; Schulz and van Harreveld 1996; Hobson 1997; Vincent and Hobson 1998; Stuetz *et al.* 1999; Gostelow *et al.* 2001). In part, this has resulted from a general migration from the cities to rural areas (Schulz and van Harreveld 1996). This increased urbanisation has put additional pressures on sewage treatment works, in conjunction with increased environmental legislation. New treatment facilities have also been needed in greenfield sites adjacent to local housing (Hobson 1997; Vincent and Hobson 1998) and in many situations rationalisation schemes (particular in sludge treatment) have exchanged many relatively low odour sources to a few high odour sources (Vincent and Hobson 1998). The increased use of land surrounding sewage treatment works can also be explained (in England and Wales) by the reorganisation of the water industry in 1974, which removed sewage treatment from local authority control, thereby separating planning and sewage treatment roles.

The general expansion of wastewater treatment facilities has therefore had the effect of exposing more people to sewage odours, which increases the probability that more people will complain about odour pollution. Additionally, the increased awareness of and expectation for the local environment has resulted in the public being more willing to complain about environmental issues (Schulz and van Harreveld 1996; Hobson 1997; Vincent and Hobson 1998). It is also though that the increased presence of pressure or protest groups has stirred more public awareness of the role of privatised water companies (Vincent and Hobson 1998).

The significant decline in the numbers of odour complaints in England and Wales after 1995/96 (Figure 1.4) indicates that the incidences of odours being released from agricultural practices and industrial processes have decreased. The larger reduction in the number of complaints against industrial processes most likely reflects the greater impact that the introduction (in England and Wales) of the Environmental Protection Act (EPA) 1990 had on the control of nuisance odours. This new legislation gave Environmental Health Officers the power to serve an abatement notice for odour nuisance, but more importantly it has forced both agricultural practices and industries to re-think their strategies for odour prevention and control. At wastewater treatment works, the control of odours has become an important consideration in the design and gaining of planning consent for new works, and the solving of odour problems at existing facilities has become more critical (Vincent and Hobson 1998). Additionally, water utilities are increasingly concerned about their public image in relation to environmental issues and are particularly aware of the high standards that the public now expect from privatised water utilities.

The increasing number of complaints about odour pollution and the introduction of new legislation has also stimulated scientific interest in the techniques used to assess the impact that an odorous emission can have on a local community (see chapters 5–13). Additionally, considerable progress has also been made in the management and development of technologies to treat odours (see chapters 14–20). The greater decline in the number of complaints against industrial processes compared with agricultural practices (Figure 1.4) suggests that the installation of abatement equipment at these facilities can directly reduce the emission of annoyance odours. Figure 1.5 supports this and shows that with a greater understanding of the sources of odours and the introduction and optimisation of abatement systems to control odours at inland and coastal wastewater treatment works in Sydney, the number of complaints about odours emissions was reduced (Sydney Water 1999).

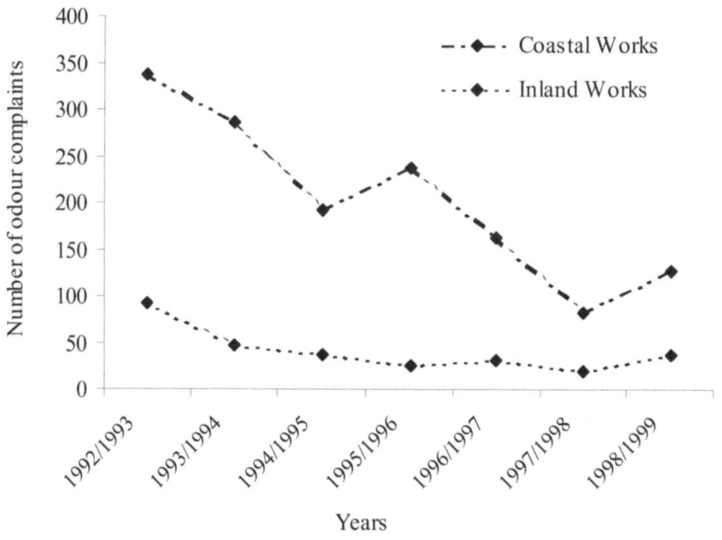

Figure 1.5. Number of odour emission complaints from coastal and inland wastewater treatment work in Sydney between 1992–1999 (Sydney Water 1999).

1.4 REFERENCES

Amoore, J.E. (1963a) The stereochemical theory of olfaction. *Nature* **198**, 271-272.
Amoore, J.E. (1963b) The stereochemical theory of olfaction. *Nature* **199**, 912-913.

Amoore, J.E. (1964) Current status of the steric theory of odor. *Annal. N.Y. Acad. Sci.* **116**, 457-476.

Balling, R.V. and Reynolds, C.E. (1980) A model for evaluating the dispersion of wastewater plant odors. *J. Water Poll. Cont. Fed.* **52** (10), 2589-2593.

Bliss, P.J., Schulz, T.J., Senger, T. and Kaye, R.B. (1996) Odour measurement - factors affecting olfactometry panel measurement. *Water Sci. Technol.* **34** (3-4), 549-556.

Brennan, B. (1993) Odour nuisance. *Water Waste Treat.* **36**, 30-33.

Cain, W.S. (1980) The case against threshold measurement of environmental odors. *J. Air Poll. Cont. Assoc.* **30,** 1295-1296.

Cain, W.S., Stevens, J.C. Nickou, C.M., Giles, A., Johnston, I. and Garcia-Medina, M.R. (1995) Life-span development of odor identification, learning, and olfactory sensitivity. *Perception* **24,** 1457-1472.

Chartered Institute of Environmental Health (2000) *Annual Report on the Work of Local Authority Environmental Health Departments in England and Wales.*

Cheremisinoff, P.N. (1988) *Industrial Odour Control.* Butterworth-Heinemann, Oxford.

Davson, H. (1968) The sense of smell. In: *Principles of Human Physiology* (H. Davson and M.G. Eggleton, eds.), pp. 1413-1421, J & A Churchill, London.

Dravnieks, A. (1985) *The Atlas of Odour Character Profiles.* American Society for Testing and Materials, ASTM Data Series DS61, Philadelphia.

Dravnieks, A. and Jarke, F. (1980) Odor threshold measurement by dynamic olfactometry: significant operational variables. *J. Air Poll. Cont. Assoc.* **30,** 1284-1289.

Fortier, I., Ferraris, J. and Mergler, D. (1991) Measurement precision of an olfactory perception threshold test for use in field studies. *Amer. J. Ind. Med.* **20,** 495-504.

Frechen, F.-B. (1988) Odour emissions and odour control at wastewater treatment plants in West Germany. *Water Sci. Technol.* **20**, 261-266.

Frechen, F.-B. (1994) Odour emissions of wastewater treatment plants - recent German expereinces. *Water Sci. Technol.* **30** (4), 35-46.

Gardner, J.W. and Barlett, P. N. (1999) *Electronic nose: principles and applications.* Oxford University Press, New York.

Gostelow, P., Parsons, S.A. and Stuetz, R.M. (2001) Odour measurements for sewage treatment works. *Water Res.* **35**, 579-597.

Griep, M.I., Mets, T.F., Vercruysse, A., Cromphout, I., Ponjaert, I., Toft, J. and Massart, D.L. (1995) Food odour thresholds in relation to age, nutritional and health status. *J. Gerontology* **50A**, B407-B414.

Griep, M.I., Mets, T.F., Collys, K., Vogelaere, P., Laska, M. and Massart, D.L. (1997) Odour perception in relation to age, general health, anthropometry and dental state. *Arch. Gerontology Geriatrics* **25,** 263-275.

Hobson, J. (1997) Odour potential. *Water Quality Internat.* (July/August), pp. 21-24.

Koe, L.C.C. and Brady, D.K. (1986) Sewage odors quantification. *J. Environ. Eng.* **112** (2), 311-327.

Laska, M. and Hudson, R. (1991) A comparison of the detection thresholds of odour mixtures and their components. *Chem. Senses* **16,** 651-662.

Leonardos, G. (1980) Selection of panelists. *J. Air Poll. Cont. Assoc.* **30,** 1297.

Ohloff, G. (1994) *Scent and fragrances.* Springer-Verlag, Berlin.

Patterson, M.Q., Stevens, J.C., Cain, W.S., and Commeto-Muniz, J.E. (1993) Detection thresholds for an olfactory mixture and its three constituent compounds. *Chem. Senses* **18,** 723-734.

Schulz, T.J. and van Harreveld, A.P. (1996) International moves towards standardisation of odour measurements using olfactometry. *Water Sci. Technol.* **34** (3-4), 541-547.

Stuetz, R. M., Fenner, R.A. and Engin, G. (1999) Assessment of odours from sewage treatment works by an electronic nose, H_2S analysis and olfactometry. *Water Res.* **33**, 452-461.

Sydney Water (1999) *Annual Environmental and Public Health Report.* Sydney Water Corporation, Sydney.

Wilson, G.E., Huang, Y.C. and Schroepfer, W. (1980) Atmospheric sublayer transport and odor control. *J. Environ. Eng. Div., Proc. Am. Soc. Civil Eng.* **106**, 389-401.

Vincent, A. and Hobson, J. (1998) *Odour Control.* CIWEM Monographs on Best Practice No. 2, Terence Dalton Publishers, London.

Wright, R.H. (1982) *The Sense of Smell.* CRC Press, Boca Raton.

2

Regulations and policies

Franz-Bernd Frechen

2.1 INTRODUCTION

Odour emissions can cause serious annoyance in the neighbourhood of the emission source. Thus, especially in densely populated areas, odour is becoming increasingly a subject of national and even international interest.

It is accepted that Article 8 of the European Convention of Human Rights applies where there is severe environmental pollution affecting the well-being of individuals even when their health is not seriously damaged. Also, the World Health Organisation defines: "Health is a state of complete physical, mental, and social well-being and not merely the absence of disease or infirmity". This is a pretentious definition. We thus have to face the fact that odour annoyance, although the odour itself does not act toxically or as a direct cause for diseases, may affect human health indirectly.

However, when discussing odour problems, effects other than annoyance, which may be summarised under toxic effects, are not to be considered. In the

 Odours in Wastewater Treatment: Measurement, Modelling and Control edited by Richard Stuetz and Franz-Bernd Frechen. ISBN: 1 900222 46 9

context of this book odour is looked at as a possible source of annoyance, not as a source of toxic effects or direct cause for diseases. If the gases present have these effects, then other legal frameworks and directives are appropriate.

It is not the aim of this chapter to present a complete description of the laws and regulations applicable in Europe, in a specific European country or in any other country. It is necessary to discuss some general valid principles which will allow one to approach existing laws and regulations or which allow the establishment of a new system of laws and regulations. However, examples will be given for a better understanding of the possibilities.

2.2 COMPONENTS OF THE PROBLEM

An odorant is a material that can cause what mankind recognises as "odour". Usually, odorants are gases and thus behave like gases. This is a very simple, a very important, a very basic and a very often misunderstood fact. For instance, many papers are written concerning a special dispersion calculation for odours. This is misleading, as odorants disperse like other gases do. There is no such thing as a special dispersion of gases which may cause odour sensations which is different to the dispersion behaviour of non-odorant gases. What is meant is that the assessment of the impact substantially differs from that required by other pollutants.

What are the basic components of the problem? We have:

- The stimulus, which is associated with the presence of a certain amount of odorants – these odorants can be measured analytically, if known, and the odour concentration in odour units per cubic metre (ou/m^3) can be measured.
- The response as an assessment of the stimulus, which can indicate an annoyance.

Processes that happen between these two anchor points, as e.g. physiological and psychophysical processes, are also presented in the book. When discussing regulations and policies, they are of minor interest.

The second bullet point involves three parts, which must be clearly separated: There must be an annoyance present, the annoyance must exceed a certain limit, and this issue must be assessed properly.

This means that in general laws, regulations and directives must set standards concerning the extent of annoyance which are feasible and measurable with an appropriate effort.

Returning to the "dispersion calculation for odours", the misleading statment above mainly results from the fact that the "odour" dispersion models usually include some kind of annoyance assessment to a certain extent, using a more or less complex set of basic data with which the "model" was "calibrated". In fact, if ever, the assessment is calibrated, not the dispersion model.

2.3 WHAT TYPE OF STANDARD?

2.3.1 Overview

It is clear that odours that are offensive and unpleasant for human beings can only exist when human beings are present. No man, no odour. Or better: no man, no annoyance.

It also is clear that if offensive odours are present in ambient air, measures at the recognised sources of the emissions that cause the annoyance are necessary. Thus, as usual in environmental protection, prevention from impact needs measures – and in most cases also standards – at the emission sources.

The total process includes three parts:

Emission ⇒ Impact ⇒ Annoyance

All three parts must be considered, and the links between them are important, as standards in general, according to their motivation, should arise from the (maximum allowable) nuisance and then look back to the (maximum allowable) emission. So, we have the difficulty of a physiological/psychophysical link between annoyance and impact, and we do have a meteorological link between impact an emission.

When starting to think about laws and regulations for the prevention from annoyance caused by malodours a layman may begin like "there are no odours allowed". This of course does not include the offensiveness of malodours and interdicts even pleasant odours. The next attempt would include the word of "unpleasant", "annoying" or similar.

The next problem to deal with would then be to define the allowable extent of nuisance. A very strict standard would be that unpleasant odours are not allowed "at any time" and "for any person". Reference to duration of impact and/or to percentage of people affected is introduced here.

At this point we have the two indispensable components of an odour impact standard:

- extent of impact and
- duration of impact

These two parameters strongly influence the *extent of nuisance*, which of course may be affected by other circumstances like age, social status, health, etc, which will not be regarded here.

All laws and regulations that deal with malodour prevention should – direct or indirect – take care of the two constituents mentioned. However, depending upon their position, laws and regulations will give more or less detailed information concerning these two parameters.

Several types of standards can be found, and one may distinguish between older ones and newer ones or simpler ones and more sophisticated ones. The following two types obviously belong to the older and simpler ones:

- Minimum distance standards (MDS): Based upon practical experience, this type of regulation usually takes type and size of plant into consideration. Nevertheless, it usually does not regard the type of sensitive vicinity. This is one of the oldest types of regulation and can be accepted today as a rule of thumb in very simple cases, but does not meet today's consideration of the annoyance.
- Maximum emission standards (MES): Either based upon general accepted experience concerning the impact resulting from the allowed emission, or even neglecting the impact, this type of regulation in most cases will not meet today's needs, even when distinction is made between type and size of the plant or magnitude of emission.

According to the growth of knowledge, newer types of standards recognise the annoyance part of the whole process more and more. This, of course, makes it more difficult to stipulate appropriate values to be met by the emitting facilities. These newer type of standards may be characterised as:

- Maximum impact standards (MIS): Impact in the relevant vicinity or at the site boundary is limited. Emission is limited indirectly due to the measured (existing plants) or expected (plants under design, atmospheric dispersion calculation needed) impact resulting from the operation of the plants.
- Maximum annoyance standards (MAS): To be collected via questionnaires, the level of satisfaction of the population, concerning several environmental impacts such as odours, noise, dust, is the key value which indicates whether in a specific area action against odours has to be taken.

The most common approach today is the MIS. For a sound stipulation of values it will be necessary to substantiate knowledge concerning the correlation between impact and nuisance, as the nuisance would be reduced by limiting the maximum allowed impact.

In case of the MAS it is necessary – or stipulation is needed – to know the correlation between emission and annoyance, thus including meteorological dispersion.

2.3.2 Extent of nuisance

The extent of nuisance has several aspects that are worth to be discussing briefly. There are two viewpoints we have to deal with – the stimulus viewpoint and the receptor viewpoint. However, the main problem is the correlation between these two processes. This correlation is not comprised by any known formula or law, and this makes it very difficult to clue from one to the other phenomenon.

2.3.2.1 The "stimulus" viewpoint

There are several properties of the stimulus that are important or are said to be important concerning the nuisance caused:

- hedonic odour tone,
- strength of the odour perceived, which may be given in terms of odour concentration or an odour intensity number,
- kind of odour,
- several time-dependent characteristics, example
  - total duration of impact,
  - rhythm of impact,
  - frequency of impact,
  - time of the day / of the week / of the year of impact.

It is generally accepted that nuisance is connected to the hedonic odour tone, determination of which is described by guidelines, e.g. bilingual VDI Guideline 3882, part 2 (VDI 3882, part 2, 1994) in Germany. No European guideline covering this topic exists up to now. The hedonic odour tone ranges from –4 (extremely unpleasant) to +4 (extremely pleasant). Vanilla is an example for a pleasant odour and should be rated by a suitable test person between +1.9 and +2.9.

It is evident and thus is also recognised by the VDI guideline mentioned that the hedonic odour tone is connected with the odour concentration. With increasing concentration of an odorant that is generally recognised as unpleasant the – negative – hedonic odour tone becomes worse (i.e. more negative), and with a generally pleasant odour it often can be observed that with very high concentrations the positive hedonic odour tone drops and can even fall below zero.

As the odour concentration itself is connected to the odour intensity sensed, measurement of which is described for example in the bilingual VDI Guideline 3882, part 1 (VDI 3882, part 1, 1992) in Germany, the three parameters

- hedonic odour tone,
- odour concentration and
- odour intensity

play an important role with the nuisance generated by an odour. Although it thus in theory would be necessary to measure all three parameters, in practice this will not be done very often, as it is not feasible to measure all three parameters.

Owing to the fact that odour concentration and odour intensity are connected via laws as for example the Weber-Fechner-law or Stevens law, it is accepted that one of the two parameters can be omitted which would increase feasibility, decrease cost and would not affect the relevance of the statement seriously.

The next step in reducing the number of parameters to be measured is to accept the hypothesis that odours which should be minimised, generally are unpleasant odours. In consequence, usually the hedonic odour tone is presumed to be below zero for those odours with which regulations and policies have to deal.

So, finally just one out of the three parameters – the odour concentration – is left, presuming unpleasant odours. But this parameter (as well as the total set of all three parameters) would not provide information on the extent of nuisance if the dimension of time were not included.

It is commonly accepted that an annoying condition which is present only for a very short time would not reach by far the nuisance level of a repeated, enduringly annoying condition. This introduces the dimension of time, which has four often discussed aspects, that is to say the total duration of the respective event, the rhythm of its appearance, the frequency of its appearance and the time-of-day/time-of-week/time-of-year dimension. As, however, the rhythm or the frequency or the time-of-day, etc. are easy to talk over but hard to cover in exact numbers with associated effects, the dimension of time is reduced to the parameter of total duration during a set period of time.

This is not only an easy to use and easy to measure approach, but it also offers excellent conditions for using atmospheric dispersion calculations which are essential especially when it is necessary to predict odour impacts from facilities that are under design or construction. Atmospheric dispersion calculations are able to calculate impact concentrations and the duration of this state, and by ordering the impact concentrations by magnitude and summing up the duration one can present the cumulative frequencies of the odour impact concentrations and thus can easily tell which impact concentration is exceeded for which total duration for the time period basing the calculation. This basic time period is often represented as one year, but the meteorological data usually are averages over a longer time period, e.g. 10 years. Figure 2.1 shows an example from a real case where some 100 receptor points, i.e. points where the odour impact had to be assessed, were calculated.

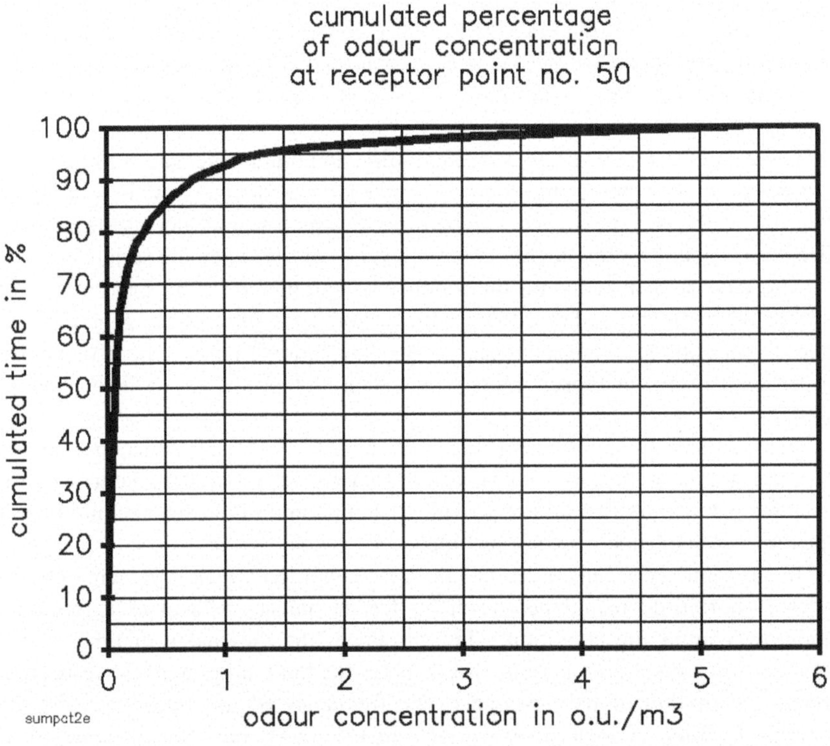

Figure 2.1. Sample sum frequencies for total duration vs. impact odour concentration.

From this evaluation, for example, it can be seen that an impact concentration of 0.5 ou/m^3 is exceeded for 15% of the time, an impact concentration of 1 ou/m^3 is exceeded for 8% of the time and the maximum impact concentration is 6 ou/m^3.

Usually, with odour, averages are not important. The average impact concentration from the example given in Figure 2.1 was not even calculated, as with odour the peak situations are relevant, and these peak situations occur during short periods of time. Thus, values for the percentages of time without odour (odour perception, clear odour perception, perception of annoying odour, perception of identifiable facility odour ...) may be found which can be in the range of 85% or 90%, as is the case in Germany, with special additional assessment due to "short-time-effect" of odours (see Both 1995). Also, percentages of 95%, 98%, 99% or even 99.5% can be found. These values must not be compared to each other without regarding all further circumstances and prerequisites applicable in each case, see the example below.

In conclusion, it can be established that the "stimulus viewpoint" describes and assesses the extent of nuisance in a *two-component standard*, i.e. mostly in the form of a given impact odour concentration, which must not be exceeded during a set duration in time per time, e.g. hours per year. From the example given in Figure 2.1 it can be directly derived that a standard stipulating that an impact concentration greater than 0.5 ou/m^3 must not be exceeded for more than 15% of time is equivalent at this receptor point to a standard stipulating that an impact concentration of greater than 1 ou/m^3 must not be exceeded for more than 8% of the time.

Finally, the parameter "kind of odour" has to be discussed. Although the kind of odour is not suitable for any direct regulatory approach, it is most important, as the task of all efforts must be the reduction of the odour emission. Thus, it is essential to identify the cause of malodours. This is easily possible when taking the kind of odour into account. Thus, for field inspections it is always essential to record the perceived kind of odour. Only when regarding the kind of odour will it be possible to apply the "polluter pays principle". So the two component standards must be extended with the constraint that the odours the respective source and cause must be identifiable.

2.3.2.2 The "annoyed population" viewpoint

The stimulus viewpoint does not care about the people living in the area that is subject to a malodour impact. Even if there were no-one living in that area, a standard following the stimulus viewpoint would be possible.

However, standards only make sense if they fulfil a protective aim. Thus, as we excluded problems like direct toxicity, etc. from the discussion about odour, only the presence of a potentially annoyed population justifies the establishment of impact standards.

If we have a population living in the relevant area, then the protection from nuisance is the motivation for standards, and thus it is consequent to use the response of local residents for an assessment of the situation and for decisions on whether there is a legal or an illegal situation in a specific case.

It is postulated here that it is generally accepted that total absence of any nuisance is an aim which is impossible to reach in practice due to technical as well as economic reasons. Thus, discussion is reduced to the question on how much annoyance may be acceptable. For example, the German Federal Protection Act for Ambient Air (BImSchG, Bundes-Immissionsschutzgesetz) distinguishes between insubstantial and substantial annoyance. If an annoyance is insubstantial, then no demand for protection against the malodour exists.

Methods to measure the extent of the annoyance must incorporate panellists, as annoyance is not measurable with any technical or analytical instrument. Psychometric methods are required.

Panellists may be members of a test person panel experienced in assessing odours, but most commonly local residents are used to gather information on the extent of odour annoyance in an actual case. The most valuable tool is population questioning, and in bilingual VDI Guideline 3883, part 1 (VDI 3883, part 1, 1997) and bilingual VDI Guideline 3883, part 2 (VDI 3883, part 2, 1993) the performance of different questioning methods are described. Also here, no European standard will be available in the near future.

Usually, questionnaires include a question concerning the extent of annoyance, expressed in an annoyance category or an indication, e.g. on a thermometer scale, ranging from 0 (no annoyance) to 10 (extremely annoying). This, unfortunately, again gives a two-parametric distribution, for example percentage of people vs. annoyance category, and thus often a method is required to reduce the information to just one number representing the total result. As an example, the questioning with annoyance categories uses weights for each of the four categories of annoyance ("slightly annoying", "annoying", very annoying" and "extremely annoying") and then can calculate the annoyance index I. Although this reduction of information results in one convenient number, the content and value of information of course decreases.

Stipulations regarding the annoyed population viewpoint must set standards that can be reviewed by the questioning method used. If, for example, category assessment is done including "slightly annoying", "annoying", very annoying" and "extremely annoying", then a stipulation can combine maximum percentages for each category ("less than 5% of extremely annoyed people", "and"/"or" "less than 15% annoyed"...) or can combine results for categories ("less than 15% very or extremely annoyed").

2.4 ENVIRONMENTAL PROTECTION POLICY

Of course, the type of standard strongly depends on the environmental protection policy. Two main directions can be distinguished – the emission limiting policy, known as "the objective approach" and the impact limiting policy, known as the "subjective approach".

2.4.1 The objective approach: emissions principle

Everyone is equal before the law. This should apply to every company that wants to build new facilities and every city discharges used water into that rivers, etc.. Thus, emission standards which apply to everyone in the same manner are fair. Problems would arise if different countries "offer" different emission standards, as this would give a competition to the disadvantage of the environment.

Most legal systems follow this principle. This gives a lot of certainty for all parties, as the conditions of acting inside such a system are reliable and do not change every day.

2.4.2 The subjective approach: impact principle

Everyone has the right to be protected against offensive environmental impacts. If someone does affect (or affect substantially) any other person, then the legal system must have a possibility and the power to correct this.

In specific cases, consequently, an emitting facility, wastewater treatment plant or whatever else must face the fact that owing to the occurrence of annoyance resulting from their emissions, requirements concerning the emission of the facility may be requested that exceed what is accepted as emission standard. These requirements will aim at an upgrading and operation of the facility which will ensure that no odours will be emitted that generate nuisance in the vicinity of the plant.

2.4.3 The optimal approach: combination

Both principles can be combined to one approach where basic rules are formulated that demand a certain minimum standard concerning emissions prevention, and additional requirements, which exceed the standard set of the basic demands.

Looking into the EU's newer legislation in general (legislation concerning odour control or annoyance is not found yet in the EU), it seems that this type of regulation is used increasingly. The emission approach, which was the main tool for a long time, is more and more complemented with an impact part, thus, both the above principles are put into practice now in combination.

2.4.4 What can be found today? Some examples

It is not possible nor useful to present all possible regulations of many different countries here. The idea with this chapter was to provide some useful basic information which should be expanded by one's own experience and should lead to a better understanding and a deeper discussion of some backgrounds that are relevant with odour nuisance. However, some examples should be enumerated here which may contribute to this aim.

2.4.4.1 Germany

In Germany, environmental protection legislation is comparatively old, as the rapid economic development after World War II together with a high population density demanded this. However, besides a minimum distance regulation and some more or less vague regulations of the types of MES (TA Luft) and a vague form of an MIS (Gem.Rd.Erl NRW), see Frechen (2000), the laws did not give too much advice on how to handle the problem until ten years ago.

Jurisdiction had to decide over several cases where annoyed persons sued the owners of odour emitting facilities, and in general it can be said that it was more and more recognised that the residents must not be annoyed substantially.

For ten years now the State of Northrhine-Westphalia, which is the most populated and industrialised part of Germany, developed and tested a new regulation according to the MIS-type, backed up and calibrated by annoyance surveys and field inspections. This "Directive on Odour in Ambient Air" is explained by Both (1995). It sets an impact odour concentration of 1 ou/m^3 as the limit impact concentration, and then limits the time percentage during which a higher impact concentration is tolerable ("insubstantial annoyance"). Time percentages are 15% for industrial areas and 10% for residential areas. Although 10% or even 15% may seem to be very high percentages, indicating a very serious impact, it must be considered that the limit concentration of 1 ou/m^3 is formed regarding the short time effect of odours. This means, for example, that when using the standard dispersion model, which calculates hourly averages of impact concentrations, one has to multiply the hourly average by a factor of ten. Thus, an impact concentration of 0.1 ou/m^3 (hourly average resulting from the calculation) equals 1 ou/m^3 in the sense of the directive.

Frechen (2000) gives more information concerning the situation in Germany today.

Recently it has been discussed whether to transfer the "Directive on Odour in Ambient Air" into the federal laws, which would mean that it would become relevant throughout Germany.

2.4.4.2 Switzerland

It is stated that "too high impacts" are not allowed. Impact is "too high" if a "relevant portion of the population" is "significantly annoyed". The determination of the annoyance uses the method of questioning. Most important response is the thermometer value of annoyance given in a thermometer scale from 0 to 10.

The following scheme applies:

Annoyance	thermometer value	Percentage of strongly annoyed (≥ 3)	Measures
strong	> 5	> 25%	immediate measures
medium	3–5	10–25%	long term measures
reasonable	< 3	< 10%	no special measures

Emission standards concerning odour concentration are not stipulated, although emission standards are set for about 150 substances which can cause odour, anticipating that no serious annoyance will occur if these standards are met.

2.4.4.3 The Netherlands

Policy has the objective to keep the population as free from annoyance as possible. It was aim that in 2000 not more than 12% of the population are annoyed by industrial odours (annoyed here stands for "perceive sometimes or even often annoying odour"). The percentage of people strongly annoyed by industrial odours should then drop below 3%.

The percentage results are renewed every year by questioning. Supplemental to the yearly questioning a telephone questionnaire was developed.

2.4.4.4 United Kingdom

Control over odour annoyance in UK law is found in the Environment Protection Act. Salter (2000) gives a comprehensive overview over "The legal context of odour annoyance" in the UK. No general valid emission standards concerning odour are set. The regulations do not incorporate impact odour concentrations or percentages of time, but retracts to the more general statements concerning the odour nuisance.

2.4.4.5 Belgium

There is no general law except the statement that "it is forbidden to cause unacceptable annoyance by dust, smoke, odours, fumes ...". This is of course a very broad stipulation but sometimes it is used on an ad hoc basis in court.

There are no detailed odour regulations per sector. No specific regulation is found for wastewater odours. There are "minimum distances" between intensive animal farming houses and residential areas, and there are general emission limits for inorganic and organic compounds (similar to TA-Luft). This sometimes helps to prevent odour nuisance but often it doesn't.

In general, problems are handled through the working licence of the companies. In this respect there is one criterion defined as 98-percentile corresponding to 1 su/m^3 (su=sniffing unit). The methodology to control this criterion is described in the working licence and approved by the environmental inspection.

Mostly, if there are complaints, the factory is forced to make a study to describe the situation and to evaluate the impact on the neighbourhood. If the impact is significant, a plan is required in which technical and other (management) measures to reduce problems are described together with a time plan for the implementation of the measures.

In general, as can be seen, odour policy in Belgium is more or less a case-to-case policy.

2.4.4.6 United States of America

In the USA there are no federal regulations for odours. Odours are regulated by the states or sometimes by local governments.

In Massachusetts, as an example, there exists a draft odour policy for composting facilities that requires that new or expanding facilities not exceed 5 Dilutions/Threshold (D/T) at the property line.

Other states in the USA have a variety of odour regulations. Some have hydrogen sulphide limits, some have D/T limits and many have general nuisance language. Such language requires that odours do not cause a nuisance, which is defined in different ways. One definition of nuisance is to unreasonably interfere with the comfortable enjoyment of life and property (which is close to the WHO health definition) or the conduct of business.

2.4.4.7 EU directive for WWTPs

The EU will issue directive 12255 for wastewater treatment plants over 50 population equivalent. Part 9 of the directive is devoted to "Odour and Ventilation", and here some very general advice and hints are given. The directive deals with the emissions source, i.e. the WWTP, and gives some hints concerning the impact, but does not give a complete standard for annoyance prevention.

2.5 SOME CONCLUSIONS

In Brazil ten years ago the author was asked to give an expert opinion over a WWTP treating industrial water from the oil and refinery industry with high amounts of solvents. The complaints of residents came from an area which was more than 5 km from the plant. First conclusion: Offensive odours can become an issue not only in the "first world". Second conclusion: although big in this case, "distance only" is no solution, as strong odours can travel a long way as can be seen from this example.

In Germany in 1960 a newly built mechanical composting works caused odours that were so annoying to the nearby residents that it was almost decided to tear down the new plant. Conclusion: even if there are no regulations (and there was no regulation in Germany at the end of the 1950s concerning odour annoyance), sometimes the case is clear that massive effort is needed to ameliorate the situation.

Concerning the design of regulations and the policy to best fight annoyance due to malodours, it may be stated that a policy that leads to an emission reduction should combine a maximum impact standard (favourably derived from annoyance observations) with an annoyance questioning or at least a field inspection (like the type that is described in the bilingual VDI Guideline 3940 (VDI 3940, 1993) in Germany) after the respective measures are in effect.

Some key statements may finally be presented:

- "No odour" is not a choice: WWTPs always will have some odour emitted, and even deodorization plants have a measurable odour concentration in the off gas.
- "No annoyance" is not a choice: no nation has enough money to keep away even the smallest extent of annoyance from its citizens. Although a question of money and desired comfort, at least a remnant risk of nuisance, sometimes called "insubstantial annoyance", will be present. The question is where the border between insubstantial and substantial annoyance may be.

2.6 ACKNOWLEDGEMENTS

Information on the current regulations in Belgium and the US were contributed by Herman Van Langenhove and Thomas Mahin.

2.7 REFERENCES

Both, R. (1995) Odour regulations in Germany – A new directive on odour in ambient air. *Proc. International Specialty conference Air Waste Management Association, Odours: Indoor and Environmental Air,* September 1995.

Frechen, F.-B. (2000) Odour measurement and policy in Germany. *Water Science and Technology,* **41** (6), 17-24.

Salter, J. (2000) The legal context of odour annoyance. *Proc. International Meeting on Odour Measurement and Modelling, Odour 1, Cranfield University.*

VDI-guideline 3882 part 1 (1992) Olfactometry - Determination of odour Intensity, *VDI-handbook on Air Pollution Prevention,* Vol. 1.

VDI-guideline 3882 part 2 (1994) Olfactometry - Determination of hedonic odour tone, *VDI-handbook on Air Pollution Prevention*, Vol. 1.

VDI-guideline 3883 part 1 (1997) Effects and assessment of odours – Psychometric assessment of odour annoyance – questionnaires, *VDI-handbook on Air Pollution Prevention,* Vol. 1.

VDI-guideline 3883 part 2 (1993) Effects and assessment of odours - Determination of annoyance parameters by questioning – repeated brief questioning of neighbour panellists, *VDI-handbook on Air Pollution Prevention*, Vol. 1.

VDI-guideline 3940 (1993) Determination of odorants in ambient air by field inspections, *VDI-handbook on Air Pollution Prevention,* Vol. 1.

Part II

ODOURS ASSOCIATED WITH WASTEWATER TREATMENT

3

Odour formation in sewer networks

Thorkild Hvitved-Jacobsen and Jes Vollertsen

3.1 INTRODUCTION

The safe and efficient collection and conveyance of wastewater to treatment and discharge is traditionally the main function of a sewer network. This simplified concept only indirectly takes into account that the sewer is also a reactor for chemical and biological processes. From a biological point of view, the redox conditions are crucial for the function of the sewer. Aerobic conditions will ensure that odour, health and corrosion problems are minimized. However, such conditions may also create problems for a subsequent advanced treatment because readily biodegradable substrate needed for denitrification and biological phosphorus removal will be degraded under aerobic transport in the sewer.

Anaerobic conditions in a sewer may give rise to a number of problems in the network itself and are typically identified by malodours, health risks and corrosion. Such problems – generally considered to be associated with the formation of hydrogen sulphide – have been addressed in early publications (Pomeroy and Bowlus 1946; Thistlethwayte 1972; Boon and Lister 1975; Pomeroy and Parkhurst 1977). Going back in time, only a few studies have focused on the transformations of the organic matter during transport in the

 Odours in Wastewater Treatment: Measurement, Modelling and Control. Edited by Richard Stuetz and Franz-Bernd Frechen. ISBN: 1 900222 46 9

sewer and typically just in terms of removal of BOD or COD (Stoyer 1970; Koch and Zandi 1973; Pomeroy and Parkhurst 1973; Green *et al.* 1985).

In-sewer processes take place in a complex system. They proceed in one or more of the four phases: the suspended water phase, the biofilm, the sewer sediments and the sewer atmosphere and by exchange of relevant substances between these phases. Processes that proceed in the sewer system therefore affect other parts of the urban system, i.e. the urban atmosphere, wastewater treatment plants and local receiving waters (Figure 3.1).

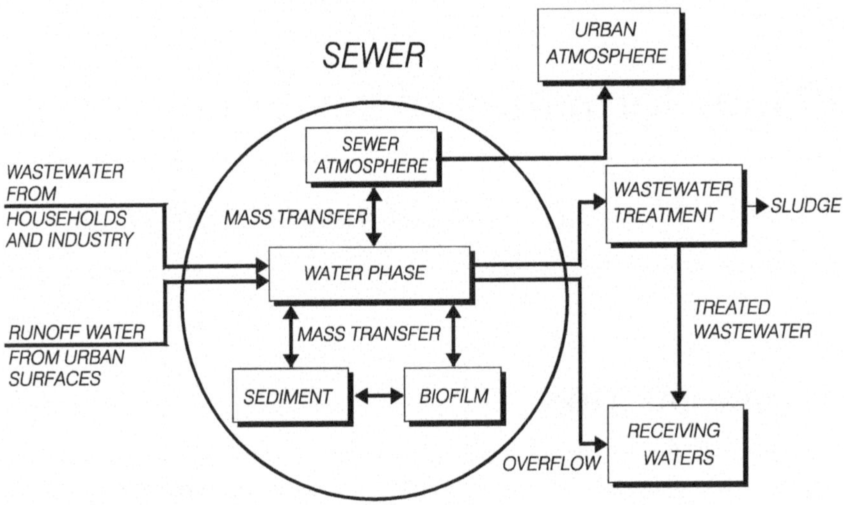

Figure 3.1. Wastewater transport and processes within and related to the urban wastewater system.

Although the odour phenomenon and associated control strategies for many years have been of concern when dealing with wastewater collection and treatment, the detailed pathways and conditions for odour formation in sewer networks have not been addressed at a detailed level. No conceptual basis for a general description of odour formation and corresponding formation modelling exists. This lack of understanding is considered crucial.

In this chapter odour formation is addressed from a fundamental microbiological point of view and related to results from experimental studies. In this context, possibilities for the prediction of odour formation in sewer networks will be assessed. For this purpose a newly developed sewer process model will be considered (Hvitved-Jacobsen *et al.* 1998a, b; Hvitved-Jacobsen and Nielsen 2000).

3.2 MICROBIAL PROCESSES IN SEWERS RELATED TO ODOUR FORMATION

Several phenomena make a sewer a complex system for microbial processes:

- Wastewater is a matrix including a great variety of microorganisms and with a great number of substrates varying in time and space.
- The microbial processes proceed in different subsystems of the sewer: the suspended wastewater phase, the biofilms, the sediments and the solid surfaces in contact with the air phase.
- The microbial processes in the sewer interact across the boundaries of these subsystems and often take place under changing aerobic and anaerobic conditions. Exchange of substrate (electron donors as well as electron acceptors) and biomass between these subsystems proceed.

Microbial production of odorous compounds in the different subsystems of a sewer network should therefore be assessed with a focus on the redox conditions established.

3.2.1 Fundamentals

Microbial transformations of organic matter include what are basically considered to be biochemical processes, i.e. changes of chemical components initiated by living organisms. A sewer system is dominated by the activity of heterotrophic – or more correctly termed chemoheterotrophic – microorganisms (bacteria). The presence of the heterotrophic biomass in wastewater, biofilms and sediments of a sewer is central for these biochemical processes. The heterotrophic biomass makes use of the organic matter in the wastewater for two fundamental purposes: the organic matter is the carbon source for formation of cell materials and it is – as an electron donor – the energy source needed for the growth process and for maintenance (Figure 3.2). The anabolic processes provide the substances necessary for growth of new biomass. The catabolic processes provide the energy needed for production of new cell biomass and for the maintenance of the fundamental functions of the existing biomass.

The energy accumulated in the organic matter is made available for the microorganisms by the catabolism, i.e. a "degradation" process performed by oxidation of the organic matter. The organic matter is thereby the electron donor. The corresponding external electron acceptor, which is reduced, is either dissolved oxygen (DO) (aerobic process), nitrate (anoxic process) or sulphate (anaerobic process). These energy-producing reactions are termed respiration processes and only proceed under the presence of an external compound that can serve as the terminal electron acceptor of the electron transport chain.

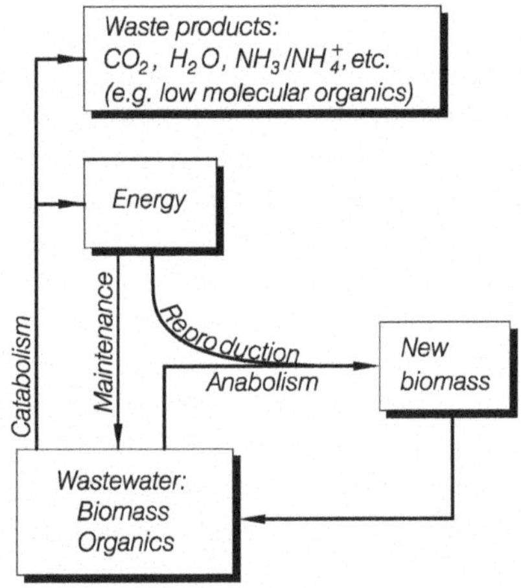

Figure 3.2. Main pathways of organic matter in wastewater of a sewer system.

3.2.2 Aerobic and anoxic heterotrophic microbial processes

The complex organic molecules – the electron donors – are in the aerobic respiration process broken down by passing electrons to oxygen, which is reduced by the formation of H_2O, CO_2 and inorganic substances. The organic carbon is thereby simultaneously transformed into inorganic carbon and released as CO_2. Although ammonia (NH_3) is produced by aerobic ammonification, it does not result in the formation of major odorous problems. The reason is that NH_3 has a relatively high recognition threshold value (about 40 ppb) and a relatively low tendency to be emitted from wastewater at a typical neutral pH value. As a consequence, the end-products of the aerobic heterotrophic processes are generally considered non-odorous substances. Transport of wastewater under aerobic conditions in sewers is therefore not important as far as odour problems are concerned.

The aerobic respiration with DO as the terminal electron acceptor is an efficient process for energy metabolism. Under DO non-limited conditions, the oxygen uptake rate (OUR) may, however, for wastewater vary considerably depending on the density and the activity of the bacteria and the biodegradability of the wastewater. Typically, values of OUR have been measured in the range 2–20 $gO_2/m^3/h$ (Boon and Lister 1975; Matos and de Sousa 1996; Hvitved-Jacobsen and Vollertsen 1998). The most biodegradable molecules, which are also often the most volatile, e.g. volatile fatty acids (VFAs), are therefore potentially subject to biodegradation. Biodegradable, volatile substances discharged into and produced in a sewer network are therefore often efficiently removed.

Anoxic conditions require absence of DO and presence of nitrates. Such conditions are typically only found when artificially implemented. The aerobic and anoxic pathways of organic matter degradation are rather identical and anoxic conditions therefore do not generally create specific odour problems. Addition of nitrate to wastewater is widely used as a control measure to avoid anaerobic conditions in sewers.

3.2.3 Anaerobic heterotrophic microbial processes

In terms of the formation of odour compounds, it is from a theoretical point of view important to evaluate the extent and occurrence of the anaerobic heterotrophic microbial processes in a sewer.

Under anaerobic conditions both respiration and fermentation processes may proceed to support the energy requirement of the organisms (Figure 3.2). Contrary to respiration, fermentation does not require the participation of an external electron acceptor. In this case, the organic substrate undergoes a balanced series of oxidative and reductive reactions, i.e. organic matter reduced in one step of the processes is oxidized in another.

As a result, the partial breakdown of the organic matter by fermentation yields organic products with a low molecular weight, e.g. VFAs, along with CO_2. Compared with the aerobic respiration, the fermentation is inefficient, however, the fermentation products can to some extent – and in addition to fermentable substrate – be used by the sulphate reducing bacteria that make use of sulphate as terminal electron acceptor (Nielsen and Hvitved-Jacobsen 1988). In the absence of sulphate, the methanogenic bacteria utilize the low molecular weight fermentation products to obtain energy, producing methane (CH_4) as an end-product. Some of the methanogenic bacteria, the chemoautotrophic, use CO_2 and H_2. Furthermore, under changing aerobic/anaerobic conditions in the sewer, low molecular organics produced under anaerobic conditions may be degraded in the aerobic parts of the system.

Fermentation may take place in the three major microbial subsystems of a sewer, i.e. the wastewater, the biofilm and the sediments (Figure 3.3). Sulphate reducing bacteria (SRB) are slow growing and are therefore primarily present in the biofilm and in the sediments where sulphate from the wastewater may penetrate (Nielsen and Hvitved-Jacobsen 1988; Hvitved-Jacobsen *et al.* 1998b; Bjerre *et al.* 1998). However, as a result of biofilm detachment, sulphate reduction may to some minor extent take place in the wastewater. Methanogenic microbial activity normally require absence of sulphate and will therefore mainly take place in deeper parts of the sediments and not in the biofilm that typically is fully penetrated by sulphates.

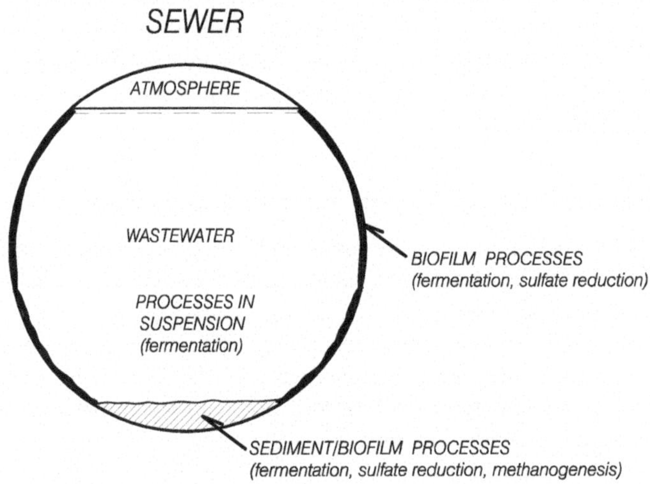

Figure 3.3. Outline illustrating the subsystems and occurrence of processes in a gravity sewer under anaerobic conditions.

In sewer networks without considerable amounts of sediments, the anaerobic processes are therefore dominated by the acidogenic production of VFA and CO_2 and by the sulphate reduction (hydrogen sulphide production). The methanogenic phase can therefore normally be excluded as being of minor importance. These facts were verified by Tanaka and Hvitved-Jacobsen (1999) and Tanaka (1998) under sewer conditions in a number of laboratory experiments and in the field.

As generally shown, both anaerobic respiration (sulphate respiration) and fermentation produce odorous substances and they will typically proceed simultaneously. The sulphate respiration has as its end-product hydrogen sulphide, well known as a component causing odour problems. Depending on the type of substrate and the microorganisms present, the fermentation pathways and the products produced may vary considerably. Figure 3.4 is just an example of fermentation of sugars, illustrating that a broad range of volatile organic compounds (VOCs) – and potentially odorous compounds – may be produced.

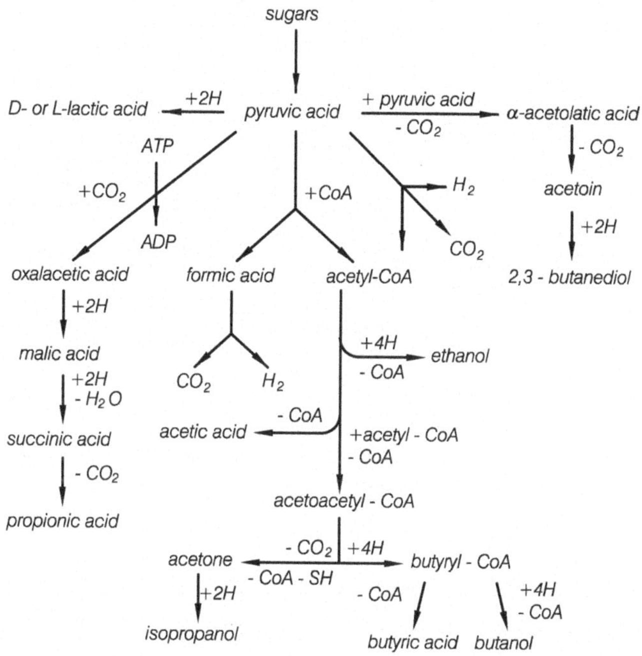

Figure 3.4. Illustration of some major pathways and end-products of the bacterial fermentation of sugars from pyruvic acid (Stanier *et al.* 1986).

The interaction between the aerobic and anaerobic transformations of wastewater in a sewer network is outlined in Figure 3.5. The concept outlined is the basis for a process model description of the dominating heterotrophic in-sewer processes.

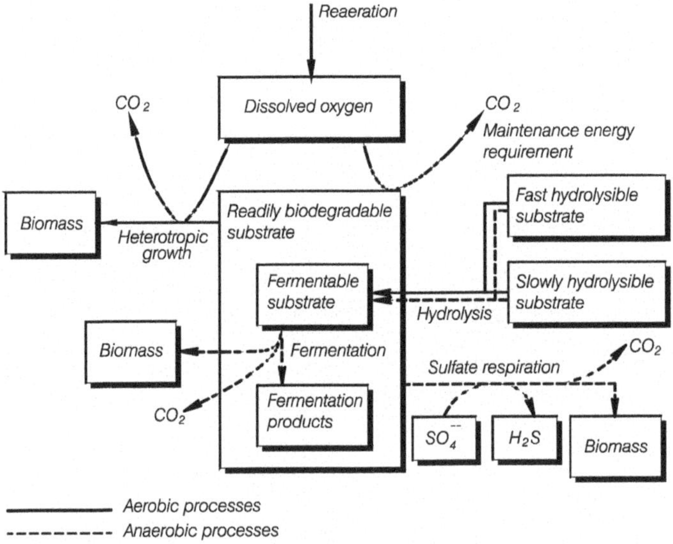

Figure 3.5. A simplified integrated aerobic and anaerobic concept for transformation of organic matter and sulphur components of wastewater in sewers.

3.3 VOLATILE ORGANIC COMPOUNDS PRODUCED UNDER ANAEROBIC CONDITIONS IN SEWERS

The anaerobic microbial processes responsible for the formation of odour-causing substances produce both inorganic gases and VOCs. The malodorous inorganic gases are primarily ammonia (NH_3) and hydrogen sulphide (H_2S).

In addition to odour, hydrogen sulphide in the gas phase causes both human health and corrosion problems. It can be detected by the human sense of smell at a concentration level about 0.001 ppm (threshold odour concentration). It has at 10–50 ppm sublethal effects (nausea and eye, nose and throat irritation) and at about 100 ppm it can cause serious breathing problems. At 300–500 ppm death may occur within a few minutes. It is important to notice that hydrogen sulphide loses it characteristic smell at about 50 ppm and above this level it cannot be detected directly by the human sense of smell (ASCE 1989).

As already depicted in Figure 3.4, a great number of VOCs can be produced. The most common organic substances associated with odours and produced from wastewater organic matter are shown in Table 3.1. Many organic compounds known as potential odour-producing substances have been identified in domestic wastewater (Raunkjær *et al.* 1994; Hvitved-Jacobsen *et al.* 1995; Hwang *et al.* 1995). Generally VFAs are known as anaerobic decomposition products of carbohydrates, e.g. starch. Mercaptans are primarily produced from proteins. Several of the compounds shown in Table 1.1 arise from the anaerobic decomposition of organic matter containing sulfur and nitrogen.

Only few studies have been concerned with measurements of specific odorous compounds in sewer systems. Hwang *et al.* (1995) have in a study of malodorous substances in wastewater at different steps of sewage treatment analysed the influent wastewater (Table 3.1).

Table 3.1. Sulfur and nitrogen containing odorous compounds in the influent wastewater at a treatment plant (Hwang *et al.* 1995).

Compound	Average concentration (μg/l)	Range of concentrations (μg/l)
Hydrogen sulphide	23.9	15 - 38
Carbon disulphide	0.8	0.2 – 1.7
Methyl mercaptan	148	11 - 322
Dimethyl sulphide	10.6	3 - 27
Dimethyl disulphide	52.9	30 - 79
Dimethylamine	210	-
Trimethylamine	78	-
n-propylamine	33	-
Indole	570	-
Skatole	700	-

Although the results in Table 3.1 only represent examples, they are interesting for several reasons. First of all the table shows that several of the odorous compounds may appear in wastewater from a sewer system in relatively high concentrations, especially when compared with H_2S. It should also be noticed that the concentrations are those observed in wastewater. What appears in the air phase may be quite different, an aspect which will be dealt with in the following example and in more detail in section 3.4.

Thistlethwayte and Goleb (1972) have reported investigations of the composition of sewer air. The main part of the samples was taken in a sewer transporting mixed municipal wastewater with a maximum residence time around 4 hours. BOD_5 of the wastewater was ranging from 300–350 g/m^3 and the temperature was typically about 24 °C. The authors divided the components in the sewer air into four groups. These groupings were determined not only by the chemical nature of the components but also according to their respective concentrations. Ammonia was not included in this scheme:

(1) Carbon dioxide: CO_2;
(2) Hydrocarbons and chlorinated hydrocarbons;
(3) Hydrogen sulphide: H_2S;
(4) Odorous gases and vapours such as mercaptanes, amines, aldehydes and VFAs.

The typical composition of sewer air reported by Thistlethwayte and Goleb (1972) is shown in Table 3.2. This investigation did not distinguish between components from inlets, e.g. industrial sources, and components produced in the sewer.

Group 1 (CO_2) indicates that microbial degradation of wastewater organic matter takes place in the sewer. In terms of odour, the other groups 2–4 are relevant. In spite of the fact that the investigation did not include the sources of the components found in the sewer atmosphere, group 2 probably is a result of inputs to the system. The components included in the groups 3 and 4, however, are interpreted as a result of anaerobic processes.

The results reported by Thistlethwayte and Goleb (1972) indicate that the concentrations of the constituents of groups 3 and 4 tend to be related. That is to say the constituents of group 4 (a, b and c) tend to vary according to the levels of the H_2S concentrations roughly in the ratio of 1:50 to 1:100. They conclude that this observation suggests that although the H_2S concentration alone may not be a sufficient measure of potential sewer air odour levels, H_2S concentration measurements probably are sufficient for most studies of sewer gases.

3.4 EMISSION OF ODOURS FROM SEWERS

3.4.1 Fundamental aspects

The design characteristics and operation mode of a sewer network determine to a great extent if anaerobic conditions and thereby related potential odour problems may arise. Table 3.3 gives an overview of major sewer system characteristics typically associated with odour formation. If neither oxygen nor nitrates are available, strictly anaerobic conditions occur and sulphate is the potential external electron acceptor. In addition to the respiration processes outlined in Table 3.3, fermentation under anaerobic conditions plays a major role for the VOC formation.

Release of odorous compounds from the wastewater into the overlaying atmosphere is a fundamental process for odour problems. As long as an odorous compound remains in the water phase, odour problems do not exist.

Table 3.2. Typical composition of sewer air reported by Thistlethwayte and Goleb (1972). The composition corresponds to dry weather conditions and anaerobic conditions in the sewer.

Group no. and components	Order of concentration range by volume
1. Carbon dioxide, CO_2	0.2–1.2%
2. Hydrocarbons and chlorinated hydrocarbons	
a. Hydrocarbons, mainly aliphatics C6-C14 and mostly C8-C12 (petrol)	up to 500 ppm
b. Chlorinated hydrocarbons, mostly trichlorethylene with ethylene dichloride and some carbon tetrachloride	10–100 ppm
3. Hydrogen sulphide, H_2S	0.2–10 ppm
4. Odorous gases and vapours	
a. Sulphides (mostly methyl mercaptan and dimethyl sulphide; some ethyl mercaptan)	10–50 ppb
b. Amines (mostly trimethylamine and dimethylamine; some diethylamine)	10–50 ppb
c. Aldehydes (mostly butyraldehyde)	10–100 ppb

It is important that odorous substances may undergo degradation in the wastewater phase under aerobic conditions. Sulfur compounds seem fast degradable whereas this is not the case for nitrogen compounds (Hwang *et al.* 1995).

For sewer systems, two major transport phenomena for odorous compounds must therefore be dealt with:

- The water-air transfer process for odorous compounds describing the release from wastewater into the sewer atmosphere.
- The ventilation of the sewer system transporting the sewer atmosphere into locations where malodours should not be accepted.

The following basic conditions and phenomena are important for the transport processes and for the amount of volatile odorous compounds transferred from the wastewater phase into the sewer atmosphere and subsequently into the city atmosphere:

- pH and temperature of the wastewater.
- Turbulence of the wastewater.
- Ventilation of the sewer system.
- Chemical and microbial processes, e.g. on the sewer walls, affecting the amount of odorous compounds in the sewer atmosphere.

Table 3.3. Electron acceptors and corresponding conditions for microbial redox processes in sewer systems, cf. text.

Process conditions	Electron acceptor	Typical sewer systems characteristics
Aerobic	+ oxygen	Partly filled gravity sewer Aerated pressure sewer
Anoxic	- oxygen + nitrate	Pressure sewer with addition of nitrate
Anaerobic	- oxygen - nitrate + sulphate (+ CO_2)	Pressure sewer Full flowing gravity sewer Gravity sewer with low slope

Typically, non-equilibrium conditions exist for the emission of odorous compounds because of dilution of the sewer atmosphere by e.g.ventilation affecting the water–air transfer of volatile compounds. The partial pressure of an odorous compound in the sewer atmosphere is therefore normally lower than corresponding to equilibrium. However, a simple understanding of the potential for release of odorous components from wastewater is based on their behaviour under equilibrium conditions.

3.4.2 Water–air equilibrium conditions for odour compounds

A first approach in this respect is described by the distribution coefficient K_A for a more or less volatile compound, A, between the gas phase and the water phase:

$$K_A = \frac{y_A}{x_A} \tag{3.1}$$

Where:
K_A = distribution coefficient or partition coefficient (-),
y_A = mole fraction of A in the gas phase (moles /(total moles)),
x_A = mole fraction of A in the water phase (moles /(total moles)).

The concepts of mole fraction of a component used in equation 3.1 is a convenient measure of concentration when dealing with trace quantities and dilute solutions, often experienced in environmental systems. Especially in case of transport phenomena and equilibrium between phases it results in simple quantitative expressions. The phenomena being of interest when dealing with release of odour components from wastewater into a sewer atmosphere is in this respect a relevant example.

The fundamentals of mole fraction can be illustrated for a binary system consisting of two components, A and B. The mole fraction of A (considering a gas phase) is defined as follows:

$$y_A = \frac{N_A}{(N_A + N_B)} \tag{3.2}$$

Where:
N_A =moles of A,
N_B = moles of B.

The mass balance (in the actual case for the gas phase) is therefore:

$$y_A + y_B = 1 \tag{3.3}$$

For trace quantities of A in air, y_A can be expressed corresponding to a situation where 1 "mole" of air (mainly consisting of N_2, O_2, Ar and CO_2) has a "molar volume" of approximately 22.4 l/mole at 0°C and 1 atm and a "molar weight" of 29 g/mole.

$$y_A = \frac{c_{1A}}{1/224} = \frac{c_{1A}}{0.0446} \tag{3.4}$$

Where:
c_{1A} = molar concentration of component A in air (moles/l).

Correspondingly, for dilute aqueous solutions, 1 mole of water equals 18 g, i.e. x_A can be expressed as follows:

$$x_A = \frac{c_{2A}}{1000/18} = \frac{c_{2A}}{55.56} \tag{3.5}$$

Where:
c_{2A} = molar concentration of component A in water (moles/ l).

Equation (3.1) expresses that the ratio of the concentrations of A in the gas phase and the water phase, respectively, is a constant at equilibrium. This constant is temperature dependent but independent of the quantity of A as long as dilute solutions are dealt with.

The relative volatility, α, is a different constant which under equilibrium conditions can be used to express the distribution of a volatile compound between a gas phase made of A and water vapour and a water phase containing A. This constant is for a component A defined as follows:

$$\alpha_A = \frac{y_A / y_{water}}{x_A / x_{water}} \quad (3.6)$$

Where:
α_A =relative volatility (-),
y_{water} = mole fraction of water vapour in the gas phase (moles/(total moles)),
x_{water} = mole fraction of water in the water phase (moles/(total moles)).

For a dilute solution of water, which is a reasonable approximation for normal wastewater, x_{water} is approximately equal to 1 and equation 3.6 is therefore reduced to:

$$\alpha_A = \frac{y_A / y_{water}}{x_A} \quad (3.7)$$

The most widely used and still simple theoretical approach for description of a gas–liquid equilibrium for a volatile compound A is expressed by Henry's law:

$$p_A = y_A P = H_A x_A \quad (3.8)$$

Where:
p_A = partial pressure of a component A in the gas phase (atm),
P = total pressure (atm),
H_A = Henry's law constant for A (atm/(mole fraction)).

Henry's law defines under equilibrium conditions and at constant temperature the relative amount of a volatile compound in the gas phase as a function of the relative concentration in the water phase; i.e. Henry's law quantifies the degree of tendency of a volatile compound to escape. The law applies for dilute solutions of A, i.e. for solutions where x_A is close to 0 provided that A does not dissociate or react in the water phase. Table 3.4 gives some examples of Henry's law constants and boiling points for selected odorous and non-odorous compounds. The list includes values for hydrocarbons that frequently appear in sewer networks from e.g. industrial sources and street runoff.

Table 3.4. Gas-liquid equilibrium of selected odorous compounds in water at 25°C. Other compounds are included for comparison (Thibodeaux 1979; Sander 2000).

Substance	Compound	Boiling point at atmospheric pressure (°C)	Henry's law constant, H_A (atm (mole fraction)$^{-1}$)
Volatile sulphur compounds (VSCs)	Methyl mercaptan	6	200
	Ethyl mercaptan	35	200
	Allyl mercaptan	69	
	Benzyl mercaptan	195	
	Dimethyl sulphide	37	110
	Dimethyl disulphide	110	63
	Thiocresol		
Nitrogenous compounds	Methylamine	-6.4	0.55
	Ethylamine	17	0.55
	Dimethylamine	7	1.3
	Pyridine	115	0.5
	Indole	254	
	Scatole	265	
Acids (VFAs)	Acetic	118	0.063
	Butyric	162	0.03
	Valeric	185	0.025
Aldehydes and ketones	Acetaldehyde	21	5.88
	Butyraldehyde	76	6.3
	Acetone	56	1.9
	Butanone	80	2.8
Inorganic gases	Hydrogen sulphide, H_2S	- 59.6	563
	Ammonia, NH_3	- 33.4	0.843
Selected non-odorous compounds	Nitrogen, N_2	- 195.8	86,500
	Oxygen, O_2	- 183	43,800
	Carbon dioxide, CO_2	- 78.5	1,640
	Methane, CH_4	- 161.5	40,200
Hydrocarbons	Pentane, C_5H_{12}	36	70,400
	Hexane, C_6H_{14}	69	80,500
	Heptane, C_7H_{16}	98	46,900
	Octane, C_8H_{18}	125	19,400

As previously mentioned, the simple equilibrium approach requires that the relevant odorous compounds exist in a non-dissociated molecular form in the water phase. For several odorous compounds this is not the case. Hydrogen sulphide is in this respect an important example with sulphide chemical species related according to the following equilibrium:

$$\underset{(gas)}{H_2S(g)} \overset{water-air\ transfer}{\Leftrightarrow} \underset{(aqueous)}{H_2S(aq)} \overset{pK_{a1}=7.0}{\Leftrightarrow} \underset{(ion)}{HS^-} \overset{pK_{a2}=14}{\Leftrightarrow} \underset{(ion)}{S^{2-}} \qquad (3.9)$$

Where:

The equilibrium constants, K_{a1} and K_{a2}, determine the ratio between the concentrations, C, at equilibrium:

$$K_{a1} = \frac{C_{H^+} \cdot C_{HS^-}}{C_{H_2S}(aq)} \qquad (3.10)$$

$$K_{a2} = \frac{C_{H^+} \cdot C_{S^{2-}}}{C_{HS^-}} \qquad (3.11)$$

The release to the atmosphere is thereby strongly dependent on the pH because only the molecular form and not the dissociated forms can be emitted. For example at a pH of about 7, an equal amount of H_2S and HS^- exists in the water phase. Increase of the pH will therefore at equilibrium conditions and at a constant total sulphide concentration reduce the hydrogen sulphide concentration in the overlying sewer atmosphere (Figure 3.6). Therefore, when applying Henry's law, equation 3.8, only the non-dissociated molecular form, H_2S, should be taken into account.

The non-dissociated H_2S (aq) form can be determined based on the equations 3.9-3.11. The sulphide ion, S^{2-}, only exists in measurable amounts above a pH of about 12. In the case of wastewater, only the equilibrium between H_2S (aq) and HS^- in equation 3.9 is therefore relevant and equation 3.12 determines the ratio between C_{HS^-} and $C_{H_2S}(aq)$ at the actual pH:

$$pH = pK_{s1} + \log \frac{C_{HS^-}}{C_{H_2S}(aq)} \qquad (3.12)$$

Where:

$pK_{s1} = 7.0$ at 25 °C.

The curves shown in Figure 3.7 are the combined result of Henry's law, equation 3.8, and the pH-dependent dissociation of H_2S (equation 3.12). Although the curves are describing equilibrium conditions, they are essential for the evaluation of the potential risk for odour problems. Both turbulence in the wastewater, e.g. caused by drops, and the degree of ventilation in the sewer are important for establishment of the equilibrium conditions shown in Figure 3.7. Release of H_2S into the atmosphere at e.g. pumping stations and hydraulic jumps may be rather high. Such conditions are described by Matos and de Sousa (1992) who have developed a model for prediction of hydrogen sulphide build-up in the atmosphere of a gravity sewer pipe.

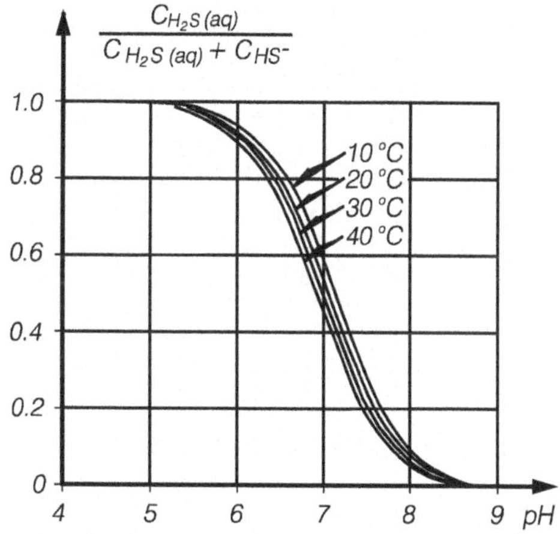

Figure 3.6. Equilibrium conditions for H_2S in aqueous solution (Melbourne and Metropolitan Board of Works 1989).

In addition to hydrogen sulphide, a number of other odorous compounds in wastewater, e.g. NH_3 and VFAs, exist in both a molecular and an ionic form. Conditions similar to those described for H_2S therefore exist for these compounds.

3.4.3 Water–air transport processes for odour compounds

From a theoretical point of view, the transport process of a volatile component across the water–air interface is depicted in Figure 3.8. The figure shows a concept of understanding that concentration gradients in both phases exist and that the total resistance for mass transfer is the sum of the resistance in each phase.

Although mass transfer across the water-air interface is difficult in terms of its application in a sewer system, it is important to understand the concept theoretically. The resistance to the transport of mass is expected mainly to reside in the thin water and gas layers located at the interface, i.e. the two films where the gradients are indicated (Figure 3.8). The resistance to the mass transfer in the interface itself is assumed to be negligible. From a theoretical point of view, equilibrium conditions therefore exist at the interface. Because of this conceptual understanding of the transport across the water–air boundary, the theory for the mass transport is often referred to as "the two-film theory" (Lewis and Whitman 1924).

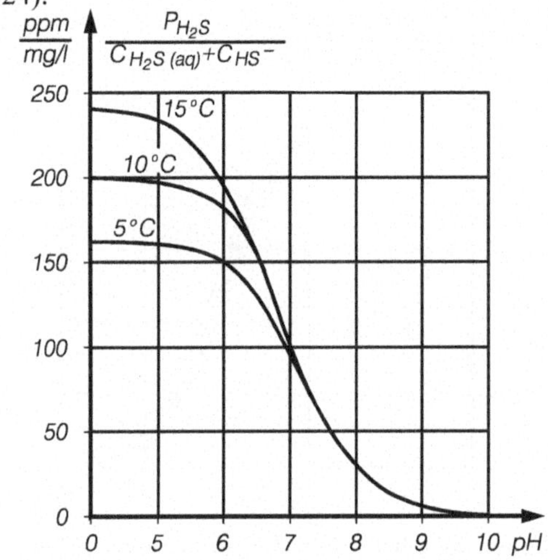

Figure 3.7. Partial pressure of H_2S measured in ppm on a volumetric basis in the atmosphere in equilibrium with a water phase of sulphide, cf. equations 3.8 and 3.9. The curves show the equilibrium partial pressure in the atmosphere per unit concentration in the water phase.

The two-film theory considering molecular diffusion through stagnant liquid and gas films is the traditional way of understanding mass transfer across the water-air boundary. Other theories exist, e.g. the surface renewal theory (Danckwerts 1951). However, the two-film theory gives an understanding of fundamental phenomena that may lead to simple empirical expressions for use in practice. Further details on water–gas mass transfer can be found in e.g. Thibodeaux (1979) and Stumm and Morgan (1981).

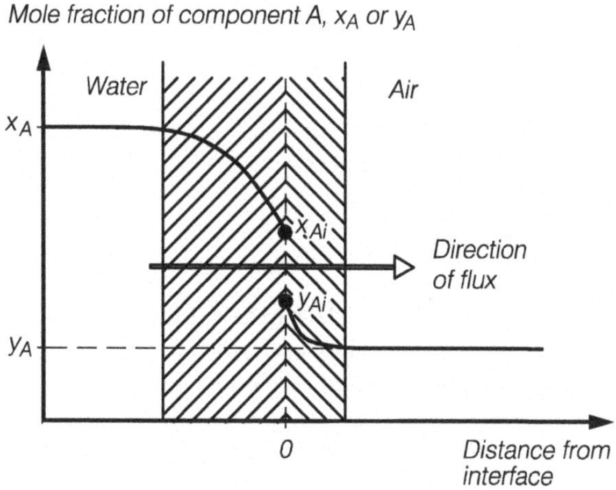

Figure 3.8. Fundamentals for the transport of a volatile component A across the water–air interface.

It is according to the two-film theory appropriate to consider the transport of the volatile components from the water phase to the air phase in two steps: From the bulk water phase to the interface and from the interface to the air. The driving force for the transfer of mass per unit surface area from the water phase to the interface and from the interface to the air phase is determined from the difference between the actual molar fractions, x_A and y_A, and the corresponding equilibrium values, x^*_A and y^*_A:

$$J_A = -k_{2A}(x^*_A - x_A) \tag{3.13}$$

$$J_A = k_{1A}(y^*_A - y_A) \tag{3.14}$$

Where:
J_A = flux rate of component A (moles/(total moles)/s/m^2),
k_{1A} = gas phase mass transfer coefficient (/s/m^2),
k_{2A} = water phase mass transfer coefficient (/s/m^2).

Which of the equations 3.13 or 3.14 is the most important depends on which part of the boundary has the major resistance to the mass transport. If, as an example, the major resistance exist in the water film of the boundary, i.e. $k_{2A} < k_{1A}$, equation 3.13 is the relevant description of the flux rate.

The two equilibrium values of the molar fraction, x^*_A and y^*_A, are fictitious, however, each determined from Henry's law, i.e.:

$$y^*_A = \frac{H_A}{P} x_A \tag{3.15}$$

$$y_A = \frac{H_A}{P} x^*_A \tag{3.16}$$

Each of the equations 3.15 and 3.16 can be substituted in either equation 3.13 or equation 3.14. In case of a dominating resistance in the water film, equation 3.13 is reformulated by substitution of equation 3.15:

$$J_A = k_{2A}\left(x_A - \frac{y_A}{H_A / P} \right) \tag{3.17}$$

Each of the mass transfer coefficients k_{1A} and k_{2A} can be interpreted as a molecular diffusion coefficient, D, divided by a film thickness, z, for the gas phase and the water phase, respectively, i.e. $k = D/z$. However, this interpretation has in practice no meaning because of the lack of knowledge on the thickness of the two films.

General expressions for the water–air mass transfer can be derived by solving the two equations, 3.13 and 3.14, for x^*_A and y^*_A, respectively, and substituting the results in each of the two equations 3.15 and 3.16. The following two expressions are thereby obtained:

$$J_A = \frac{k_{1A} \cdot k_{2A}}{\frac{k_{1A} \cdot P}{H_A} + k_{2A}} \left(-\frac{y_A \cdot P}{H_A} + x_A \right) = K_L \left(-\frac{y_A \cdot P}{H_A} + x_A \right) \tag{3.18}$$

$$J_A = \frac{k_{1A} \cdot k_{2A}}{k_{1A} + \frac{k_{2A} \cdot H_A}{P}} \left(x_A \cdot \frac{H_A}{P} - y_A \right) = K_G \left(x_A \cdot \frac{H_A}{P} - y_A \right) \tag{3.19}$$

Where:
K_G = overall mass transfer coefficient referring to the gas phase (/s/m^2),
K_L = overall mass transfer coefficient referring to the water phase (/s/m^2).

From the equations 3.18 and 3.19, the following two expressions are derived:

$$\frac{1}{K_L} = \frac{1}{k_{1A}} + \frac{P}{H_A \cdot k_{2A}} \tag{3.20}$$

$$\frac{1}{K_G} = \frac{1}{k_{2A}} + \frac{H_A}{P \cdot k_{1A}} \tag{3.21}$$

The two pairs of equations 3.18, 3.19 and 3.20, 3.21 are equally valid but typically the two corresponding equations 3.18 and 3.20 are applied.

Equation 3.20 expresses that the total resistance to mass transfer across the water-air boundary is equal to the sum of the resistances across the liquid film and the gas film. The importance of the magnitude of Henry's constant is in this respect evident. For high values of H_A, e.g. exemplified by O_2, the resistance mainly exist in the water film and turbulence in a sewer will therefore enhance the water-air transfer process. The importance of turbulence in the water phase is reduced for odorous components with a relatively low H_A value and turbulence in the air phase will correspondingly increase the release rate (Table 3.4). As seen from the equations 3.20 and 3.21, these facts also depend on the k_{1A}/k_{2A} ratio that varies according to system characteristics.

Liss and Slater (1974) have, based on the value of H_A, evaluated which type of mass transfer resistance exists. They propose the following criteria valid for most systems, cf. Table 3.4:

- Flow through the liquid film controls the mass transfer if $H_A > 250$ atm/(mole fraction).
- The resistance in both the water and the air film may be of importance if H_A is between 1 and 250 atm/(mole fraction).
- The flow conditions is controlled by the air film if $H_A < 1$ atm/(mole fraction). This situation corresponds not only to compounds with a relatively low volatility but also to compounds which are reactive in the water phase, e.g. like NH_3.

As can be seen from Table 3.5, all three situations are relevant for odorous compounds.

A major problem in the quantification of water–air transport phenomena in terms of the rate expression, equation 3.17, is to find appropriate values for K_L. As far as sewer systems are concerned, the most well established knowledge concerning water-air mass transfer is on the oxygen transfer (re-aeration). Based on theoretical considerations and empirical knowledge, a number of equations have been developed for determination of the re-aeration in pipes.

When considering sewer pipes, re-aeration is traditionally dealt with using an approach that, compared with equation 3.18 and 3.19, is formulated in different units:

$$F = K_L a\left(S_{OS} - S_O\right) = K_{LO_2}\left(S_{OS} - S_O\right) \tag{3.22}$$

Where:
F = rate of oxygen transfer (g/m^3/s),
$K_L a = K_{LO_2}$: overall oxygen transfer coefficient (/s),
S_{OS} = dissolved oxygen saturation (equilibrium) concentration of wastewater (g/m^3),
S_O = oxygen concentration in bulk water phase (g/m^3).

The overall oxygen transfer coefficient is defined as follows:

$$K_L a = K_L^{'} \cdot a = K_L^{'} \, A/V = K_L^{'} \, d_m^{-1} \tag{3.23}$$

Where:
$K_L^{'}$ = oxygen transfer velocity (m/s),
a = water-air surface area, A, to volume of water, V (/m),
d_m = hydraulic mean depth of the water phase (m).

Different empirical expressions have been developed for determination of K_{LO_2} (Krenkel and Orlob 1962; Parkhurst and Pomeroy 1972; Tsivoglou and Neal 1976; Taghizadeh-Nasser 1986). The following expression by Jensen and Hvitved-Jacobsen (1991) and Jensen (1994) is developed and validated under field conditions for prediction of reaeration in sewer pipes:

$$K_{LO_2} = 0.86\left(1 + 0.2\,F_r^2\right)\left(s\,u\right)^{\frac{3}{8}} d_m^{-1}\,1.024^{T-20} \tag{3.24}$$

Where:
F_r = u g$^{-0.5}$ d$_m^{-0.5}$, Froude number (-),
u =mean velocity of flow (m/s),
g = gravitational acceleration (m/s^2),
s = slope (m/m),
T = temperature (°C).

Approaches have been suggested for the determination of K_L values for odorous compounds based on knowledge on the molecular diffusion coefficient and the experience gained from air–water oxygen transfer in terms of K_{LO_2} values. It has been mentioned that the mass transfer coefficient, k, according to the two-film theory is equal to D/z for each of the two films. Contrary to this theory, the surface renewal theory implies that $k = D^{0.5}/z$. The value of n in the following expression 3.25 is therefore not well defined from a theoretical point of view.

$$\frac{K_L}{K_{LO_2}} = \left(\frac{D_L}{D_{LO_2}}\right)^n \tag{3.25}$$

Furthermore, the resistance to oxygen transfer across the air–water interface almost only exists in the water film. Therefore, equation 3.25 should only be applied to compounds that are comparable to oxygen, i.e. according to Liss and Slater (1974) have an H_A value greater than about 250 atm (mole fraction)$^{-1}$.

Although both theoretical and practical constraints exist for predicting the water–air mass transport of odorous compounds, the theoretical knowledge on the behaviour of these compounds is highly valuable. This knowledge can be applied when evaluating odour problems and when considering development of empirical equations.

3.5 PREDICTION OF HYDROGEN SULPHIDE IN SEWER NETWORKS

Thistlethwayte and Goleb (1972) made an important, however somewhat dubious statement when they concluded that although the H_2S concentration alone may not be a sufficient measure of potential sewer air odour levels, H_2S concentration measurements probably are sufficient for most studies of sewer gases. Their statement was based on rather limited measurements in sewers. However, it corresponds to the theoretical considerations concerning anaerobic microbial processes and investigations of these processes under sewer conditions that have been outlined.

It is, according to the existing theoretical and practical knowledge, not realistic to establish a general applicable model for prediction of the emitted odours from a sewer network. Even a more limited prediction of hydrogen sulphide emission from sewers is very difficult. The realistic approach is therefore to let hydrogen sulphide in the wastewater of a sewer network be an indicator of the potential risk for odour problems.

From a practical point of view concerning prediction of odours originating from sewers, these corresponding theoretical and practical facts are important. If hydrogen sulphide can be adopted as a convenient indicator for odour problems in sewers, the lack of possibility for modelling the formation of all relevant odorous components makes it interesting to consider a formation model for hydrogen sulphide in sewers as a substitute for odour prediction. Knowledge on hydrogen sulphide formation in sewers and its prediction is therefore important when considering odour problems related to wastewater collection.

A great number of processes and sinks for the sulphur cycle in a sewer affect the extent to which extent hydrogen sulphide is an odour problem. Figure 3.9 outlines the major pathways. Although not all aspects can be easily quantified, they should be included in an evaluation of odour problems associated with sewage transport.

Details related to the different phenomena and processes shown in Figure 3.9 will not be dealt with in this context although a brief description – especially related to the formation of sulphide in sewer networks – will be given in section 3.5.2. Further information is available in the literature, e.g. in terms of overviews in USEPA (1974), ASCE (1982), USEPA (1985), ASCE (1989), Melbourne and Metropolitan Board of Works (1989) and Hvitved-Jacobsen and Nielsen (2000).

3.5.1 Criteria for evaluation of odour problems

The hydrogen sulphide concentration level in the wastewater phase can be considered a first and relevant estimate of odour potential related to a specific sewer network. The actual hydrogen sulphide concentration in the atmosphere as a result of anaerobic processes in wastewater is of course a more correct, however also complicated, indicator of odour problems. Taking the existing possibilities for model simulation of water-air hydrogen sulphide transfer, sulphide oxidation and sewer ventilation into account, system and operational characteristics for a sewer must be extremely well known if the concentration in the air phase should be the basis for odour assessment.

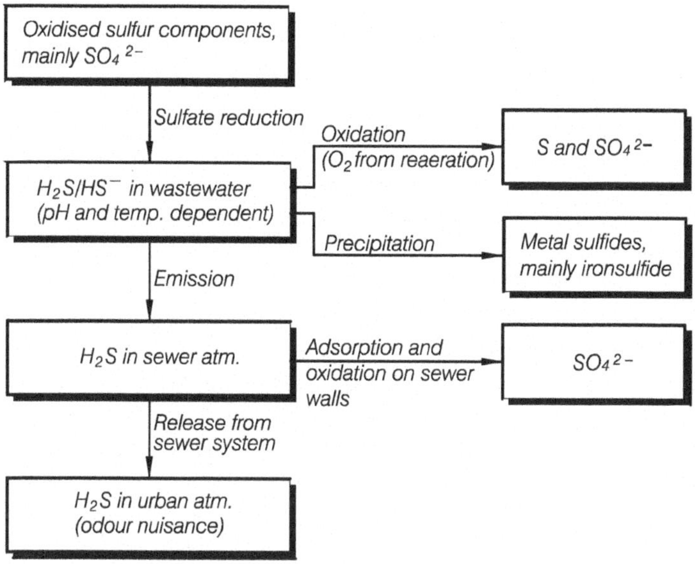

Figure 3.9. Main processes and sinks for the sulfur cycle in a sewer network associated with odour problems.

The partial pressure of H_2S on a volumetric basis in the atmosphere in equilibrium with a water phase of sulphide (H_2S + HS^-) is at a pH of 7 equal to about 100 ppm/(mg/l) (Figure 3.7). It is therefore clear that under equilibrium conditions very low sulphide concentrations in the wastewater will produce an unpleasant smell compared with the threshold odour value. This fact is also evident when considering the relatively high value of Henry's constant for H_2S, H_A = 563 atm (mole fraction)$^{-1}$ (Table 3.5). However, under real conditions such situations rarely exist and concentrations of hydrogen sulphide in wastewater of sewer systems should typically exceed 0.5 mgS/l before problems are identified. Sulphide concentrations of 0–0.5, 0.5–3 and 3–10 mgSl^{-1} may be considered as low, moderate and high, respectively, in terms of problems that are typically reported (Hvitved-Jacobsen and Nielsen 2000). Such values are especially important as criteria in case of model simulations for prediction of odour problems related to wastewater collection.

3.5.2 Factors affecting the formation and presence of hydrogen sulphide in sewer networks

Anaerobic conditions, i.e. the absence of DO and nitrate, are required for sulphate reduction. Under such conditions, the most important factors determining sulphate reduction rates will be outlined. These factors are important to consider when using models for sulphide prediction.

3.5.2.1 Presence of sulphate

Sulphate is typically found in all types of wastewater in concentrations greater than 5–15 mgS/l, i.e. in concentrations that are not limiting the sulphide formation rate in relatively thin biofilms (Nielsen and Hvitved-Jacobsen 1988). In sewer sediments, however, where sulphate may penetrate into the deeper sediment layers, the potential for sulphate reduction may increase with increasing sulphate concentration in the bulk water phase. Under specific conditions, e.g. in the case of industrial wastewater, it is important that sulphur components (e.g. thiosulphate and sulphite) other than sulphate may act as sulphur sources for sulphate reducing bacteria (Nielsen 1991).

3.5.2.2 Quantity and quality of biodegradable organic matter

Biodegradable organic matter is available in wastewater as a substrate for sulphate reduction. However, in wastewater from, for example, food industries with a relatively high concentration of readily biodegradable organics preferred by the sulphate-reducing bacteria, the sulphate reduction rate may be higher than in wastewater from households. However, also in domestic wastewater, COD may be high in certain areas owing to a shortage or reuse of water, leading to a higher potential for sulphide formation. Several specific organics, e.g. formate, lactate and ethanol, have been identified as particularly suitable substrates for sulphate reducing bacteria (Nielsen and Hvitved-Jacobsen 1988).

3.5.2.3 Temperature

The temperature dependency of sulphate reduction for sulphate reducing bacteria is high corresponding to a temperature coefficient of about 1.13 per degree Celsius, i.e. a change in the rate with a factor Q_{10} = 3.0–3.5 per 10 °C of temperature increase. Full scale studies have shown that the temperature coefficient will be reduced to about 1.03 (Nielsen *et al.* 1998).

3.5.2.4 pH

Sulphate-reducing bacteria mainly exist between pH 5.5 and 9. A significant inhibition of the sulphate-reducing bacteria will, however, not take place below a pH of about 10.

3.5.2.5 Area:volume ratio in pressure mains

Sulphide is primarily produced in the biofilms. The corresponding water phase concentration of the sulphide therefore relates to the area:volume ($A{:}V$) ratio of a sewer pipe. Relatively low sulphide concentrations in wastewater from large diameter pipes therefore exist compared with small diameter pipes.

3.5.2.6 Flow velocity in pressure mains

The potential production of sulphide depends on the biofilm thickness. If the flow velocity in the pipe is 0.8–1 m/s, the corresponding biofilm is rather thin, typically 100–300 μm. However, high velocities also reduce the thickness of the diffusional boundary layer and thereby the resistance against transport of substrates and products across the biofilm-water interface.

3.5.2.7 Anaerobic residence time in a sewer network

The anaerobic residence time of the wastewater during transport is a factor that affects the level of sulphide concentration in the wastewater. The residence time is determined by the magnitude of wastewater inflow compared with the water volume of the pipe. The level of sulphide formation – especially in a pressure pipe – is therefore subject to the diurnal variation of the inflowing wastewater and to the precipitation pattern in combined sewered catchments.

3.5.2.8 Sinks for hydrogen sulphide

As depicted in Figure 3.9, a number of sinks for hydrogen sulphide may reduce the actual concentration in the water phase. Emission to the sewer atmosphere (also affected by the sewer ventilation), oxidation in especially gravity sewers caused by the reaeration process and precipitation of heavy metal sulphides, mainly iron sulphide, are important sinks to consider.

3.5.3 Prediction of the formation of hydrogen sulphide using empirical equations

A number of simple empirical equations have been developed to predict sulphide formation in both gravity sewers and pressure mains. These equations will not be dealt with in this chapter. An overview of these models is given in Hvitved-Jacobsen and Nielsen (2000).

3.5.4 Integrated model for the formation of hydrogen sulphide

Until now only empirical models for the prediction of hydrogen sulphide in sewer networks have been available. The sewer process model approach published by Hvitved-Jacobsen *et al.* (1998a,b) and Hvitved-Jacobsen and Nielsen (2000) add a new dimension to combine anaerobic transformations of organic matter with the formation of sulphide (Figure 3.5). Although this concept is rather crude compared with the complexity of formation of odorous substances, it includes a link between the quality of the wastewater, the processes occurring and the formation of sulphide. The process concept developed is outlined in Table 3.5. Further details, e.g. concerning model formulation and process descriptions, may be found in Hvitved-Jacobsen *et al.* (1998b) and Hvitved-Jacobsen and Nielsen (2000).

Table 3.5. Integrated aerobic and anaerobic process model concept for transformations of organic matter and sulphur components of wastewater in sewers.

	S_F	S_A	X_{S_1}	X_{S_2}	X_{Bw}	S_{H_2S}	$-S_O$	Process rate*
Aerobic growth in bulk water	$-1/Y_{Hw}$				1		$(1-Y_{Hw})/Y_{Hw}$	Eq. a
Aerobic growth in biofilm	$-1/Y_{Hf}$				1		$(1-Y_{Hf})/Y_{Hf}$	Eq. b
Maintenance energy requirement	-1						1	Eq. c
Aerobic hydrolysis, fast	1		-1					Eq. d, n=1
Aerobic hydrolysis, slow	1			-1				Eq. d, n=2
Anaerobic hydrolysis, fast	1		-1					Eq. e, n=1
Anaerobic hydrolysis, slow	1			-1				Eq. e, n=2
Fermentation	-1	1						Eq. f
Hydrogen sulphide production						1		Eq. g
Reaeration							-1	Eq. h

* The formulation of the process rates is found in Hvitved-Jacobsen *et al.* (1998b) or Hvitved-Jacobsen and Nielsen (2000).

When considering the sulphide formation, the sewer process model is an attempt to include a conceptual understanding of relevant processes. Compared with the empirical hydrogen sulphide prediction models, there are also further advantages. One major advantage is that the model is designed to simulate changing aerobic and anaerobic conditions. The model is therefore able to simulate quality changes in a sewer network with arbitrary gravity and pressure pipe sections.

3.5.5 Odour formation modelling based on sewer processes

The sewer process model described and outlined in Figure 3.5 and Table 3.4 has been developed based on laboratory and pilot scale investigations. It has been validated in both gravity sewers and pressure pipes for its ability to simulate transformations of organic matter and sulphide formation in sewers (Tanaka *et al.* 1998; Tanaka and Hvitved-Jacobsen 2000). However, until now the model has not been used as a tool for evaluation of odour problems in sewer networks. For this purpose a number of case studies still remain to be performed.

Based on theoretical considerations, the sewer process model, however, possesses fundamental characteristics to predict odour formation in sewer networks. This statement is supported by experimental results produced as a background for the formulation and validation of the model. The following are major characteristics supporting this statement:

- Fermentable organisms and fermentation products are substrates for the sulphate respiring biomass; i.e. malodorous substances produced by fermentation may therefore appear simultaneous with H_2S. Often the production rates will be limited by fast biodegradable organic substrates produced from hydrolysis and fermentation.
- Hydrogen sulphide is a respiration product with a low threshold odour value around 0.5–1 ppb. This threshold value is of the same order of magnitude as many malodorous VOCs produced by fermentation.
- Hydrogen sulphide is a component with a relatively high Henry's law constant. It has therefore a high tendency to be emitted from the wastewater phase and occurs as a malodorous substance.
- The sewer process model proposed for odour modelling is designed from a conceptual point of view. It includes the quality transformation of wastewater organic matter under both aerobic and anaerobic conditions integrated with sulphide formation. Quality aspects of wastewater in terms of its biodegradability, which are considered crucial for the formation of all odours, are thereby included as a basis for sulphide formation.

- The sewer process model has the ability to simulate changing aerobic/anaerobic processes in both gravity sewers and pressure mains. The model is therefore applicable under varying and realistic conditions.

Because the sewer process model has a conceptual background, it possesses the ability to be used for design purposes. In this respect it is superior to existing purely empirical models for hydrogen sulphide prediction.

As a part of the evaluation process for the sewer process model, it is of major importance to assess which criteria must be put into operation to distinguish between different levels of odour problems. The criteria discussed in section 3.5.1 are in this respect considered sound and realistic.

3.6 EXAMPLE OF SIMULATIONS WITH THE SEWER PROCESS MODEL

The purpose of the following example is to give an impression of simulation results and performance of the sewer process model.

The model has been used for simulation of transformations of organic matter and the formation of sulphide in a 50 km sewer line to be implemented in the Emscher area, Germany. A complexity of problems has been the cause for changes of the existing sewer system. Odour and corrosion problems in the sewer catchment and improvements for treatment of the wastewater in terms of biological nitrogen and phosphorus removal at the subsequent treatment plant were focussed on.

Different scenarios of sewer systems, gravity sewers and pressure mains, have been compared. The geography of the Emscher area is relatively flat and allows only for the construction of a gravity sewer with a corresponding low slope. A gravity sewer of this type will be subject to aerobic/anaerobic changing conditions. As a part of the planning process, the possibility of simultaneous aerobic/anaerobic processes has been ascertained and quantified by model simulations. An example of simulation with the sewer process model is shown in Figures 3.10 and 3.11 (Hvitved-Jacobsen and Vollertsen 1998; Hvitved-Jacobsen *et al.* 1999).

The example shows that sulphide formation can be simulated under varying conditions and for different types of sewer systems. As previously mentioned the present lack of information on the ability of the model for simulation of odours is an important point to overcome.

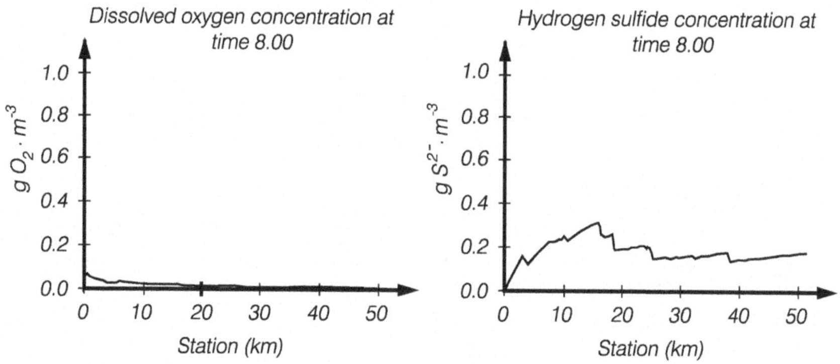

Figure 3.10. Results from simulations with the sewer process model for a 50 km gravity sewer pipe with a slope less than 0.13%. Variations are shown in the dissolved oxygen concentration (DO) and the hydrogen sulphide concentration at 8:00., cf. text.

3.7 CONTROL OF ODOURS FROM SEWERS

Methods and procedures for sulphide control in sanitary sewers are well established and a great number and varieties of control methods exist. These methods are generally well described from both a theoretical and a practical point of view in the literature. The principles of several of these methods are found in e.g. Thistlethwayte (1972), USEPA (1974), ASCE (1982), USEPA (1985), ASCE (1989), Melbourne and Metropolitan Board of Works (1989) and Boon (1995). Table 3.6 is in this respect just an overview of common control methods.

There is often no clear distinction whether odours in sewers originate from sulphate reduction or fermentation processes in the wastewater. In this respect it is important that not all the methods outlined in Table 3.6 are generally suitable for control of odours. As an example, chemical precipitation of sulphide may have no influence on the anaerobic production of VOCs. From a general and theoretical point of view the methods mentioned in Table 3.6 under point 1 are the most suited for control of septicity and thereby odour formation. However, it is important to note that the suitability of any control method should be assessed from a site-specific point of view.

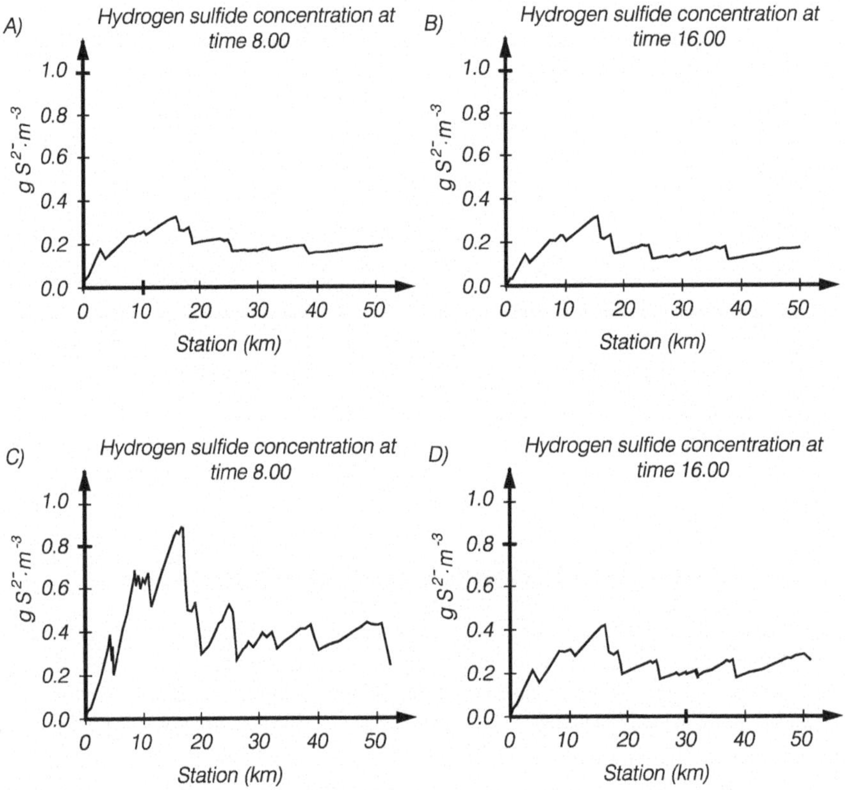

Figure 3.11. Simulated hydrogen sulphide profiles in a gravity sewer (A and B) and a pressure main (C and D), cf. text.

Table 3.6. Methods for control of sulphide in sewer systems.

General principle of the method	Specific measure
1. Prevention of sulphate reducing conditions	Addition to the wastewater of: - air - pure oxygen - nitrate
2. Prevention of adverse effects	Chemical precipitation of sulphides by: - iron (II) sulphate - iron (III) chloride
3. Methods aiming at specific effects on the biological system	- alkaline substances increasing pH - chlorine - hydrogen peroxide - ozone
4. Mechanical methods	- flushing - ball for detachment of biofilm
5. Other methods	- reduction of turbulence - protective coatings of corrosion-resistant materials - control of ventilation

3.8 REFERENCES

ASCE (1982) Gravity sanitary sewer design and construction, ASCE (American Society of Civil Engineers) manuals and reports on engineering practice **60** or WPCF (Water Pollution Control Federation) manual of practice **FD-5**, pp. 275.

ASCE (1989) Sulphide in wastewater collection and treatment systems, ASCE (American Society of Civil Engineers) manuals and reports on engineering practice **69**, pp 324.

Bjerre, H.L., Hvitved-Jacobsen, T. Schlegel, S. and Teichgräber, B. (1998) Biological activity of biofilm and sediment in the Emscher river. Germany, *Water Sci. Technol.* **37**(1), 9-16.

Boon, A.G. and. Lister, A.R (1975) Formation of sulphide in rising main sewers and its prevention by injection of oxygen. *Prog. Water Tech.* **7** (2), 289-300.

Boon, A.G. (1995) Septicity in sewers: causes, consequences and containment. *Water Sci. Technol.* **31**(7), 237-253.

Dague, R.R. (1972) Fundamentals of odor control. *J. Water Poll. Control Fed.* **44**, 583-595.

Danckwerts, P.V. (1951) Significance of liquid-film coefficient in gas adsorption. *Industrial and Engineering Chemistry* **43**(6), 1460.

Green, M., Shelef, G. and Messing, A. (1985) Using the sewerage system main conduits for biological treatment. *Water Res.* **19**(8), 1023-1028.

Hvitved-Jacobsen, T., Raunkjær, K. and Nielsen, P.H. (1995) Volatile fatty acids and sulphide in pressure mains, *Water Sci. Technol.* **31**(7), 169-179.

Hvitved-Jacobsen, T., Vollertsen, J. and Nielsen, P.H. (1998a) A process and model concept for microbial wastewater transformations in gravity sewers. *Water Sci. Technol.* **37**(1), 233-241.

Hvitved-Jacobsen, T., Vollertsen, J. and Tanaka, N. (1998b) Wastewater quality changes during transport in sewers - an integrated aerobic and anaerobic model concept for carbon and sulfur microbial transformations. *Water Sci. Technol.* **38**(10), 257-264 (read text pp. 249-256) or errata in *Water Sci. Technol.* **39**(2), 242-249.

Hvitved-Jacobsen, T. and Vollertsen, J. (1998) An intercepting sewer from Dortmund to Dinslaken, Germany, report submitted to the Emschergenossenschaft, Essen, Germany, pp. 35.

Hvitved-Jacobsen, T., Vollertsen, J. and Tanaka, N. (1999) An integrated aerobic/ anaerobic approach for prediction of sulphide formation in sewers. Proc. CIWEM and IAWQ joint International Conference on Control and Prevention of Odours in the Water Industry, London, September 22-24, 1999, 27-36.

Hvitved-Jacobsen, T. and Nielsen, P.H. (2000) Sulfur transformations during sewage transport. In: *Environmental Technologies to Treat Sulfur Pollution - principles and engineering* (P. Lens and L.H. Pol, eds.), IWA Publishing, London, pp. 131-151.

Hwang, Y., Matsuo, T., Hanaki, K., and Suzuki, N. (1995) Identification and quantification of sulfur and nitrogen containing odorous compounds in wastewater. *Water Res.* **29**(2), 711-718.

Jensen, N.Aa. and Hvitved-Jacobsen, T. (1991) Method for measurement of reaeration in gravity sewers using radio tracers. *J. Water Poll. Contr. Fed.* **63**(5), 758-767.

Jensen, N.Aa. (1994) Air-water oxygen transfer in gravity sewers. *Ph.D. dissertation*, Environmental Engineering Laboratory, Aalborg University, Denmark.

Koch, C.M. and Zandi, I. (1973) Use of pipelines as aerobic biological reactors. *J. Water Poll. Contr. Fed.* **45**, 2537-2548.

Krenkel, P.A. and Orlob, G.T. (1962) Turbulent diffusion and the reaeration coefficient, *J. Sanit. Eng. Div.* **88**(SA2), 53.

Lewis, W.K. and Whitman, W.G. (1924) Principles of gas adsorption. *Industrial and Engineering Chemistry* **16**(12), 1215.

Liss, P.S. and Slater, P.G. (1974) Flux of gases across the air-sea interface. *Nature* **247**, 181-184.

Matos, J.S. and de Sousa, E.R. (1992) The forecasting of hydrogen sulphide gas build-up in sewerage collection systems. *Water Sci. Technol.* **26**(3-4), 915-922.

Matos, J.S. and de Sousa, E.R. (1996) Prediction of dissolved oxygen concentration along sanitary sewers. *Water Sci. Technol.* **34**(5-6), 525-532.

Melbourne and Metropolitan Board of Works (1989) Hydrogen sulphide control manual - septicity, corrosion and odour control in sewerage systems, Technological Standing Committee on Hydrogen Sulphide Corrosion in Sewerage Works, vol. 1 and 2.

Nielsen, P.H. and Hvitved-Jacobsen, T. (1988) Effect of sulphate and organic matter on the hydrogen sulphide formation in biofilms of filled sanitary sewers. *J. Water Poll. Contr. Fed.* **60**, 627-634.

Nielsen P.H. (1991) Sulfur sources for hydrogen sulphide production in biofilm from sewer systems. *Wat. Sci. Tech.* **23**, 1265-1274.

Nielsen, P.H., Raunkjaer, K. and Hvitved-Jacobsen, T. (1998) Sulphide production and wastewater quality in pressure mains. *Water Sci. Technol.* **37** (1), 97-104.

Parkhurst, J.D. and Pomeroy, R.D. (1972) Oxygen Absorption in streams, *J. Sanit. Eng. Div.* **98**(SA1), 101.

Pomeroy, R.D. and Bowlus, F.D. (1946) Progress report on sulphide control research. *Sewage Works Journal* **18** (4).

Pomeroy, R.D. and Parkhurst, J.D. (1973) Self-purification in sewers, Advances in Water Pollution Research. Proc. 6th International conference, Pergamon Press.

Pomeroy, R.D. and Parkhurst, J.D. (1977) The forecasting of sulphide buildup rates in sewers. *Prog. Water Techn.* **9** (3), 621-628.

Raunkjær, K., Hvitved-Jacobsen, T. and Nielsen, P.H. (1994), Measurement of pools of protein, carbohydrate and lipid in domestic wastewater, *Water Res.* **28**(2), 251-262.

Sander, R. (2000), Henry's law Constants. In: *Chemistry WebBook*, (W.G. Mallard and P.J. Lindstrom (eds.), NIST Standard Reference Database Number 69, National Institute of Standards and Technology, USA, http:// webbook.nist.gov/chemistry.

Stanier, R.Y., Ingraham, J.L., Wheels, M.L. and Painter, P.R. (1986) *The Microbial World*, Prentice-Hall, Englewood Cliffs.

Stoyer, R.L (1970) The pressure pipe wastewater treatment system. Presented at the 2nd Annual Sanitary Engineering Research Laboratory Workshop on Wastewater Reclamation and Reuse, Tahoe City, CA, USA.

Stumm, W. and J.J. Morgan (1981) *Aquatic Chemistry: An introduction emphasizing chemical equilibria in natural waters.* John Wiley and Sons, New York.

Taghizadeh-Nasser, M. (1986) Gas-liquid mass transfer in sewers (in Swedish); Materieöverföring gas-vätska i avloppsledningar. Chalmers Tekniska Högskola, Göteborg, Publikation 3:86 (Licentiatuppsats).

Tanaka, N. (1998) Aerobic/anaerobic process transition and interactions in sewers. Ph.D. dissertation, Environmental Engineering Laboratory, Aalborg University, Denmark.

Tanaka, N., Hvitved-Jacobsen, T., Ochi, T. and Sato, N. (1998) Aerobic/anaerobic microbial wastewater transformations and reaeration in an air-injected pressure sewer. Proc. 71st Annual Water Environment Federation Conference & Exposition, WEFTEC'98, Orlando, Florida, USA, October 3-7, **2**, 853-864.

Tanaka, N. and Hvitved-Jacobsen, T. (1999) Anaerobic transformations of wastewater organic matter under sewer conditions. In: Proceedings of the 8th International Conference on Urban Storm Drainage (I.B. Joliffe and J.E. Ball, eds.), Sydney, Australia, August 30 - September 3, 1999, 288-296.

Tanaka, N. and Hvitved-Jacobsen, T. (2000) Sulphide production and wastewater quality - investigations in a pilot plant pressure sewer. Proc. 1st World Congress of the International Water Association (IWA), Paris, France, July 3-7, 2000, pp 8.

Thibodeaux, L.J. (1979) *Chemodynamics – Environmental Movement of Chemicals in Air, Water and Soil*, John Wiley & Sons, pp. 501.

Thistlethwayte, D.K.B. (ed.) (1972) *The Control of Sulphides in Sewerage Systems*, Butterworth, Sydney.

Thistlethwayte, D.K.B. and Goleb, E.E. (1972) The composition of sewer air. Proc. 6th International Conference on Water Pollution Research, Israel, June 1972, 281-289.

Tsivoglou, E.C. and Neal, L.A. (1976) Tracer measurement of reaeration. III Predicting the reaeration capacity of inland streams. *J. Water Pollut. Control Fed.* **48**(12), 2669.

USEPA (1974) Process design manual for sulphide control in sanitary sewerage systems, USEPA (US Environmental Protection Agency) Technology Transfer, Washington, D.C., USA.

USEPA (1985) Design manual – odor and corrosion control in sanitary sewerage systems and treatment plants, USEPA (US Environmental Protection Agency) publication **EPA 625/1-85/018**, Washington, D.C., USA.

Vincent, A. and Hobson, J. (1998) Odour control, CIWEM Monographs on Best Practices **2**, Terence Dalton, London.

4

Sources of odours in wastewater treatment

Alison J. Vincent

4.1 INTRODUCTION

Wastewater is a mixture of constituents from domestic and industrial sources, often diluted with groundwater from infiltration and run-off water where the system is partially combined. Fresh wastewater smells, as do all its products. The degree to which it smells and the extent to which the smell will cause a problem is dependent on the original components, the way in which the wastewater and its products are treated and handled and the extent to which they are exposed to the atmosphere and to potential complainants. This chapter looks at the compounds associated with wastewater odours, their source and the factors affecting their release.

edited by Richard Stuetz and Franz-Bernd Frechen. ISBN: 1 900222 46 9

4.1.1 Identified chemicals associated with wastewaters

Analysis of gases, using gas chromatography (GC) and other techniques, show a wide range of chemicals present at different stages of wastewater and sludge treatment and handling, including compounds not normally associated with odour problems (Van Langenhove *et al.* 1985; Zeman and Koch *et al.* 1983; Koe and Tan 1987).

The main groups detected are:

- A wide range of aliphatic, aromatic and chlorinated hydrocarbons, termed volatile organic compounds or VOCs,
- Hydrogen sulphide,
- Organic sulphur compounds,
- Aldehydes and ketones,
- Lower molecular weight fatty acids,
- Ammonia and amines.

The main odour causing chemicals are summarised in Table 1.1. Some of the compounds are hazardous to health and may have occupational exposure limits (Health and Safety Executive 1997). Hydrogen sulphide, the most common of the gases, is lethal at concentrations over about 300 ppm (v/v), a concentration commonly present below covers of enclosed processes at wastewater treatment works (World Health Organisation 1987).

The link between measurement of individual chemicals and odour, as measured by olfactometry, may be greater or lesser than the additive effect of the components of an odour as calculated from the threshold odour concentrations of individual chemicals. The strength of the link will depend on the number and type of components, the strength of the odour and the process stage from which the sample was taken (Laing *et al.* 1994; Patterson *et al.* 1993; Laska and Hudson 1991). The relation between a "marker compound" such as hydrogen sulphide and odour will also vary between process stage and site (Koe 1985). In the future, the development of an "electronic nose" technology may enable wastewater odours to be characterised and quantified (Stuetz *et al.* 1999).

4.2 SOURCE OF ODOURS IN WASTEWATER AND SLUDGE

The compounds contributing to the smell of wastewaters and their products come from the original components of the sewage, the biochemical changes that take place and to the chemicals that may be added as part of the treatment process.

4.2.1 Odours due to the original components of the wastewater

The smell of fresh wastewater results from its components, which consist of a mixture of discharges from toilets, baths, sinks, dishwashers and washing machines, and industrial wastes. Fresh wastewater rarely causes a major odour problem unless potential complainants are located very close to an open discharge point or to the air ventilation system of a pumping station or if there are significant odorous industrial discharges. The mixture of odorous chemicals typically contains:

- a wide range of aliphatic, aromatic and chlorinated hydrocarbons derived from cleaning agents used in the home (such as toluene, limonene, aromatic benzene derivatives, saturated aliphatic hydrocarbons C9 to C14, xylene, phenol),
- solvents (such as chlorinated hydrocarbons),
- petrol derivatives (such as benzene),
- odours associated with human waste such as urea and ammonia from urine and skatole and indole (breakdown products of tryptophan) from faeces.

Most volatile organic carbons (VOCs), originating from discharges of solvents or petrochemicals, have relatively low solubility and, therefore, are partially stripped from solution in the sewerage system, at pumping stations and during aeration. Some VOCs will be adsorbed onto primary sludge and may be released during subsequent mesophilic anaerobic digestion and other heated processes. The presence of hydrocarbons in ventilated air may have an impact on the required design of odour treatment plant; for example VOCs are adsorbed on activated carbon and may reduce its capacity.

Other components of the wastewater may also increase the potential for generation of odours in downstream treatment processes. Adverse discharges are:

- Strong, putrescible wastes from the food industry,
- Warm wastewater,
- Sulphate-containing wastewater or infiltrated ground water,
- Seawater infiltration (due to its high content of sulphate and the presence of essential trace nutrients),
- Toxic discharges.

4.2.2 The impact of biochemical changes during transport and treatment on odours

The majority of the chemicals associated with odour problems develop in wastewater and wastewater sludges when they have become anaerobic or septic, that is when all dissolved oxygen (DO) and nitrates have been used. The rate at which microorganisms consume DO is fairly constant until the concentration reaches about 0.2 to 0.4 mg/l. In wastewater the respiration rate of microorganisms is about 3 to 15 mg/l/h and the respiration of bacterial slimes on submerged walls is about 700 mg/m^2/h at 15 °C (Boon and Lister 1975*)*. As the DO becomes rate limiting, any oxidised nitrogen present in the wastewater will provide an alternative electron acceptor for the anoxic dissimilation of organic matter. Under such conditions, the microorganisms will continue to 'respire' and oxidise substrate but at a slower rate than the aerobic rate. The rate of respiration under anoxic conditions is approximately 40% of the aerobic respiration rate. These steps are accompanied by a decrease in the redox potential of the wastewater (Figure 4.1). Under anoxic conditions, the redox potential will decline from about +50 mV to about -100 mV (Eh).

In primary sludges, depletion of residual DO occurs very rapidly because the numbers of microorganisms in the sludge are several orders of magnitude higher than in wastewater and the availability of substrate per unit volume is also much greater. Surplus activated or humus sludges, derived from aerobic biological treatment processes, may contain DO and nitrate with only residual substrate and will become septic less rapidly unless co-settled with primary sludge.

Under increasingly anaerobic conditions, the following odour-producing biochemical reactions take place:

- Fermentation (hydrolysis, acidogenesis and proteolysis) of fats, polysaccharides and proteins leading to the production of fatty acids, alcohols, aldehydes, ketones, ammonia, amines, mercaptans and sulphides (section 4.2.2.1),
- Utilisation of sulphate as an electron receptor producing hydrogen sulphide (section 4.2.2.2).

Other biochemical reactions will reduce odours:

- Under aerobic and anoxic conditions malodorous compounds are oxidised (section 4.2.2.3),
- Within the anaerobic digestion process, methanogenic bacteria break down volatile fatty acids to produce odourless methane (section 4.2.2.4).

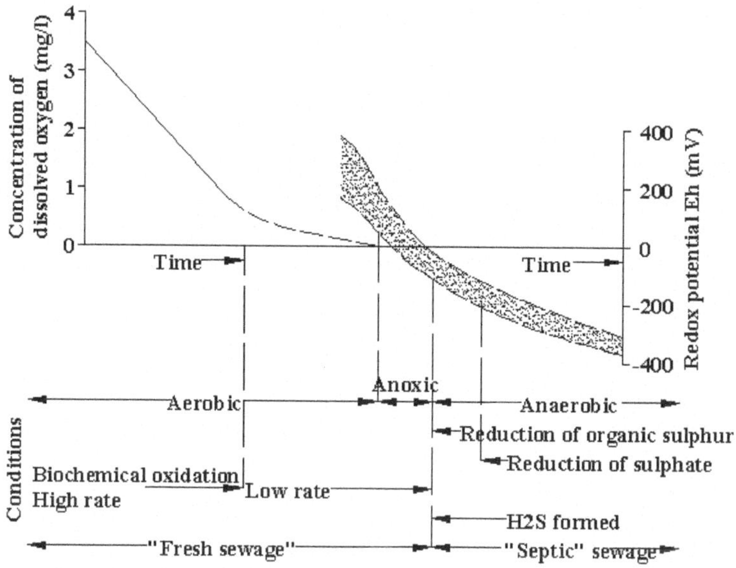

Figure 4.1. Variations in condition of sewage in relation to concentration of dissolved oxygen and redox potential (Boon 1995).

The extent to which all biochemical reactions take place will be affected by environmental factors including retention time, temperature, pH value, redox potential, concentrations of substrate and nutrients, the presence of toxic chemicals, salinity and the composition of wastewater or sludges (particularly the concentration of organic material and suspended solids).

These steps are described in more detail below.

4.2.2.1 Fermentation processes

Under anaerobic conditions, fermentation of fats, polysaccharides and proteins, occurs. In the fermentation process, these compounds are hydrolysed first to fatty acids, shorter chain saccharides, amino acids and peptides and then to shorter chain compounds. Within a heated anaerobic digester, fermentation is the 'acid forming' stage of the process and volatile fatty acids (VFAs) are rapidly converted to methane.

Hydrolysis of proteinaceous material (which contains the sulphur-containing amino acids cysteine, cystine and methionine) and organic sulphur compounds leads to the production of hydrogen sulphide and organic sulphides and disulphides. Domestic wastewater normally contains about 3 to 6 mg/l of organic sulphur in proteinaceous matter, and additional organic sulphur in the form of sulphonates (about 4 mg/l), derived from household detergents (Boon 1995).

The bacteria responsible for hydrolysis with the production of sulphides are anaerobic or facultative anaerobic species, for example *Proteus* spp., *Bacteroides* spp. and some *Clostridium* spp. (Crowther and Harkness 1975) and are active at a higher redox potential than those which subsequently reduce sulphate to form H_2S.

Many of the products of fermentation processes are volatile and odorous with low odour thresholds and include:

- ammonia,
- amines,
- alcohols,
- aldehydes,
- ketones,
- carbon dioxide,
- short chain VFAs such as butyric, propionic, lactic, acetic acid
- organic sulphides such as ethyl mercaptan (ethanethiol), dimethyl sulphide, methyl disulphide and hydrogen sulphide.

In crude or settled wastewater, the impact of the formation of fermentation products on odours is relatively low compared to the impact of hydrogen sulphide. However, fermentation products may be the main source of odours from stored sludges and sludge liquors resulting from thickening or dewatering of sludges. During storage of primary sludge, significant concentrations of VFAs and other fermentation products may develop, increasing with increasing retention time under anaerobic conditions. The production of VFAs in sludges also leads to a reduction in pH value with values as low as 5.5 found in septic sludges. Acid conditions enhance the release of malodorous organic sulphides as well as hydrogen sulphide. Fermentation will also occur when waste-activated or humus sludges are stored for prolonged periods under anaerobic conditions.

Examples of locations where fermentation processes take place are:

- sludge retained in primary sedimentation tanks,
- raw sludge storage,
- gravity thickening,
- anaerobic digestion,
- secondary (biological) sludge storage,
- VFAs generator associated with biological phosphorus removal.

4.2.2.2 Reduction of sulphate

The reduction of sulphates by sulphate-reducing bacteria (SRB) with the production of hydrogen sulphide is the most important of the odour-generating reactions because hydrogen sulphide is always present when there are odours due to septicity even when it is not the main cause of odour. Hydrogen sulphide generation is also a common cause of corrosion of concrete, iron and mortar, particularly when sulphide is biochemically oxidised to sulphuric acid. Because of this, the generation and control of hydrogen sulphide has been studied by numerous research workers (Boon and Lister 1975; Hvitved-Jacobsen *et al.* 1999) and several design guides for odour and corrosion control in sewerage systems have been written (Bowker *et al.* 1989; EPA 1974; Pomeroy 1990). An empirical equation for the estimation of sulphide generation in a rising-main sewer is given below (Boon and Lister 1975).

$$Cs = K_C\, t\, \mathrm{COD}[(1+0.004D)/D]1.07^{(T-20)} \quad (4.1)$$

Where:
Cs = concentration of sulphide (mg S/l)
K_C = constant, usually taken to be 0.00152
t = anaerobic retention time (minutes)
D = diameter of rising main (cm)
T = temperature of wastewater (°C)
COD = COD of wastewater (mg/l)

SRB are heterotrophic bacteria which 'respire' sulphate to provide energy for the dissimilation of organic matter and release sulphide into solution (Postgate 1959, 1984). They are strict anaerobes operating at a lower redox (below -200 mV) than fermentation processes, which occur at the same time. They grow at a slower rate than aerobic microorganisms. While oxygen or nitrate is present, SRB cannot function. However, they can survive adverse conditions of temperature, aerobicity, salinity and pressure, and are ready to become active whenever local conditions become anaerobic (Lens *et al.* 1995). SRB are present within wastewater, sludges, bacterial slimes on the submerged surfaces of sewers and holding-tanks, undisturbed sediments of sewers (Schmitt and Seyfried 1992), rivers and estuaries and in anaerobic digestion processes.

The total amount of sulphide that can be produced by SRB is limited by the initial sulphate content of the wastewater or sludge (present as inorganic sulphate and sulphonates) and the presence of nutrients and fermentation products (Hvitved-Jacobsen *et al.* 1999). The concentration of inorganic sulphate in wastewater can vary greatly from area to area, depending on the hardness of ground water and potable water supply, the method of potable water treatment and the composition of any industrial wastewater. In inland areas of the UK, concentrations generally exceed 10 mg/l (as S), with typical concentrations about 20mg/l (as S). Where there is seawater infiltration or sulphate-containing industrial wastewaters, concentrations may be much higher and very high concentrations of sulphide may develop during sludge storage or anaerobic digestion. Some sulphide will be naturally precipitated as insoluble sulphide by metal salts present in the wastewater or sludge. The addition of iron salts to precipitate sulphide has been used as an odour control technique.

Examples of locations where sulphate reduction takes place are:

- rising main sewers,
- sediments and slimes within tanks, grit channels, and wet wells,
- primary sedimentation tanks,
- slimes in high rate or overloaded biological filters or rotating biodiscs,
- storm or tidal storage tanks,
- sludge storage tanks,
- gravity thickeners,
- upflow anaerobic sludge blanket process,
- anaerobic digesters.

4.2.2.3 Action of bacteria under aerobic conditions

Under aerobic or anoxic conditions, bacteria oxidise organic matter. This will occur in wastewater, in slimes on wetted surfaces, in the mixed liquor or biofilm of aerobic secondary treatment processes or in sludges (Einarsen *et al.* 1999). Maintaining aerobic conditions will also have the effect of:

(1) inhibiting sulphate-reducing bacteria and
(2) chemically and biochemically oxidising malodorous chemicals previously formed under anaerobic conditions to less odorous compounds, including sulphuric acid, nitrates and carbon dioxide. The oxidation of sulphide in wastewater takes place at a rate of between 1 and 15 mg/l/h (Bowker *et al.* 1989), typically 2.5 mg/l/h. Oxidation will proceed much faster in biofilms and mixed liquor.

The oxidation reaction of hydrogen sulphide to sulphuric acid by autotrophic thiobacilli is utilised in the removal of odour by means of an aerobic biofilter or bioscrubber, where a bacterial support media such as shells, peat, coir, heather or a plastic media is used. The use of the activated sludge process as an odour 'bioscrubber' to directly treat odorous ventilation air by feeding it to the blowers has also been tried successfully at a number of sites.

Aeration of septic wastewater or mixed liquor, either by turbulence or using an aeration system, will also result in the release of VOCs and dissolved odorous gases. In a sewerage system this may cause odour problems at manholes, house connections and discharge points particularly if the wastewater from upstream systems is already septic. Hydrogen sulphide released to the sewer atmosphere and adsorbed onto slimes that grow on exposed damp surfaces out of direct contact with the flow of wastewater, will also be oxidised to sulphuric acid, which is a common cause of corrosion in sewers.

Examples of locations where aerobic, odour-reducing, processes take place are:

- gravity sewers,
- biological filters,
- activated-sludge plants,
- submerged biological aerated filters,
- sequencing batch reactors.

Examples of locations where anoxic processes take place are:

- rising main sewer with added or recirculated nitrates,
- aerated or anoxic selector zone upstream of an activated sludge plant.

4.2.2.4 The action of methanogenic bacteria

The methanogenic bacteria operate in the same redox range as the SRB. They act to convert VFAs to methane, and in doing so significantly reduce both the odour level and the obnoxious nature of the smell of raw sludges. SRB operate in competition with methanogens and biogas may contain significant concentrations of hydrogen sulphide as well as significant levels of other organic sulphides such as dimethyl disulphide (Winter and Duckham 2000), Table 4.1.

Table 4.1. Composition of biogas.

Compounds	Proportion in digester gas
Methane	65–70% (v/v)
Carbon dioxide	25–30% (v/v)
Hydrogen, water vapour, nitrogen and other gases	About 5% (v/v)
Hydrogen sulphide	0.03–0.3% (v/v)
Other volatile compounds including organic sulphides and VOCs	10–30 ppm (v/v)

Digested sludges still smell, particularly during or immediately after discharge from the digester, due to residual digestion activity or the emission of biogas entrained in the sludge. Odours will also be released during dewatering and during storage and disposal of sludge cake. Sludge liquors from dewatering of digested sludge will contain ammonia and other reduced nitrogenous compounds as well as hydrogen sulphide. Odour problems are often associated with start-up or difficulties with a digester. Loss of methanogenesis can result in the formation of high concentrations of VFAs and sulphides with a significant risk of creating odour problems.

Biogas from the digestion process is typically burned in boilers to provide heat for the digestion process, or in combined heat and power engines during which odours are thermally oxidised. Excess gas is burnt in flare stacks or burners. Thermal oxidation is time/temperature dependent and residual sulphide from flare stacks or boilers may cause problems where initial levels were very high or if temperatures are insufficiently high. Intermittent release of biogas from pressure-release valves may cause an odour problem.

4.2.3 Odours associated with chemicals used in the treatment process

Chemicals are routinely used in wastewater treatment for a range of purposes, including enhanced removal of suspended solids and BOD and for sludge conditioning and stabilisation. These may have an impact on odours, particularly on the release of odorous chemicals already present within the wastewater or sludge. For example:

- Addition of iron salts to wastewater for phosphate removal or to aid sedimentation of suspended solids and BOD removal will precipitate sulphides, reducing the potential for odour release. However, high dose rates of iron salts (such as ferric chloride) can result in a lowering of pH value, and consequent sulphide release.
- Addition of lime, to aid sedimentation or for the stabilisation of sludges, will inhibit the release of hydrogen sulphide, due to an increase in pH value, but it will increase significantly the release of ammonia and other odorous reduced nitrogenous compounds.

Other physico-chemical processes for wastewater treatment are potential sources of odour, in particular the use of lime and ammonia stripping. The use of membrane treatment has less impact on the potential generation of odours, other than that which may be associated with the storage of unstabilised wastewaters or sludges

Chemicals are used for odour control both as additives to wastewater and sludge and in odour treatment systems for vented air. There may be residual odours in treated air, for example residual chlorine from a hypochlorite scrubber, ozone from an ozone scrubber or the smell of peat from a peat-bed biofilter. These odours may be of importance when demonstrating compliance with an odour standard or if the discharge of treated air is close to receptors.

4.3 RELEASE OF ODOURS TO THE ATMOSPHERE

Odorous compounds have an impact only when they are transferred into the atmosphere and to a potential complainant. Odorous chemicals that remain dissolved and are subsequently oxidised chemically or biochemically while in solution will not create odour problems. The extent to which odours released from wastewater or its products will cause a problem depends on:

- the specific odorous chemicals that are released,
- the amount of odour released i.e. the flow rate of gas multiplied by its odour content. Thus, a very strong odour (e.g. from a grit skip) may have a limited impact compared with a high flow-rate of less-contaminated air (e.g. at the weirs of a primary sedimentation tank),
- the volume of air in which the odour is dispersed,
- the proximity of receptors, who may become 'sensitised' to an odour, detecting it at lower concentration than the 'non-sensitised' receptor,
- the frequency, duration and time of day of odour release, for example: occasional removal of stored sludge from a storage tank, regular release of strong odours from sludge dewatering operations and continuous release of low concentrations of odour from discharge of septic wastewater. In general, the most distress is caused by regular or continual release of odours.

4.3.1 Factors affecting release of odours

Gas laws describe the amount of an odorous gas transferred from liquid to gas phases. The rate of transfer is described by mass transfer theory:

$$r_v = K_L a\,(C\text{-}C^*)V \qquad (4.2)$$

Where:
r_v = rate of volatilisation (mg/h)
K_L= mass transfer coefficient (m/h)

a = specific interfacial area for mass transfer (1/m)
C = concentration of volatile compound in water (mg/m^3)
C^* = concentration of volatile compound in water in equilibrium with the gas phase (mg/m^3)

Henry's Law describes the maximum concentration in the gas phase:

$$P = Hp \tag{4.3}$$

Where:
P = mole fraction in gas phase
p = mole fraction of dissolved gas in liquid phase
H =Henry's constant (483 atm/mol fraction at 20 °C for hydrogen sulphide) (Tchobanoglous and Burton 1991)

The factors affecting the amount of odorous gases released are therefore:

- The solubility of the dissolved gases.
- The concentration of the compound in the gas and liquid phases
- The overall volumetric mass-transfer coefficient (K_La), which combines mass transfer coefficient and interfacial area. The rate of release at points of turbulence is very much greater than from quiescent surfaces.
- Temperature: the solubility decreases and rate of transfer increases with increasing temperature .
- pH value which affects the concentration of dissolved gas as only the un-ionised form of odorous compounds are available for transfer to the atmosphere. Low pH values favour the emission of H_2S (Table 4.2), mercaptans and volatile fatty acids, while high pH values favour the emission of ammonia and reduced nitrogenous compounds. Ammonia is about 100% un-ionised at pH 11. Changes in pH values have no effect on non-polar compounds.

Table 4.2. Proportions of H_2S and HS^- in dissolved sulphide (Pomeroy 1990).

pH value	Proportion of un-ionised H_2S	Proportion of HS^-
5	0.99	0.01
6	0.91	0.09
7	0.50	0.50
8	0.09	0.91
9	0.01	0.99

Concentrations of H_2S in the gas phase rarely reach the equilibrium values predicted by Henry's Law due to the dynamic nature of most systems (Table

4.3). However, in enclosed systems they may rapidly exceed concentrations that are a risk to health, particularly under the covers of enclosed sludge storage tanks.

Table 4.3. Examples of typical liquid and equivalent gas phase concentrations of hydrogen sulphide in sewerage systems (Cranny 1994; Matos and Aires 1994).

Sulphide concentration in sewage (mg/l)	Maximum predicted concentration of H_2S in sewer atmosphere (20 °C, pH 7.0) ppm (v/v)	Measured concentration of H_2S in sewer atmosphere ppm (v/v)
Hunters Green Pumping station		
10	1357	225
11	1493	185
15	2036	200
9	1222	40
Costa do Estoril		
10	1357	300
1.5	204	60

Above open quiescent surfaces, such as lagoons, the movement of air above the surface lowers the concentration of odorous chemicals, providing a positive driving force for the progressive emission of odours. The amount of odour released in this case may be proportional to the wind speed.

The bulk of odour release occurs at points of turbulence such as weirs and discharge points or where wastewater/sludge is mixed or aerated vigorously. The movement of air can provide a positive driving force with almost complete stripping of odorous gases possible. These locations are also at greatest risk of corrosion to concrete or iron work. A significant reduction in odour release can be achieved by reducing the height of drops over weirs and into tanks and channels or by selective covering at these locations.

4.3.2 Quantification of odour release and its impact

The value of odour emission equals the flow-rate of vented air multiplied by the odour content measured using olfactometry or a marker such as hydrogen sulphide concentration. This can be measured easily on a vent stack and less easily on area sources.

Tests to measure the potential for odour release of liquors at different stages in the treatment process have been described by a number of workers (Koe and Tan 1990; Frechen 2000a; Hobson 1995). In this test, a known volume of air is recirculated through a given volume of liquor and the odour content of the off-gas is measured by olfactometry. Examples of odour potentials are given in Table 4.4 (Hobson 1995) which show the impact of activated sludge treatment and digestion on the reduction in odours. An indication of the amount of odour release from certain processes can be made from a 'mass balance' or by using empirical equations. Yang and Hobson (1998) use the value of odour potential in an equation for odour release at weirs.

$$OE = 7.16\ 10^{-4}\ OP\ F_{weir}\ hK_{pH} \qquad (4.4)$$

Where:
OE = odour emission rate per unit length of weir (ou/sm)
OP = odour potential of the liquid flowing over the weir (ou/m^3)
F_{weir} = weir loading rate (m^2/h)
h = height of drop of liquid flow at weirs (m)
K_{pH} = pH correction coefficient, takes a value of 1.17 at pH 7

Table 4.4. Examples of odour potential (ou/m^3).

Odour source	Odour potential (ou/m^3)
Raw wastewater sludge	100,000 - >2,500,000
Digested wastewater sludge fresh from digester	300,000
after storage	10,000
Digested sludge filtrate	2000
Septic wastewater from rising main	1000,000
Mixed liquor	620
Storm tank (sub surface)	305,650
Gravity thickener overflow (maximum)	4,000,000
Oxidation ditch selector tank	2000

Hydrogen sulphide can be also be used to indicate the potential for odour problems. Sulphide has been measured along the process stream at problem sites in the UK to identify locations of odour generation and release (Table 4.6).

In these examples, odours at Site A were due to generation and release of odours at the primary settlement stage. No problems were associated with sludge handling or digestion due to containment of sludge and an effective gas handling system. Odours at Site B were generated in primary sludges and released at the aerated sludge-holding tank. Odours at Site C were present in sludges and released during mechanical dewatering.

For greenfield sites, values of odour emission or predictive equations can be used to derive odour emission rates for use in dispersion modelling (Boon *et al.* 1998; Yang and Hobson 1998; Witherspoon *et al.* 2000). Equations describing the release of VOCs are given in the literature (Melcer *et al.* 1994) and are available in commercial computer models such as TOXCHEM. Examples of emission rates from area sources are given by Frechen (1992, 2000b).

Table 4.6. Sulphide concentrations in wastewater and sludges (mg/l as S).

	Site A	Site B	Site C
Inlet	0.03	0	0.06
Primary tank before weir	4.4	-	-
Primary tank after weir	1.1	0.5	0.11
Mixed liquor	0.2	0.3	-
Final effluent	0	-	-
Primary/co-settled sludge	50.1	16	48
Sludge pump sump	-	14	-
Sludge storage tank	-	2 (aerated)	38
Thickened sludge	57.6	-	-
Sludge liquors	-	-	7
Digested sludge	23.3	-	-

The impact of a measured or estimated odour emission can be determined using dispersion modelling. A rough estimation of the effect of an odour emission can be made using an empirical equation (Keddie 1980), which relates flow-rate and odour content of the gas to the radius from an odour source in which complaints may be expected.

$$\mathrm{OR} = (2.2E)^{0.6} \tag{4.5}$$

Where:
OR = odour complaints radius (m) and
E = odour emission rate = flow rate of air (m^3/s^1). odour content (ou/m^3)

The measure of uncertainty is given as the range $(0.7E)^{0.6}$ to $(7E)^{0.6}$

Examples of locations where there may be significant release of odours are:

- discharge point of rising main sewers,
- primary tank weirs,
- primary tank sludge bellmouths,
- free drops of sludge into open holding tanks or over weirs,
- mechanical sludge thickening and dewatering plant,
- discharge points of septage or imported sludges,
- discharge point of sludge liquors.

4.4 DESIGN TO MINIMISE ODOUR PROBLEMS ASSOCIATED WITH WASTEWATER TREATMENT PROCESSES

Odour control techniques look at both the prevention of generation of odours and the minimisation of their release. Often it is not possible to prevent septicity but in many cases release can be reduced. Process selection can significantly affect the likely risk of odours at a site. An extended aeration plant, treating crude wastewater from a gravity sewerage system with sludges tankered from site would not be expected to require odour control provision. In comparison, a works treating pumped wastewater with primary sedimentation, high-rate biological filtration, imported sludge and sludge treatment facilities would be expected to require significant odour control provision. However, almost any stage of treatment or handling of wastewater or sludge can cause odour problems with odours generated in one process being released at a number of downstream stages.

A considerable reduction in the risk of odour problems can be achieved by design (Vincent and Hobson 1998). The principles for successful design are:

- minimise odour generation, for example by minimising the time of storage of wastewater and sludges under anaerobic conditions,
- minimise odour release, for example by avoiding turbulent flow or large hydraulic drops (e.g. over weirs or into storage tanks) of crude or settled wastewater, sludges or sludge liquors,
- minimise the effect of odour release by location of odorous stages away from sensitive site boundaries.

4.4.1 Potential for odour release at different process stages

The potential for odour generation and release at common stages in the wastewater treatment process and methods of minimising potential problems are briefly described below.

4.4.1.1 Sewerage systems

In rising main sewers respiration of wastewater and slimes rapidly depletes any dissolved oxygen or nitrates. Sulphate reduction, together with fermentation, takes place within the body of wastewater and on the slimes present on the submerged sewer walls. Odour release occurs at the discharge point and may cause a problem if the receiving sewer is relatively small diameter or if there are house connections close to the discharge point at the sewerage system or downstream pumping station or wastewater treatment plant (WWTP).

In gravity sewers the velocity of wastewater generated by the slope should ensure that the rate at which oxygen dissolves from the atmosphere in the sewer exceeds the respiration rate of the microorganisms in the wastewater and in slimes on the submerged surfaces. Downstream of a rising main sewer, re-aeration of the wastewater may allow oxidation of dissolved sulphides. The re-aeration rates in gravity sewers are reviewed by Boon (1995). Odour generation may occur in gravity sewers if the slope is shallow and sediments can accumulate. Sewer gas from gravity sewers will vent at the WWTP.

Design to minimise odours should:

- minimise the length of pumped sewers
- ensure adequate slopes in gravity sewers
- avoid hydraulic drops or sharp bends in gravity sewers
- ensure that odours cannot escape outside the sewerage system possibly by sealing manholes
- use chemical addition (e.g. nitrate, oxygen or iron) to prevent or precipitate sulphides if a problem develops

4.4.1.2 Wastewater pumping stations

Odours including hydrogen sulphide and VOCs may be present from upstream sources and released at channels, screens and sumps. These may pose a health and safety hazard and adequate ventilation should be available if personnel require entry. Pumping stations close to houses may have a risk of causing an odour problem. Design to minimise odours should:

- reduce the height of hydraulic drops into sumps,
- avoid turbulence of flow in channels,
- minimise operational volume of sumps
- provide sufficient slopes in sumps so that there is no accumulation of sediments or rags,
- enable removal of fat and grease.

4.4.1.3 Storm storage

Storm storage may be provided within the sewerage system or at the inlet to a WWTP. Odours are generated during the storage of storm sewage, sludges and accumulated debris. Odour emission occurs during filling and emptying, particularly if vigorous mixing or jet-cleaning systems are used. There will also be odours released at the discharge point as it is returned to the main flow of sewage. Design to minimise odours should:

- minimise retention of storm sewage,
- keep tank clean of debris,
- keep inlet and discharge points at a low level to minimise splashing.

4.4.1.4 Inlet works

Raw wastewater inlet channels can be a source of an odour problem, particularly if receiving septic sewage, returned storm sewage, imported sludges, sludge liquors or septic tanks wastes. Gravity sewers will also vent into the atmosphere at the inlet of the treatment works. Odours can be released from the discharge points, channels, screening and grit removal (particularly aerated grit channels). Screenings and grit will smell during storage and transfer, particularly if not washed after separation. Design to minimise odours should:

- avoid accumulation of grit,
- provide screenings washing,
- avoid hydraulic drops or sharp bends in gravity sewers,
- minimise height of discharge points, if possible odorous discharges should be below water level.

4.4.1.5 Primary sedimentation tanks

Sewage entering the primary sedimentation stage may already be septic. Some increase in septicity in sewage and sludges during sedimentation is unavoidable except where chemical additives are used. Co-settlement of primary and surplus activated sludge can increase the speed at which the sludge deteriorates. Odour release occurs at the stilling chamber, the top-water horizontal surface, and mostly at the settled sewage overflow weir and channels, desludging chambers and bell-mouth discharges. Design to minimise odours should:

- allow tanks to be removed from operation at times of low flow to maintain retention close to design values,
- minimise the height of drops over effluent weirs,
- provide close coupled automatic desludging,
- remove sludges at a low concentration (about 2%) to avoid excessive retention.

4.4.1.6 Aerated secondary treatment processes

Odours are removed from wastewater by adsorption of anaerobic compounds onto the sludge floc and biochemical oxidation. However, the aeration system will strip odours from the mixed liquor and the off-gases have a characteristic odour, with higher rate plants having a greater odour emission than low rate, nitrifying plants. Greater stripping of odours with a mechanically aerated plant than a fine bubble diffused air plant has been reported (Melcer et al, 1994). Odours may be released at the inlet of the aeration tank if the incoming sewage is septic or there is a discharge of sludge liquors. Odour emissions are significantly less than those from the primary settlement stage. Design to minimise odours should:

- ensure adequate aeration and mixing,
- use diffused air aeration rather than mechanical surface aeration if the site is sensitive with respect to odours,
- discharge sludge or other odorous liquors below water level.

4.4.1.7 Biological filters

Biological filters remove odours in wastewater by adsorption of anaerobic compounds onto the biofilm and biochemical oxidation. Low rate, nitrifying plants have lower potential odours than higher rate treatment plants. Odours may be stripped from the wastewater at the surface of the filter. The natural ventilation of the filter, which draws air up through the filter when the temperature of the wastewater is greater than ambient temperature, can exacerbate this. Biological filters can be a source of odour generation if overloaded, affected by toxic discharges, or if media have deteriorated and areas of ponding occur. High rate filters are often a source of odours as anaerobic conditions can develop within the thick slimes that develop at high loading rates. Any odours developing within the filter are then emitted in the ventilation air. Design to minimise odours should:

- provide covers, forced ventilation and odour control equipment for high rate filters, with ventilation drawn through the filter from the top to the bottom,
- minimise the drop between distributor and media surface, avoid spray distributors,
- use recirculation if there are signs of ponding,
- ensure adequate ventilation.

4.4.1.8 Sludge and septage import and export facilities

Odorous air will be displaced and vented from tankers and at the discharge point during emptying and filling operations. Odour will also be released if sludges in the tanker or the tank are mixed before pumping. Design to minimise odours should:

- ensure discharges are at low level or close coupled,
- cover the reception chamber or tank,
- if there is a problem, tanker vents can be connected to odour treatment units,
- locate tanker discharge point away from sensitive site boundaries.

4.4.1.9 Raw and co-settled sludge storage and gravity thickening

The amount of hydrogen sulphide and fermentation products generated will increase significantly with time of storage. The strength of sludge liquors will also increase markedly with time. Strong odours can be emitted during filling and draining of tanks, from the surface and weirs of full tanks, during mixing, and during subsequent treatment of sludges and liquors. Design to minimise odours should:

- minimise sludge storage capacity prior to thickening, digestion and dewatering stages to minimise odour generation in sludges and liquors,
- provide covers with venting to odour treatment (N.B. toxic levels of hydrogen sulphide will develop below covers),
- discharge sludges and sludge liquors at low level to avoid splashing,
- mix at low, rather than high, speed.

4.4.1.10 Mechanical thickening and dewatering

Major emission of odour from sludges and sludge liquors can occur during thickening and dewatering. The intensity of the odour will depend on the length of time that the sludge has been retained in primary tanks and sludge storage tanks before the thickening or dewatering stage. As mechanical thickening is generally carried out in enclosed equipment or an enclosed building, ventilation and odour control is often provided. Failure of mechanical thickening plant can increase the risk of odour problems downstream e.g. if it results in storage of raw sludges for long periods with consequent high emissions of odour when sludge thickening restarts. Design to minimise odours should:

- ensure more than adequate capacity is provided so that raw sludge does not ‘back up’ in the system,

- ensure that the raw sludge is as fresh as possible before dewatering,
- contain and treat odours released at the equipment and subsequent discharges to storage facilities,
- discharge sludges and sludge liquors at low level to avoid splashing.

4.4.1.11 Anaerobic digestion

Odours may be released at overflow weirs and discharges to secondary digester tanks. Problems can occur at start up and during operational difficulties with the digester. Some odours from biogas may be released from pressure relief valves and from waste gas burners if there is a delay in ignition. Residual odours from biogas may be present after flaring and in the vent from boiler or CHP unit. Design to minimise odours should:

- ensure waste gas burners operate reliably,
- minimise drop of digested sludge into secondary digestion tanks,
- iron salt addition may be used if hydrogen sulphide levels are very high and are causing problems.

4.4.1.12 Thermal treatment processes and sludge drying

Odour release from thermal treatment of wastewater and sludges occurs because of the volatilisation of compounds in the liquid phase due to the increase in temperature, and also because breakdown of cells releases more organic matter including ammonia into the liquid phase. The degree to which the products of the thermal treatment process (liquid and gas phase) cause problems depends on the degree of containment of gas and liquid phases of the particular proprietary system and the subsequent handling and treatment of the thermally treated liquors.

4.4.2 Most common causes of odour problems

The main causes of odour problems identified at 26 sewerage and WWTPs are listed in Table 4.7 (Vincent 1998). Thirteen of these were greenfield sites, six had problems due to commissioning difficulties and three had problems that had developed due to encroachment by housing.

Table 4.7. Main causes of odour problems at 26 sites.

Perceived cause of odour problem	Number
Rising mains	12
Gravity sewerage	2
Industrial waste	4
Storm/balancing tanks or sumps	4
Primary settlement tanks	10
High rate filters	1
Sludge thickening/dewatering	13
Tankering/sludge export	3
Sludge digestion (heated and cold stages)	6
Total	55

4.5 REFERENCES

Bonnin, C., Laborie, A. and Paillard, H. (1990) Odour nuisances created by sludge treatment: problems and solutions. *Water Sci. Technol.* **22** (12), 65-74.

Boon, A.G. (1995) Septicity in sewers: causes, consequences and containment. *Water Sci. Technol.* **31** (7), 237-253.

Boon, A.G., Vincent, A.J., and Boon, K.G. (1998) Avoiding the Problems of Septic Wastewater. *Water Sci. Technol.* **37** (1) 223-231.

Boon, A.G. and Lister, A.R. (1975) Formation of sulphide in a rising-main sewer and its prevention by injection of oxygen. *Prog. Wat. Technol.* **7** (2), 289-300

Bowker. D. G. Bowker, J. M. Smith and Webster, N. A. (1989) *Odour and Corrosion Control in Sanitary Sewerage Systems and Treatment Plants.* Hemisphere Publishing Corporation.

Cranny, P. (1994) Stripping and volatilisation in wastewater facilities. *Proc. Water Environment Federation Speciality Conference Series*, Jacksonville.

Crowther, R.F. and Harkness, N. (1975) Anaerobic bacteria. In: *Ecological Aspects of Used Water Treatment* Volume 1, The Organisms and their Ecology (C.R. Curds and H.A. Hawkes, eds.) Academic Press, London.

Einarsen, A.M. Æsøy, A., Rasmussen, A-I., Bungum, S. and Sveberg, M. (1999) Biological Prevention and removal of hydrogen sulphide in Sludge at Lillehammer Wastewaster Treatment Plant. *Water Sci. Technol.* 41 (6), 175-182.

EPA (1974) *Process Design Manual for Sulfide Control in Sanitary Sewerage Systems*, EPA-625/1-74-005.

Frechen, F-B. (1992) Odor emissions of large WWTPs: source strength measurement, atmospheric dispersion calculation, emission prognosis, countermeasures - case studies. *Water Sci. Tech.* **25** (4-5), 375-382.

Frechen, F.-B, (2000a) Sampling methods for odour analysis *Proc. International Meeting on Odour Measurement and Modelling, Odour 1*, Cranfield University.

Frechen, F-B. (2000b) Overview of olfactometric emission measurements at wastewater treatment plants, *IWA Specialist Group on Odours and Volatile Emissions* Newsletter No 3 (September).

Health and Safety Executive (1997) *Occupational Exposure Limits* EH40.

Hobson, J. (1995) The odour potential: a new tool for odour management, *J.CIWEM* **9**, 458-463.

Hvitved-Jacobsen, T., Vollertson, J., and Tanaka, N. (1999) An integrated aerobic/anaerobic approach for prediction of sulfide formation in sewers. *Water Sci Technol.* **41** (6), 107-116.

Keddie, A.W.C. (1980) Dispersion of odours. In: *Odour Control - A Concise Guide.* Published by Warren Spring Laboratory for Department of the Environment. pp 93-107.

Koe, L.C.C. (1985) Hydrogen sulphide odor in sewer atmospheres. *Water, Air Soil Pollution*, **24**, 297-306.

Koe, L.C.C. and Tan, YG (1987) GC-MS analysis of odorous emissions from the dissolved air flotation units treating surplus activated sludge at a wastewater treatment works. *Inter.J. Environ. Stud*ies **30**, 37-44.

Koe, L.C.C. and Tan, N.C., (1990) Odour generation potential of wastewaters, *Water Res.* 24 (12), 1453-1458.

Laing, D.G., Eddy, A. and Best, D.J. (1994) Perceptual characteristics of binary, trinary and quaternary mixtures and their components. *Physiology and Behaviour* **56** 81-93.

Laska, M. and Hudson, R. (1991) A comparison of the detection thresholds of odour mixtures and their components. *Chem. Senses* **16** 651-662

Lens, P.N., De Poorter, M.-P., Cronenberg, C.C., and Verstraete, W.H. (1995) Sulfate reducing and Methane Producing bacteria in Aerobic Wastewater Treatment Systems. *Water Res.* 29 (3), 871-880.

Matos, J.S. and Aires, A.M. (1994) Mathematical modelling of sulphides and hydrogen sulphide gas build-up in the Costa do Estoril Sewerage System. *Proc. IAWQ Specialised International Conference, 'The sewer as a Physical, Chemical and Biological Reactor*, May 16-18.

Melcer, H., Bell, J.P., Corsi, R.L., MacGillivray, B. and Child P. (1994) Stripping and volatilisation in wastewater facilities. *Proc. Water Environment Federation Speciality Conference Series.* Jacksonville.

Patterson, M.Q., Stevens, J.C., Cain, W.S., and Commeto-Muniz, J.E., (1993) Detection thresholds for an olfactometry mixture and its three constituent compounds. *Chem. Senses* **18** 723 -734.

Pomeroy, R.D. (1990) *The Problem of Hydrogen Sulphide in Sewers*, 2nd Ed. Clay Pipe Development Association Limited, London.

Postgate, J.R., (1959), Sulphate reduction by bacteria. *A. Rev. Microbiol.* **13**, 505 -520.

Postgate, J.R. (1984) *The Sulphate-reducing Bacteria.* Cambridge University Press.

Schmitt, F., and Seyfried, C.F. (1992) Sulfate reduction in sewer sediments. *Water Sci. Tech.* **25** (8), 83-90.

Stuetz R.M., Fenner, RA, and Engin G. (1999) Assessment of odours from wastewater treatment works by an Electronic Nose, H_2S analysis and olfactometry. *Water Res.* 33 (2), 453-461.

Tchobanoglous, G. and Burton, F.L. (1991) *Wastewater Engineering: Treatment, Disposal and Reuse,* Metcalf and Eddy Inc., McGraw-Hill Inc., New York.

Van Langenhove, H., Roelstraete, K., Schampp, N. and Houtmeyers, J. (1985). GC-MS identification of odorous volatiles in wastewater. *Water Res.* **19** 597-603.

Vincent, A. and Hobson, J. (1998) Odour Control. *CIWEM Monographs on Best Practice*, No. 2, Terence Dalton Publishing, London.

Vincent, A.J. (1998) The Management of odours. paper presented to CIWEM, East Midlands Branch, November.

Winter, P. and Duckham, S.C. (2000) Analysis of volatile odour compounds in digested wastewater sludge and aged wastewater sludge cake. *Water Sci. Technol.* 41 (6), 73-80.

Witherspoon, J.R., Sidu, A., Castleberry, J., Coleman, L., Reynolds, K., Card, T. and Daipper, G.T. (2000) Odor emission estimates using models and sampling, odour dispersion modelling and control strategies for east bay municipal utility district's (EBMUD's) collection sewerage system and wastewater treatment plant. *Water Sci. Technol.* 41 (6), 65-71.

World Health Organisation (1987) *Air Quality guidelines for Europe*. WHO Regional Publications Series No. 23, Regional Office for Europe Copenhagen.

Yang, G., and Hobson, J. (1998) Validation of the wastewater treatment odour production (STOP) model. *Proc. 2nd CIWEM National Conference on Odour Control in Wastewater Treatment*, London.

Zeman, A. and Koch, K. (1983) Mass spectrometric analysis of malodorous air pollutants from wastewater plants. *Internat. J. Mass Spectrometry and Ion Physics* **48**, 291-294.

Part III

ODOUR SAMPLING AND MEASUREMENT

5

Sampling techniques for odour measurement

John Jiang and Ralph Kaye

5.1 INTRODUCTION

The use of dynamic olfactometry can provide the basis of an effective and comprehensive approach to establishing odour strength and odour intensity levels of complex odours. When coupled with odour dispersion modelling, dynamic olfactometry can provide a particularly useful basis for odour impact assessment. The use of dispersion modelling for odour impact assessment requires the acquisition of sound and reliable source data.

In the past there have been difficulties with scientifically sound quantification of odours by olfactometry. Even today there are those who still believe the science of odour measurement is a "black art". No doubt many that hold this opinion have had experience with early attempts at the sensory evaluation of odours using earlier olfactometry techniques. Fortunately in recent

© 2001 IWA Publishing. *Odours in Wastewater Treatment: Measurement, Modelling and Control* edited by Richard Stuetz and Franz-Bernd Frechen. ISBN: 1 900222 46 9

years the science of olfactometry has advanced greatly, resulting in improved accuracy and greater repeatability of results (Wenzel 1948; Hangartner *et al.* 1985). Much of the development of olfactometry has occurred in Europe where high population density has resulted in odour generated by intensive agricultural operations, which has severely impacted on residential amenity.

The design of instruments and materials of construction used in modern olfactometry, differ greatly from those used a decade or so ago. The use of calibrated olfactometers and screened panels has greatly improved the reliability of odour concentration measurements (Jiang 1996). Most importantly, olfactometric techniques have been standardised. The draft European standard prEN 17325 (CEN 1997) is a performance based standard and defines the unit of odour measurement in terms of a butanol reference material. The forthcoming Australia and New Zealand standard (DR 99306) is based on the European standard and applies identical performance-based criteria.

Similarly, there have been developments in odour sampling techniques that have improved the science. It is now possible to transport odour samples to a laboratory without significantly affecting odour concentration or odour intensity during transport. Furthermore, the draft European olfactometry standard, referred to above, also specifies, at least in a general way, techniques for emission sampling.

5.1.1 Emission source types

The draft European olfactometry standard characterises emissions as point sources, area sources with outward flow, and area sources without an outward flow. Volume sources, which are especially important for intensive agricultural industries, are not discussed.

The sampling of an environmental parameter such as odour concentration is an extremely complex task for environmental engineers and scientists. Adequate representative samples can be taken only if the professional personnel undertaking sampling fully understand odour generation processes.

A number of industries produce odours with different emission characteristics during various phases of their operations. Some examples of such industries are listed in Table 5.1.

Table 5.1 describes odour-related industries together with some typical characterisations of emissions from the odour generation processes.

Table 5.1. Odour emissions from various industries.

Industry	Characterisation of odour generation
Sewage treatment	Large open liquid surfaces with no outflow (eg. primary sedimentation tanks, process tanks) Large open liquid surfaces with outflow (eg. aeration tanks, biofilters) Large open solid surfaces with no outflow (eg. dewatered sludge stockpiles) Point sources (eg. vent stacks, scrubbers) Volume sources (eg. sludge dewatering buildings)
Intensive agriculture (piggery, poultry, cattle feedlot, abattoirs)	Volume sources (eg. animal housing) Large open solid surfaces no outflow (eg. manure pads in open pens) Large open liquid surfaces no outflow (eg. effluent lagoons)
Food processing (biscuit, pet food)	Point sources (eg. stacks) Volume sources (eg. buildings)
Composting	Point sources (eg. scrubbers for indoor processes) Area sources with outflow (eg. biofilters to treat emissions from indoor processes, outdoor compost piles)
Industrial operations (petroleum processing plants)	Point sources (eg. stacks) Volume sources (eg. buildings) Area sources (eg. effluent treatment lagoons)
Others (fast food shops)	Point sources (eg. stacks) Volume sources (eg. building)

5.1.2 Sampling error

While the draft European standard recognises that emission rate measurements from area sources are sensitive to the characteristics of the sampling apparatus used and the selection of sampling conditions, these are not specified. The selection of inappropriate sampling apparatus and insufficient attention to sampling conditions will cause substantial errors. Such sampling errors can overshadow the errors that could potentially occur in subsequent olfactometric testing.

The possibility of introducing errors through inappropriate procedures during sample collection and transportation should be considered at an early stage in the execution of the project. The possible situations that may introduce some degree of sample error are:

- Rinsing sampling bags: Sample bags may absorb some odorous compounds and this may result in lower than expected odour concentrations. However, in some cases excess rinsing with the emission to be sampled may actually increase apparent odour concentratiions by causing extra amounts of some odorous compounds to be adsorbed onto the bag walls. These compounds may be re-released during testing. This would result in higher than expected odour concentrations.
- Sample storage materials: While Nalophan (polyterephthalic ester co-polymer) is recommended as a satisfactory storage material in the draft European standard, little research has been reported on the adsorption, diffusion, and chemical transformation characteristics of this material. A preliminary investigation on the use of Nalophan indicated that odour concentration measurements apparently increased and decreased substantially during storage times that are still within the limits allowed by the draft standard. Maximum and minimum values were observed to vary by a factor of more than three times (Pollock 2000).
- Pre-dilution: For high humidity (> 90%) and high temperature (> 50 °C) source sampling, pre-dilution should be used to prevent condensation occurring in the storage bag which may otherwise result in some degree of sample loss. Furthermore, high strength samples may require pre-dilution to prevent subsequent contamination of olfactometers. However, pre-dilution can certainly introduce some errors and should not be used unnecessarily.
- Storage time: Prolonged storage time may cause some sample losses (Schuetzle *et al.* 1975). Tedlar bags have demonstrated excellent performance in preservation of air samples (Pau *et al.* 1991) and have been recommended by the US EPA for air toxic sampling.

5.2 ODOUR IMPACT ASSESSMENT AND SAMPLING PROGRAM DESIGN

The design of an odour impact study comprises the following components: the establishment of objectives, site inspection, scheduling the sampling programs, source condition measurement, sample collection, result calculation.

5.2.1 Objectives

Before an odour impact study is implemented, it is important that the objectives of the study are fully understood. Typical main objectives are:

- Measure odour emission rates from all potential sources as primary input data for odour dispersion modelling.
- Rank odour emission sources at the facility as a step towards the preparation of an odour control strategy.
- Evaluate the efficiency of existing and potential odour control abatement technology.
- Prepare an odour emission inventory.
- Evaluate compliance of odour emission rates with company environmental management plans and external regulatory requirements.

5.2.2 Site inspection

Prior to preparing a sampling program, it is necessary for a field inspection to be carried out to determine the potential odour emission sources at the facility. Factors to be noted include:

- Operational conditions. Any changes during the processes should be discussed with on-site personnel.
- Locations of odour emissions and odour sampling points. If necessary, any extra equipment and preparation must be arranged.
- Conditions of odour emission sources. For an air temperature above 50 °C or an air relative humidity above 90%, pre-dilution of the sample is required. In such situations, a thermometer and a pitot tube should be inserted into the stack to measure air temperature and air velocity respectively. Relative humidity should be measured using a relative humidity sensor.
- Accessibility of sampling points. If required, a working platform should be built to ensure that sampling is undertaken in a safe manner. For further details, see Annex I in the European draft standard for odour measurement.
- Toxicity safety requirements. Extra care should be taken when toxic compounds are present at the sampled source. Toxic compounds should be identified and likely concentrations quantified.
- Availability of power and water. If required, a portable power generator can be used.

5.2.3 Scheduling sampling program

The sampling of an environmental parameter, such as odour concentration, is an extremely complex task for environmental engineers and scientists. Commonly, odour emissions, such as sewage lagoons and animal housing, are intrinsically heterogeneous, both spatially and temporally. Consequently, the design of an odour impact study is just as critical as the olfactometry analysis to the achievement of a rational representation of the real situation. Particular care is needed in obtaining data under highly fluctuating ambient wind conditions and at very low pollutant concentrations, sometimes at ppt (parts per trillion or 10^{-12}) levels.

Adequate representative samples can be taken only if the professional personnel undertaking sampling fully understand odour generation processes. Controllable operational factors, such as mechanical ventilation rates, as well as uncontrollable factors, such as weather conditions that may affect odour generation, should be considered before the sampling program is implemented. The first step is to identify, with the assistance of on-site personnel, the process units at the facility that generate and emit significant odour. In most cases, the odour samples should be taken in worst case conditions.

The number and location of sample points and the frequency, duration and averaging time of sampling, should reflect the temporal and spatial pattern of the particular facility being studied. It must also be kept in mind that increasing the duration and number of samples will increase the cost of a study. In some situations, it may be appropriate to take a composite sample across several sampling points, to represent an average value. For reasonably homogeneous sources, such as a continuously emitting stack, at least two samples are needed to represent the odour emission. In some cases, further samples are needed to represent the odour emission pattern. All samples must be referenced in terms of location and time.

It is particularly important that the sampling program is appropriate to the testing program adopted. The design of the odour sampling program should consider the time elapsed between sampling and testing. For odour testing, the time constraint is for sample collection, transport and testing to be carried out within 30 hours. Consequently, the sampling programme must be coordinated in advance with the selection and convening of olfactometric testing panels. It is preferred for the sampling and testing to be done on the same day. There are other considerations such as the number of samples, limitations of the sample preservation methods, geographic location of the facility site, geographic location of the testing site and the method used to transport samples.

The sampling program should include details of sampling sources, locations, frequency and any special sampling requirements such as pre-dilution. For some studies, the definition of the odour impact area, calculated using an air dispersion model or based on complaint records, will be required.

5.2.4 Economic and practical considerations

Controllable operational factors, such as mechanical ventilation rates for intensive animal production sheds, as well as uncontrollable factors, such as weather conditions, may affect odour generation and should be considered before the sampling program is implemented. Ideally, sampling should be undertaken over the range of operating conditions that are known to occur in practice.

The number and location of sample points and the frequency, duration and averaging time of sampling should reflect the temporal and spatial pattern of the particular facility being studied. However, the availability of resources, such as laboratory capacity and manpower, may impose unavoidable constraints on sampling programs. For example, in some situations, it may be necessary to represent average emission values for large area sources using composite samples across several sampling points rather than using a larger number of discrete samples.

5.3 SAMPLE COLLECTION – GENERAL PRINCIPLES

Odour samples are collected in the field using special purpose atmospheric sampling bags. The air sampling bags may be filled by either a "direct" or an "indirect" technique. For "direct" sampling the bag is filled under pressure by pumping the sample air into the bag. Because of the risk of contamination, the direct sampling approach is seldom appropriate and indirect sampling generally recommended. In indirect sampling, the bag is placed in a sealed vessel. The vessel is connected to the suction of an air pump. Sample air is drawn into the bag by reducing the pressure inside the vessel. The sampling vessel may be equipped with a clear polycarbonate lid, or a window to enable the filling of the bag to be observed during sampling. Figure 5.1 shows a recommended arrangement of apparatus for indirect sampling.

5.3.1 Materials

Only impervious materials such as Teflon™, Tedlar™, stainless steel and glass can be used in odour sampling processes that may come into contact with odorous samples. While PTFE or FEP type Teflon™ can be used as sampling lines, it is best to use FEP as it is translucent and any unexpected dust, condensation, or entrained moisture can be readily seen. Stainless steel can be used for fittings. The use of other materials such brass and rubber in fittings should be avoided as they may generate their own slight odour or react chemically with the odorants.

Clean sampling tubing should be used for each sample. Sampling lines and fittings can be reused if they have first been cleaned and rendered odourless. In cleaning sampling lines and fittings, all residues from previous samples must be removed by rinsing with clean (hot) water and dried using odour-free air in the odour testing laboratory.

Commercially sourced Tedlar™ and FEP Teflon™ atmospheric sampling bags are quite expensive and the usual practice has been to reuse bags. Bags can sometimes be cleaned for reuse by repeated flushing with odour free air in the olfactometry laboratory. This is a very labour intensive process that is not always successful. At best, bags can be reused up to 10 times and labour costs per reuse cycle are about 20 % of the cost of a new bag.

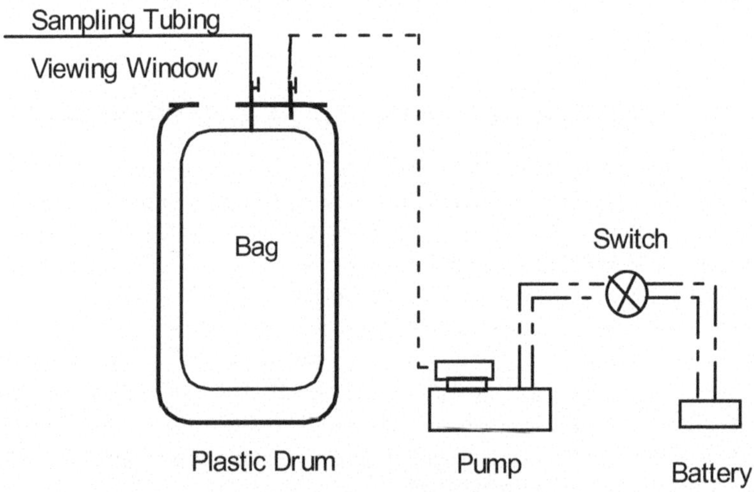

Figure 5.1. Arrangement of apparatus for indirect sampling.

5.3.2 Sample bags

Quality of construction is very important. Nothing is worse than spending a day collecting samples, only to find that the bags have collapsed prior to testing. The sample bag is a critical component that must conform to the following criteria:

- Odour free
- Does not adsorb odours or react with the odorous samples

- Sufficiently impervious to prevent any significant loss of odour components between the time of collection and the time of measurement
- Reasonably robust
- Leak-free
- Equipped with leak-free fittings which are compatible with other sampling equipment and with the olfactometer
- Sufficient volumetric capacity to enable a full test series to be completed

Sample losses from a sample bag may occur through adsorption of odorous compounds on to the bag wall, permeation through the plastic wall, condensation (where there is a temperature gradient between the sample and the ambient air), and photo-catalysed reactions between odorous gases. Figure 5.2 shows the effect of bag material on ethylbenzene recovery (Schuetzle *et al.* 1975). Tedlar™ bags have demonstrated an excellent performance in preservation of odour samples (Pau *et al.* 1991) and have been recommended by the US EPA for air toxic sampling.

From the above figure, Tedlar™ and FEP Tedlar™ are shown to be appropriate materials for sample bags. In practice, Tedlar™ is preferred because it is less fragile. However, problems have been encountered in the past with indigenous odour in commercially sourced atmospheric sampling bags and apparently, there can be "bad batches". In practice, all odour bag materials can have inherent odour caused by surplus solvents used in manufacture. Consequently, the levels of residual odour in all bags, new or unused, should be checked to determine that they are sufficiently low so as not to interfere with odour measurements. Before use, new, commercially sourced sample bags should be filled in the laboratory with odour free air and left for several hours to be checked for indigenous odour by olfactometry as required.

Because of the inherent problems with reusing sampling bags there has been a recent trend to single use Nalophan™ NA bags. Nalophan™ NA is a low cost material that is listed in the draft European standard for this purpose. However, little literature is available on the performance of Nalophan™ NA and there is some controversy regarding its use. As discussed above, there is Australian evidence suggesting that measurement anomalies may be caused by the use of Nalophan™ NA. Furthermore, samples of this material have been observed to have a slight odour.

In practice, Tedlar™ sheet is available in bulk rolls and the material can be heat-sealed. Fabrication costs for single use bags can be less than for reuse of commercially sourced atmospheric sampling bags. Samples of the Tedlar™ sheet material from the intended batches should be made up into bags and tested by olfactometry to confirm that indigenous odour levels are sufficiently low so as not to interfere with odour measurements.

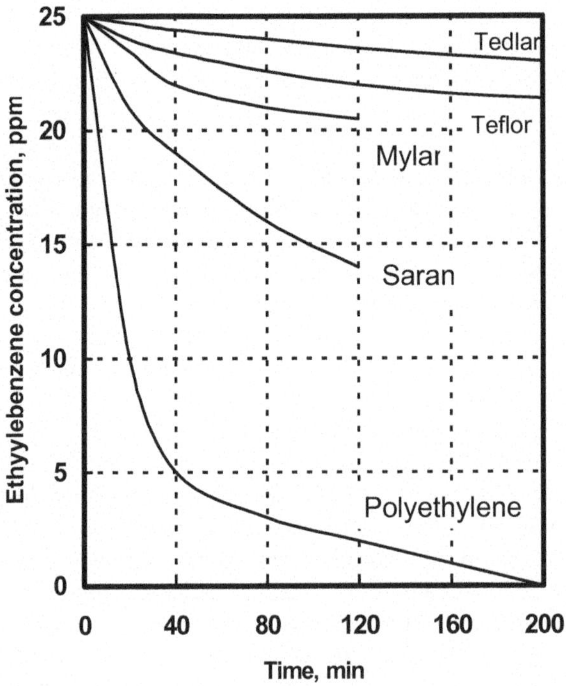

Figure 5.2. The effect of bag material on ethylbenzene recovery (Schuetzle *et al.* 1975).

5.3.3 Documentation

Sound documentation and quality control / quality assurance procedures should be strictly adhered to. It is important that the details of the sampling source characteristics including geometric dimensions, temperature, humidity, and gas velocity are recorded during the sampling. It is advised that the pre-printed forms should be used for all sample collections. A separate form should be used for each sample. Table 5.2 lists some of details that are required for each source type.

Table 5.2. Details to be recorded for odour samples.

Source	Point	Area	Volume
Client name	Yes	Yes	Yes
Job No	Yes	Yes	Yes
Sequence number	Yes	Yes	Yes
Client contact	Yes	Yes	Yes
Location	Yes	Yes	Yes
Source identification	Yes	Yes	Yes
Date	Yes	Yes	Yes
Time	Yes	Yes	Yes
Stack dimensions	Yes	-	-
Gas velocity in stack	Yes	-	-
Temperature in stack	Yes	-	-
Humidity in stack	Yes	-	-
Air velocity at the exit of wind tunnel of static sampling hood	-	Yes	-
Air temperature at exit of wind tunnel or static sampling hood	-	Yes	-
Humidity in at exit of wind tunnel or static sampling hood	-	Yes	-
Weather	Yes	Yes	Yes
Wind direction	Yes	Yes	Yes
Ambient temperature	Yes	Yes	Yes
Wind velocity	Yes	Yes	Yes
Technician signature	Yes	Yes	Yes

5.4 SAMPLE COLLECTION FROM POINT SOURCES

Typically a point source will be a stack with a known flow rate such as a discharge stack from abattoir or a vent from a processing building. Ventilation ducts that extend from buildings should generally be sampled from outside the building. Occasionally, the erection of scaffolding or the provision of a "cherry picker" lift is required to obtain safe access.

Gaseous samples should be collected from air streams with known gas flow rates or measurable air velocities and cross sectional areas. It is relatively easy to measure an odour emission rate from a point source. Samples are taken through clean Teflon tubing probes inserted into the stack or duct at the required sampling plane and the flow rate is calculated as the product of the air velocity and the cross sectional area.

5.4.1 Measurement of flow rates

Flow rate is critical in calculating odour emission rates. The accuracy of air velocity measurements greatly affects the reliability of results. Much emphasis is placed on the quality of odour concentration measurement and the accuracy of flow rate measurement requires equivalent attention. Accurate velocity measurement ideally requires measurement of a grid series of point velocities across the stack cross section. A simple way to achieve this is to divide the cross sectional area into a number of small equal sub-areas for rectangular duct or a number of annuli for a circular duct as per ISO 9096 (Figure 5.3). As a rule of thumb, a minimum of 4 measurement points for an rectangular area of up to 0.18 m^2 and 8 measurement points for a circular duct of up 0.25 m^2 are required.

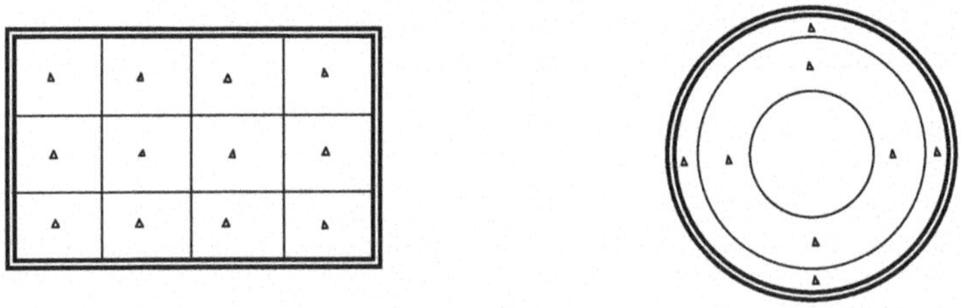

Figure 5.3. Points for flow rate measurement in rectangular and circular ducts.

The measurement plane selected should be at least two diameters upstream and eight diameters downstream of any flow disturbance. If such criteria cannot be met, the number of sampling points should be increased.

5.4.2 Selection of sampling points

Isokinetic sampling procedures are generally not required for odour sampling. However, the numbers and positions of points required for isokinetic sampling are the same as for the characterisation of the average velocity in a duct as discussed above. These are inferred to be the numbers and positions of measuring points required for collecting a composite sample from a duct.

More details concerning velocity measurement and sampling point location procedures for stacks and ducts may be adapted from: ASTM D 3464–75, "Standard Test Method for Average Velocity in a Duct Using a Thermal Anemometer"; Australian Standard AS 4323.1–1995, "Stationary source emissions, Method 1: Selection of sampling positions" ISO 10780 and ISO 9096.

5.4.3 Pre-dilution

Pre-dilution of the sample should be undertaken where samples are to be collected directly from combustion processes, where the air temperature exceeds 50 °C, where the relative humidity exceeds 90%, where the sample may be otherwise saturated with water vapour, or where the sample has an extremely high odour concentration. Pre-dilution is used to prevent condensation in the sample bag and to reduce odour concentration to a level suitable for olfactometry. Pre-dilution may be undertaken either dynamically, using an ejector, or statically, by metering an appropriate quantity of clean, dry, odour free air (or bottled nitrogen) into the sample bag prior to sample collection. It is necessary to employ a technique to measure accurately the volume of pre-dilution gas and sample collected.

5.5 SAMPLE COLLECTION FROM AREA SOURCES

Typically an area source will be a liquid or solid surface such as a primary sedimentation tank at a sewage treatment plant or a sludge stockpile. A sample collection enclosure, eg. a portable wind tunnel system, can be used for sampling to determine specific odour emission rate (SOER).

5.5.1 Wind tunnel systems – sources without outward flow

Wind tunnel systems are used to sample odour emissions from area sources such as primary sedimentation tanks at sewage treatment plants. Early wind tunnel devices such as the Lindvall hood (Lindvall 1970) and Lockyer (1984) wind tunnel system were developed to compare odour emissions from area sources under different conditions. Since 1991, further research and development at The University of New South Wales (Australia) has led to a significant improvement in aerodynamic performance (Jiang *et al.* 1995) and the experimental establishment of a relationship between chemical evaporation rate and air velocity based on boundary layer theory (Bliss *et al.* 1995).

5.5.1.1 Alternative sampling devices

An alternative to using a wind tunnel system would be to use an isolation chamber (Klenbusch 1985; Klenbusch 1986 and Gholson *et al.* 1991). Isolation chambers are sometimes referred to as “flux hoods”. In Australia, both isolation chamber and wind tunnel systems are being used to collect samples for measuring odour emission rates from area sources.

Wind tunnel and isolation chamber systems are differentiated from each other mainly in the rate of "sweep air" (i.e. carrier gas) used to transport the emission from the surface being sampled. Isolation chamber systems generally utilise "sweep air" rates of 5 to 24 l/min. Internal crossflow velocities are not usually considered. Wind tunnel systems use much higher carrier gas rates, generally more than 1800 l/min, to produce internal crossflow velocities of 0.3–1.0 m/sec.

In general, isolation chambers cannot be recommended for odour sampling purposes, because of their inappropriate mixing and aerodynamic characteristics. The use of flux hoods has been observed to cause a randomly distributed low bias to emission measurements. The inherent design and performance characteristics of static isolation flux chambers in comparison with portable wind tunnels for measuring odour emission rates have been discussed in detail (Jiang and Kaye 1997). Under field conditions measured odour emission rates from these two types of apparatuses have been observed to differ by up to 300 times in some cases (Jiang and Kaye 1997).

Emissions from area sources are of critical importance for the sewage industry as the emissions from many sewage treatment plants are dominated by these sources. Table 5.3 shows comparative total odour emission rate results for various sewage treatment plant sources using both sampling apparatuses in parallel.

Table 5.3. Comparison of total odour emission rates using flux hood and wind tunnel apparatuses.

Processing unit	Flux hood (ou/s)	Wind tunnel (ou/s)
Primary Sedimentation Tank	6780	76,076
Anaerobic tank	4690	86,539
Anoxic tanks	1697	53,428
Aeration tanks	486	40,797
Mixed liquor channel	31	10,365
Sludge dewatering	487	12,376
Sludge lagoon	4820	100,800

As discussed above, the performance of flux chambers for measuring odour emission rates varies from sample to sample under field conditions. Consequently, "correction factors" cannot be inferred from Table 5.2 for similar sources at other sewage treatment plants. The comparative performances of isolation flux chambers and portable wind tunnels for measuring emissions of pure volatile chemical compounds has also been tested (Jiang and Kaye 1997). Under controlled laboratory conditions, the degree of underprediction using flux hoods appears to be related to the Henry's Law constant for the compound in question. However, the same degree of underprediction may not occur under actual field conditions.

5.5.1.2 Wind tunnel description

The wind tunnel system is designed to simulate a simple atmospheric condition – parallel flow without vertical mixing. Odour emission from a surface takes place as odorous compounds evaporate from the known surface area sampled into the horizontal air stream (at known velocity) across the surface. Ensuring the capacity of the sampling system to collect repeatable and reproducible samples from surfaces, such as lagoons, has been a major consideration in the development of the wind tunnel sampling technique. An isometric drawing of the wind tunnel system developed at The University of New South Wales (UNSW) is shown in Figure 5.4. The system comprises several parts: extension inlet duct, connection duct, expansion section, main section, contraction section and mixing chamber. The cylindrical floats are used where the odour source is a liquid surface but removed in the cases of solid sources such as broiler litter. The extension inlet duct can be separated from the connection duct to enable cleaning and transport of the hood.

The principle of the wind tunnel system is that activated carbon filtered air is introduced at the inlet duct using a fan. The air is controlled through flat vanes in the expansion section and enters the main section via a perforated baffle. The air entering the main section forms a consistent parallel flow over a defined liquid or solid surface under the wind tunnel. A convective mass transfer takes place above the emitting surface. The odour emissions are then mixed into the bulk of the carrier air and vented out of the hood. A proportion of the mixture is drawn into a Tedlar bag via Teflon tubing.

The air velocity used in the wind tunnel is 0.3 m/s. The selection of air velocity was based on substantial odour complaint histories over 18 months around two sewage treatment plants in, one near Sydney, and the other near Perth, Australia (Jiang and Kaye 1997). It was found that most (nearly 70%) odour complaints occurred at wind speeds of 1.5 m/s or less at a height of 10 metres. The corresponding ground level wind speeds, at 0.125 metre (half wind tunnel height) would range between 0.2–0.65 m/s for various atmospheric stability classes. The aerodynamic performance at 0.3 m/s, which is the lowest reliably measurable air velocity directly inside the main section of the wind tunnel, has been validated (Jiang 1996).

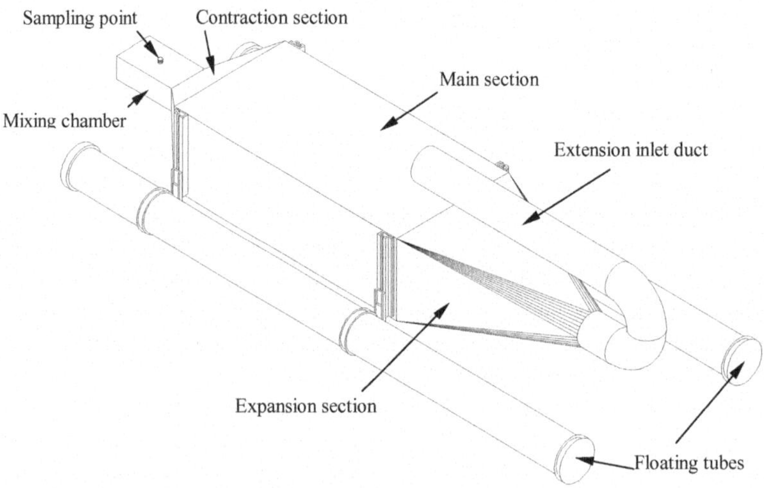

Figure 5.4. An isometric drawing of the UNSW odour emission hood.

The odour emission hood is intended to create an environment where the boundary layer is well developed. The aerodynamic performance of the hood is considered a critical parameter. Streamline flow makes it possible for the velocity measured at the mixing chamber exit to be correlated with the mean velocity in the main section. This measurement can be used in the field to confirm the velocity through the hood.

The wind tunnel system was designed so that it could be transported and manipulated in the field by a single person. In accordance with best practice, the wind tunnel is constructed entirely of stainless steel and is easily cleaned between successive samples. The design has been proven during seven years of field testing.

5.5.1.3 Sampling procedure

In practice, the wind tunnel system is disassembled for transport and assembled on site prior to use. The activated carbon filter is connected to the fan and the hood via flexible duct and secured using duct tape. Teflon sampling tubing is fitted to the hood and the sampling drum via stainless steel Swagelok fittings. Floats are not used when sampling odour emission from solid surfaces. Where multiple sources are to be sampled, the odour sources are sampled in order of increasing odour strength. The least odorous source is tested first and the most odorous source is sampled last. The sampling enclosure should be washed with clean water between samples.

The hood is placed gently on the liquid or solid surface at the desired sampling location. In ponds, the edge of the hood is submerged into the water by about 5 mm. The flexible ducts and the Teflon sampling tubing are checked to ensure they are free of kinks. Air velocity at the wind tunnel exit is checked by anemometer to ensure that it is within specification for the desired cross-flow velocity. The odour sample is taken three minutes after the fan is switched on as instrument testing with model volatile compounds has demonstrated that steady state conditions are assured after this time.

Any observation of water droplets within the Teflon sampling tubing during sampling with the wind tunnel may indicate that the mixing section of the wind tunnel has become submerged or, the sampling tubing has become detached at the wind tunnel fitting. Consequently, in the event that water droplets are observed during emission hood sampling, sample collection is terminated immediately and corrective action taken before repeating the sampling.

Figure 5.5. Odour sampling at a sewage treatment plant.

5.5.2 Static sampling enclosures – sources with outward flow

Aeration tanks at sewage treatment plants should not be regarded as simple examples of area sources with outward flow. The natural movement of ambient air causes a significant component of the emission from the surface of the liquid.

Consequently, aeration tanks are a special case that needs to be sampled using a wind tunnel system. The rate of outflow for an extended aeration process using fine bubble diffusers is estimated to be about 1.5 l/m^2/sec. The outflow rate for the area covered by the UNSW wind tunnel would therefore be only about 0.5 l/sec. This is small in comparison with a sweep air rate of 30 l/sec. The discharge of the air bubbles from the liquid surface is driven by strong buoyant forces and should not be significantly affected by the slightly higher ambient pressure inside the wind tunnel. However, it is not known how the discharge of air bubbles through the boundary layer affects the emission mechanism.

However, as a general rule, a wind tunnel system as described above, can have significant limitations when used for other area sources with outward flow such as open biofilters. In some situations, the rate of outward flow may be significant in comparison with the sweep air rate used in the wind tunnel. Furthermore, the placement of the wind tunnel may create a back pressure, limiting the flow of outward moving air into the wind tunnel and leading to an underestimation of the odour emission rate. Such sources must be sampled using static sampling hoods (i.e. no sweep air is introduced). Static sampling hoods have been developed with a capacity to balance internal and external ambient pressures.

Further caution should be used in sampling open biofilters as short-circuiting of air can occur where the media contacts the sidewalls. This can be a major source of treatment inefficiency in biofilters. Consequently, the ideal situation would be to cover the entire surface of the biofilter with sheeting material to enclose the sidewalls. The sheeting must be left open at some point to allow the air to escape and air samples are collected at this location using point source sampling apparatus. If this is not practicable and a static sampling hood is to be used, emission samples should also be collected at the sidewall perimeter using a point source sampling apparatus. The sidewall samples should be collected at a low air-pumping rate to avoid unintended dilution with ambient air. The flow rate of the fugitive emissions from the sidewall perimeter can be reasonably estimated as the difference between the total airflow rate measured at the inlet to the biofilter and the apparent total airflow rate determined using the static sampling hood.

5.6 SAMPLE COLLECTION FROM VOLUME (BUILDING) SOURCES

For intensive agricultural industries, volume sources such as chicken and pig sheds are important sources of odour emissions. However, for sewage treatment industries, volume sources such as sludge dewatering buildings are often not as significant as are area sources. Nevertheless, emissions from volume sources such as sludge dewatering buildings should be included in the sampling program. These emissions can cause serious impacts, particularly if they are situated near residences.

For a mechanically ventilated building, the exhaust air velocities and fan diameters are measured to calculate the ventilation rate. Ideally, the odour samples should be taken at the fan. Alternatively, if the odour samples cannot be taken at the fans, composite air samples are taken within the building for odour measurement and the air ventilation rate may be estimated from the mechanical specifications of the fans.

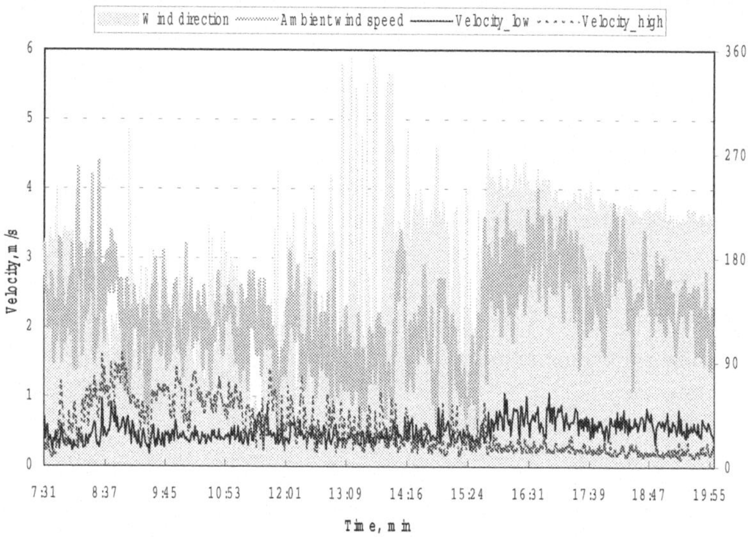

Figure 5.6. Air velocity and ambient wind speed and direction at a broiler grow-out shed.

Naturally ventilated buildings present a problem as they may have a number of openings. This type of building is one of the most difficult in which to measure the ventilation rates. Accurate estimates can only be made by using a tracer gas released at a known rate and measuring the concentration within the building. However, a simpler method of estimating the ventilation rate is by measuring air velocities at the openings. Particular attention must be paid to ambient wind direction when monitoring naturally ventilated buildings. Ideally, it is recommended that air velocity be measured continuously over 24 hours on the windward side of a naturally ventilated building. A hand-held anemometer may be used if automatic continuous monitoring cannot be arranged. Figure 5.6 shows patterns recorded for air velocity and ambient wind speed and direction measured at a broiler grow-out shed on an Australian chicken farm. Odour samples may be collected at the apertures. If this is not practical, again, composite air samples may be taken inside the building.

5.7 RESULT CALCULATION

Information concerning emission rates is required for odour impact assessments. However, olfactometry measures only the odour concentrations. In general, emission rates must be calculated using the measured odour concentrations together with other measured properties of the emission source and the sampling apparatus.

5.7.1 Point sources

For point sources, the Odour Emission Rate (OER) is calculated using the odour concentration measured by olfactometer and the measured gas flow rate:

$$OER = Q \times OC \tag{5.1}$$

Where:
OER = odour emission rate (ou/s),
Q = gas flow rate (m^3/s),
OC = odour concentration (ou/m^3).

5.7.2 Area sources without outward flow

The Specific Odour Emission Rate (SOER) may be defined as the quantity (mass) of odour emitted per unit time from a unit surface area. Area emission sources with no outward flow are sampled using a wind tunnel. Consequently, the quantity of odour emitted is calculated from the concentration of odour (as measured by olfactometry) which is then multiplied by the volume of sweep air passing through the hood per unit time. The volume per unit time is calculated from the measured velocity through the wind tunnel multiplied by the known cross sectional area of the wind tunnel. SOER is calculated by the equation:

$$SOER = \frac{Q \times OC}{A} \tag{5.2}$$

Where:
$SOER$ = specific odour emission rate (SOERs) (ou/s),
Q = flow rate through the wind tunnel (m^3/s),

OC = odour concentration (ou/m^3),
A = area covered by the wind tunnel (m^2).

Odour impact assessment requires the total emission rate for each source to be determined as the product of SOER and the total surface area of the emission source.

In the odour dispersion modelling calculation, the odour emission rates can be set as functions of wind speeds and atmospheric stability classes. In general, this is not required for point source and area source emissions with an outward flow. In these cases, internal processes determine emission rates.

Similarly, emissions from building sources at sewage treatment plants can be considered to result from internal processes. While velocities at the openings in naturally ventilated buildings are determined by ambient wind speeds, for the sake of simplicity, the emissions may be regarded as resulting from internal processes. Consequently, average odour emission rates from these building sources may be used in odour dispersion modelling calculations.

However, it is the movement of ambient air over the surface boundary layer that causes emissions from area sources without outward flow (including aeration tanks as discussed above). Consequently, wind speeds and atmospheric stability classes should be included in the atmospheric dispersion modelling calculations for these sources. Emission rates may be determined for actual ground level wind speeds corresponding with the meteorological data. The following relationship between emission rates and air velocities is derived from boundary layer theory and has been verified experimentally for the wind tunnel system (Bliss *et al.* 1995):

$$SOER_2 = SOER_1 \times \left(\frac{V_2}{V_1} \right)^{0.5} \tag{5.3}$$

Where:
$SOER_1$ = specific odour emission rate measured using the wind tunnel,
$SOER_2$ = specific odour emission rate corresponding to actual ground level wind speed,
V_1 = air velocity inside wind tunnel for sample collection (0.3 m/s in UNSW wind tunnel system),
V_2 = actual ground level wind speed.

On-site meteorological station wind sensors are usually affixed to a 10 m mast. Consequently, the ground level wind speeds at half wind tunnel height

(0.125 m) may be calculated from the 10-m height wind speeds using the following relationship:

$$U_{0.125} = U_{10} \times \left(\frac{0.125}{10} \right)^n \qquad (5.4)$$

Where:
$U_{0.125}$, U_{10} = wind speeds (m/s) at 0.125 m and 10 m heights.
(Note, where the emission source is above ground level, the actual height (rather than 0.125m) is substituted in equation 5.4.).

The wind profile exponent, *n* is assigned on the basis of the Pasquill stability class. In a recent Australian study (Kaye and Jiang 2000), median values for each of the 6 Ausplume default wind categories together with the exponent for the corresponding stability classes were used, such that for each areal emission source a 6x6 matrix of emission rates was generated (36 values for each areal source). Irwin Urban exponents of 0.15, 0.15, 0.2, 0.25, 0.4, and 0.6 are used respectively for stability classes A, B, C, D, E, and F.

In the above study, model default wind speed categories were modified to improve the resolution of the model at the low wind speed range. The new wind speed settings were selected to provide improved resolution for conditions that might be expected to generate the 98.5, 99, 99.5 and 99.9 percentile output values. These percentile output values are often used for odour impact assessment purposes. The new wind speed categories, corresponding median wind speeds (at 10 m height), and Irwin Urban exponents together with corresponding examples of emission rates are presented in Table 5.4. The emission rate calculation examples are based on a hypothetical emission rate of 100 ou/s measured at a simulated ground level wind speed of 0.3 m/s.

Emissions from "weak" sources were not included for modelling purposes. Generally, "weak" sources include secondary clarifiers and tertiary lagoons. Odour concentration measurements for such sources in the first instance may be so low as to be obscured by indigenous odours from the sampling train. Emissions from secondary clarifiers need to be considered only where they cease to be low intensity sources due to one or more of the following operating conditions:

- short sludge age (< 5 days),
- insufficient aeration capacity in the activated sludge process,
- poor oxygen transfer in the activated sludge process,
- chronic overloading of the clarifiers / rising sludge.

Table 5.4. New Ausplume wind speed settings.

Wind Speed Category	Speed Range (m/s)	Median Speed (m/s)	Stability Class A	B	C	D	E	F
			Emission Rates (ou/s)					
1	0–0.6	0.3	72	72	65	58	42	27
2	0.6–1.2	0.9	125	125	112	100	72	47
3	1.2-1.8	1.5	161	161	144	129	93	60
4	1.8-2.4	2.1	190	190	171	153	110	71
5	2.4-3.0	2.7	216	216	194	173	125	81
6	>3.0	6.5	335	335	300	269	194	125

The volume of outflow air per unit time is calculated from the measured velocity through the exhaust stack of the sampling enclosure multiplied by the known cross sectional area of the exhaust stack. In a similar fashion to other types of area sources, SOER is calculated by the equation:

$$SOER = \frac{Q \times OC}{A} \tag{5.5}$$

Where:
$SOER$ = specific odour emission rate (SOERs) (ou/s),
OC = odour concentration (ou/m^3),
Q = flow rate through the exhaust stack of the sampling enclosure (m^3/s),
A = area covered by the static sampling enclosure (m^2).

As for other types of area sources, odour impact assessment requires the total emission rate for such sources to be determined as the product of SOER and the total surface area of the emission source.

5.7.3 Building sources

For building sources, the Odour Emission Rate (OER) can be calculated from the odour concentrations measured by olfactometer and gas flow rates through the door and window openings. The following equation is applied to volume sources:

$$OER = Q \times OC \tag{5.6}$$

Where:
Q = gas ventilation rate (m^3/sec),
OC = odour concentration (ou/m^3).

For samples from sources where temperatures and pressures are significantly different from ambient conditions, the gas flow rate is calculated and adjusted to NTP (Normal Temperature and Pressure i.e. 20 °C and 101.3 kPa) conditions using the following equation:

$$Q = Q_m \frac{(273+20)}{(273+t)} \frac{p}{101.3} \tag{5.7}$$

Where:
Q = the volume flow rate at NTP conditions (m^3/s),
Q_m = the volume flow rate measured inside the vent (m^3/s),
t = air temperature inside the vent (°C),
p = the absolute pressure inside the vent (kPa).

5.8 CONCLUSIONS

Olfactometric techniques have been standardised and the use of calibrated olfactometers and screened panels has greatly improved the reliability of odour concentration measurements. However, sampling techniques, while greatly improved, still need to be standardised. The selection of inappropriate sampling apparatus and insufficient attention to sampling conditions will cause substantial errors. Such sampling errors can overshadow the errors that could potentially occur in subsequent olfactometric testing.

Flow rate is critical in calculating odour emission rates from point sources. The accuracy of air velocity measurements greatly affects the reliability of results. While isokinetic sampling procedures are generally not required for odour sampling, the numbers and positions of points required for isokinetic sampling are the same as for the characterisation of the average velocities for point sources.

Emission rate measurements from area sources are especially sensitive to the characteristics of the sampling apparatus used and the selection of sampling conditions. Emissions from area sources are of critical importance for the sewerage industry as the emissions from many sewage treatment plants are dominated by these sources.

Wind tunnel systems are used to sample odour emissions from area sources such as primary sedimentation tanks at sewage treatment plants. Research and development at The University of New South Wales has led to the design of a significantly improved portable wind tunnel system and a relationship has been established between chemical evaporation rate and air velocity based on boundary layer theory. Consequently, emission rates may be determined for actual ground level wind speeds corresponding with the meteorological data. Atmospheric dispersion modelling calculations for these sources can be set up in this way to take account of wind speeds and atmospheric stability classes. In general, adjustment of emission rates for actual ground level wind speeds is not required for point source and area source emissions with an outward flow.

5.9 REFERENCES

Bliss, P. Jiang, J.K. and Schulz, T. (1995) The development of a Sampling System for the Determination of Odour Emission Rates from Areal Surfaces: II Mathematical Model. *J. Air Waste Manage. Assoc.* **45**, 989-994.

Gholson, A. R., Albritton, J. R., Jayanty, R. K. M., Knoll J. E. and Midgett, M. R. (1991) Evaluation of an enclosure method for measuring emissions of volatile organic compounds from quiescent liquid surfaces. *Environ. Sci. Technol.* **25**, 519-524.

Hangartner, H., Hartung, J. and Voorbury, J. H. (1985) Recommendations of olfactometric measurements. *Environ. Technol. Lett.* **6**, 415-420.

Jiang, J. (1996) Odor Concentration measurement by dynamic olfactometer. *Water Enviro. Technol.* **8**, 55 -58.

Jiang, J.K., Bliss, P. and Schulz, T. (1995) The development of a sampling system for the determination of odour emission rates from area surfaces: I aerodynamic performance. *J. Air Waste Manage. Assoc.* **45**, 917-922.

Jiang, J. and Kaye, R. (1997). The selection of air velocity inside a portable wind tunnel system using odour complaint database. *Proc. Odors/VOC speciality conference*, Houston, April.

Klenbusch, M.R. (1986) Measurement of gaseous emission rates from land surfaces using an emission isolation flux chamber user's guide. EPA/600/8-86/008; U.S. Environmental Protection Agency, Las Vegas.

Lindvall,T. (1970) On sensory evaluation of odorous air pollutant intensities. *Nord. Hyg. Tidskr.,* suppl. 2. Stockholm: Karolinska Institute and National Institute of Public Health.

Lockyer, D. R. (1984) A system for the measurement in the field of losses of ammonia through volatilization. *J. Sci.. Food Agric.* **35**, 837-848.

Pau, J. C., Knoll, J. E. and Midgett, M. R. 1991. A tedlar bag sampling system for toxic organic compounds in source emission sampling and analysis. *J. Air Waste Manage. Assoc.* **41**, 1095-1097.

Schuetzle, D., Prater, T.J. and Ruddell S.R. (1975) Sampling and analysis of emissions from stationery sources I Odor and total hydrocarbons. *J. Air Poll. Cont. Assoc.* **25**, 9, 925-932.

Wenzel, B. M. (1948) Techniques in olfactometry: a critical review of the last one hundred years. *Psychological Bulletin,* **45**, 231-247.

6

Hydrogen sulphide measurement

Peter Gostelow and Simon A. Parsons

6.1 INTRODUCTION

An odour can be defined in terms of a property of a substance, or in terms of a physical sensation. This is paralleled in odour measurement where there are two broad classes of measurement. Analytical measurements are concerned with the properties of the odorous compounds (odorants) whereas sensory measurements refer to the perceived effect of the odorous compounds on the sense of smell.

Sensory measurements employ the human nose as the odour detector and hence relate to the effects of odours as experienced by humans. This is useful in terms of nuisance assessment, but is of limited use for the examination of how odours are formed, how they are emitted or how they can be controlled. For these areas, analytical measurements are required, giving information on the compounds responsible for imparting the odour.

In isolation, either class of odour measurement is of limited use. Sensory measurements give little information on the chemical composition of an odour,

 Odours in Wastewater Treatment: Measurement, Modelling and Control edited by Richard Stuetz and Franz-Bernd Frechen. ISBN: 1 900222 46 9

whereas analytical measurements give little insight on the effect of the odour on the sense of smell. Discovering links between analytical and sensory measurements is one of the major challenges in the study of odours.

The most common sensory measurement is threshold olfactometry which measures odour concentration in terms of the number of dilutions required to reduce a sample to it's threshold concentration. Unfortunately, there are many factors other than the properties of the odour sample itself that may influence the perception of the odour. These have to a large extent been addressed by the development of standards but it is still unlikely that any sensory measurement will ever approach the accuracy offered by many analytical measurements.

Analytical measurements have the advantage of objectivity, repeatability and accuracy. More importantly, they can be related directly to theoretical models relating to odorant formation or emission. Analytical measurements are not, however, without their disadvantages. Principal amongst these is the fact that most environmental odours consist of many components. Odorants may be present in very small concentrations compared with non-odorous gases, which may in turn interfere with the analysis. The analytical detection limit for many odorants is below their threshold concentration.

Although for many individual odorants a relationship between odorant concentration and perceived effect on the sense of smell can be determined, the situation for mixtures of odorants is much more complicated. Interactions between mixtures of odorants may lead to synergistic or antagonistic effects, leading to difficulties in linking analytical and sensory measurements for environmental odours.

6.2 HYDROGEN SULPHIDE

Qualification and quantification of all the odorants present in a sample is very difficult. In many cases, however, a single odorant may be dominant and can give an indication of the overall odour concentration. This is certainly the case for many sewage treatment works, as hydrogen sulphide (H_2S) is often present in higher concentrations than other odorants. Hydrogen sulphide can be measured down to low parts-per-billion levels rapidly using hand-held equipment. This allows for many measurements to be made in a short period of time, and eliminates the delay between sampling and measurement necessary for laboratory-based measurements.

Hydrogen sulphide, in common with most sewage treatment works odorants, is associated with anaerobic conditions. Hydrogen sulphide forms a good marker for odours arising from processes upstream of secondary treatment, especially where a works is fed by a septic sewage, and is also a good marker for sludge processes. It is a poor marker for odours arising from aerobic treatment, except for specific cases where these processes are overloaded.

Hydrogen sulphide is a weak dibasic acid and dissociates as shown in Figure 6.1. It is only molecular hydrogen sulphide that will lead to odour problems and at neutral pH approximately 50% of the total sulphide will be in this form. Acidic conditions will enhance hydrogen sulphide odour problems, alkaline conditions will suppress them. Hydrogen sulphide can be a poor marker where alkaline conditions exist, for example where lime dosing is employed. The presence of metal ions can lead to the formation of metal sulphides which are insoluble and do not therefore contribute to odours. If ferric dosing is employed hydrogen sulphide may be a poor marker.

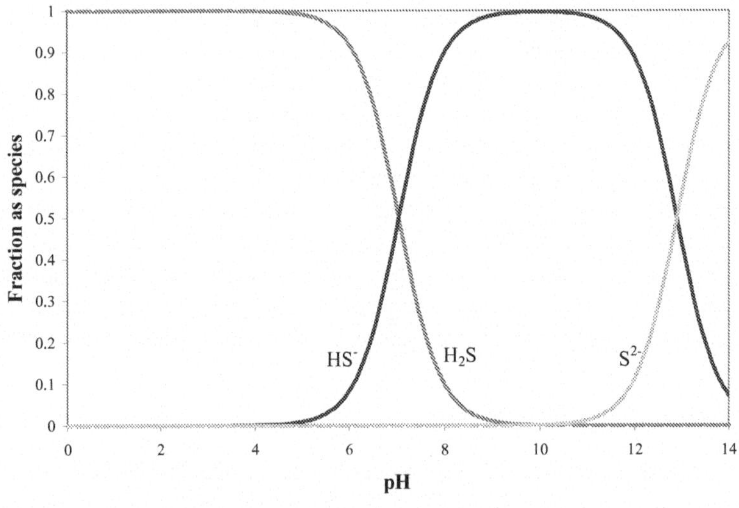

Figure 6.1. Dissociation of hydrogen sulphide.

The formation of hydrogen sulphide in sewers has been extensively studied for reasons of its toxicity and corrosive properties as well as its contribution to sewage odour. This is advantageous as it allows the liquid-phase sulphide concentration at the inlet to a works to be predicted which in turn allows theoretical emission models to be used.

6.3 HYDROGEN SULPHIDE MEASUREMENT

The fact that odorant concentrations can in many cases be measured at low concentrations in both gas and liquid phases is a principal advantage of analytical measurements. This is particularly the case for hydrogen sulphide and the development of field-portable instruments has increased the popularity of H_2S as an odour marker.

6.3.1 Liquid-phase measurement

Consideration of Figure 6.1 indicates that sulphide will exist in several states in the liquid phase. These can be characterised as:

Total sulphide:	$H_2S + HS^- + S^{2-}$ + suspended metallic sulphides
Dissolved sulphide:	$H_2S + HS^- + S^{2-}$
Un-ionised H_2S:	H_2S

Three common tests for total and dissolved sulphide are the methylene blue method, the iodimetric method and the ion-selective electrode method (APHA 1995). These are summarised in Table 6.1.

Table 6.1. Liquid-phase sulphide measurements (APHA 1995).

Method	Description
Methylene blue method	Colourimetric method utilising reaction of sulphide, ferric chloride and dimethyl-p-phenylenediamine to produce methylene blue
Iodimetric method	Titration utilising oxidation of sulphide solution by iodine
Ion-selective electrode method	Utilises silver sulphide ion selective electrode, potential related to sulphide ion activity

Total and dissolved sulphide are determined by the removal of suspended solids by flocculation or similar means, The relative species of dissolved sulphide can be determined from dissolved sulphide using the following equations:

$$\alpha_{H_2S} = \frac{[H^+]^2}{[H^+]^2 + K_{a_1}[H^+] + K_{a_1}K_{a_2}} \quad (6.1)$$

$$\alpha_{HS^-} = \frac{K_{a_1}[H^+]}{[H^+]^2 + K_{a_1}[H^+] + K_{a_1}K_{a_2}} \quad (6.2)$$

$$\alpha_{S^{2-}} = \frac{K_{a_1} K_{a_2}}{[H^+]^2 + K_{a_1}[H^+] + K_{a_1} K_{a_2}} \tag{6.3}$$

Where:
α = fraction of species,
$[H^+] = 10^{-pH}$,
$K_{a_1} = 10^{-7.04}$,
$K_{a_2} = 10^{-12.89}$.

6.3.2 Gas-phase measurement

The most common method of hydrogen sulphide measurement in the gas-phase is by the use of the gold-film monitor. These instruments utilise the change in resistance of a gold-film sensor caused by adsorption of H_2S molecules, with an output proportional to the H_2S concentration. A common gold-film monitor, the Jerome 631-X H_2S analyser (Arizona Instruments, USA) has a sensitivity of 3 ppb and can measure up to 50 ppm H_2S. Sample times range from 13–30 seconds, depending on H_2S concentration (Arizona Instrument Corporation 1997). A Jerome 631-X is shown in Figure 6.2.

Extensive testing of a Jerome 631-X was carried out by Winegar and Schulz (1998). They concluded that the analyser is capable of quantitative detection of hydrogen sulphide over a range of 2 ppb to 50 ppm with acceptable precision and accuracy. Precision was tested by repeated analysis of the same standard, with the results shown in Table 6.2. Accuracy was assessed by parallel analysis of bag samples using the Jerome 631-X and a gas chromatography (GC) method with a precision of 5%. The Jerome 631-X and GC results were in very good agreement.

Table 6.2. Jerome 631-X precision results (Winegar and Schulz 1998).

H_2S Concentration (ppm)	Relative Standard Deviation (%)
0.002	32.2
0.005	10.8
0.13	11.6
0.43	4.3
0.72	6.0
0.87	2.1
33	1.6

Figure 6.2. Jerome 631-X H_2S analyser (Courtesy of Arizona Instruments, USA).

Because the Jerome 631-X relies on adsorption, it is susceptible to interference from other reduced sulphur compounds. As these tend to be odorants, this may not be a disadvantage if an instrumental indication of odour concentration is required (Vincent and Hobson 1998). If, however, specificity to H_2S is required, any interference is a disadvantage. Winegar and Schulz (1998) addressed the issue of interference for the 631-X. Table 6.3 shows the response factors for a series of reduced sulphur compounds, which are shown as a percentage of the hydrogen sulphide response. As can be seen, the 631-X shows a significant response to many of these compounds. Winnegar and Schulz (1998) found that these compounds were present in much lower concentrations than H_2S in wastewater samples, which combined with their lower response factors allowed for quantitative detection of H_2S.

Another common instrument for gas-phase H_2S measurement is the paper-tape monitor. These utilise the reaction between hydrogen sulphide and lead acetate to produce a coloured stain on the paper tape, the opacity if which is measured optically and converted to a concentration. These instruments can be used over a similar concentration range to a gold-film monitor but have the disadvantage that sampling times can be in the order of minutes rather than seconds. This can be restrictive if a large number of samples are required in a short time period.

Table 6.3. Jerome 631-X response to reduced sulphur compounds (Winegar and Schulz 1998).

Compound	Response factor (%)
Hydrogen sulphide	100
Methyl mercaptan	45
Dimethyl disulphide	40
n-propyl mercaptan	40
Carbonyl sulphide	36
t-butyl mercaptan	35
n-butyl mercaptan	33
Diethyl sulphide	25
Diethyl disulphide	17
Tetrahyrothiophene	10
Dimethyl sulphide	7
Thiophene	0.8
Carbon disulphide	0.01

An alternative method with similar detection limits to gold-film or paper tape monitors is the UV-fluorescence type meter. These instruments actually measure sulphur dioxide (SO_2) concentrations but are applied to H_2S by first scrubbing out SO_2 from the sample and then catalytically oxidising H_2S to SO_2. The specificity to H_2S will depend on the specificity of the catalyst. These instruments are reported to be very stable and reproducible and have been applied to sewage odour measurement (McIntyre 2000).

A list of instrument manufacturers for H_2S analysers is shown in Table 6.4.

Table 6.4. List of H_2S analyser manufacturers.

Company	Models	Type
Arizona Instrument (www.azic.com)	Jerome 631-X	Gold film
Trace Technology (www.tracetechnology.com)	050 Series Portable 100 Series	Paper tape
MDA Scientific (www.zelana.com/mda/mda.asp)	Chemkey TLD SPM	Paper tape
Enviro Technology (www.et.co.uk)	API M101A	UV-Fluorescence
Interscan Corporation (www.gasdetection.com)	1000 Series Portable 4000 Series Portable	Electrochemical

6.4 LINKING H_2S AND ODOUR CONCENTRATION

Analytical measurements have many advantages, but their use is limited if they cannot be related to sensory measurements of odour. The principal barrier in linking analytical and sensory concentration measurements is the effect of mixtures. Recent work on mixtures of two to twelve odorants suggests that odorants are additive – a mixture of odorants will have a stronger odour than any of the component odorants alone (Laska and Hudson 1991; Patterson *et al.* 1993; Laing *et al.* 1994). The degree of additivity appears to vary however. The investigations of Laing *et al.* (1994) are particularly relevant to sewage treatment works (STW) odours as they investigated 2, 3 and 4 component mixtures using odorants characteristic of sewage odours. Their results suggested partial additivity, whereby the odour intensities of mixtures were less than suggested by simple summation of the individual component intensities. The intensity of the mixtures was close to that of the dominant (most intense) component implying that where, for example, H_2S is dominant, this should give a good indication of the overall odour concentration.

Koe (1985) showed that for sewage odours H_2S and odour were better correlated using an equation of the form $C_{(ou)} = mC_{(H2S)}{}^n$ where $C_{(ou)}$ is the odour concentration in ou m^{-3} and $C_{(H2S)}$ is the H_2S concentration in ppm. The values of m and n differ according to the composition of the odour. Gostelow and Parsons (2000) performed similar correlations for a number of processes at a total of 17 sewage treatment works. The resulting correlations are summarised in Table 6.5. The correlation for sludge storage and handling prior to odour tratment is shown in Figure 6.3.

Table 6.5. Summary of H_2S/odour correlations (Gostelow and Parsons 2000).

	m	n	r^2	p
Before odour treatment				
Preliminary treatment	52555	0.62	0.45	7.7×10^{-5}
Aeration tanks	14555	-0.12	0.07	0.433
Sludge storage & handling	38902	0.64	0.69	4.13×10^{-12}
After odour treatment				
Preliminary treatment	29704	0.47	0.36	8.01×10^{-4}
Aeration tanks	44465	0.60	0.35	0.093
Sludge storage & handling	48099	0.38	0.39	2.6×10^{-3}

The r^2 values for the statistically significant correlations of Gostelow and Parsons (2000) suggest that between 36–69% of the variance in odour concentration could be explained by H_2S for these samples. The strongest correlations were for sludge storage and handling and preliminary treatment before odour control. H_2S would be the dominant odorant for many of these samples. Poorer correlations were seen after odour control, which may be due to preferential removal of H_2S over other odorants. Correlations of H_2S against odour for aeration tanks were not statistically significant ($p > 0.05$) which is not unexpected as aeration tanks are not associated with H_2S odours unless overloaded.

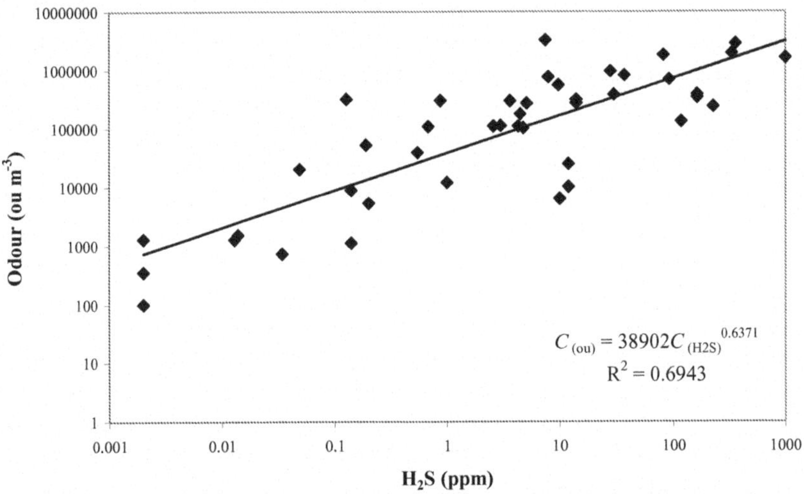

Figure 6.3. Correlation of H_2S against odour concentration for sludge storage/handling units (Gostelow and Parsons 2000).

6.5 CONCLUSIONS

The perception of odour is complicated. Until a reliable theory of olfaction exists, both analytical and sensory measurements will be necessary. Unfortunately, detailed analytical or sensory measurements are both time consuming and expensive in practice. They are very difficult to perform on-site.

Hydrogen sulphide offers an inexpensive, rapid and easy alternative to detailed analytical or sensory measurements. The use of portable instruments allows easy and rapid measurements on-site, meaning that many measurements are possible within a short period of time.

Correlations between H_2S and odour concentration suggest that H_2S is an acceptable surrogate for odour for processes where H_2S is the dominant odorant, for example sludge treatment or processes upstream of aerobic treatment. It is a poor surrogate when the H_2S content of the odour is lower, for example aeration tanks. Odour concentration can be predicted to a certain extent using correlations of the form $C_{(ou)} - mC_{(H2S)}{}^n$ with improving accuracy as the H_2S content of the odour increases.

6.6 REFERENCES

APHA (1995) *Standard Methods for the Examination of Wastewater*. American Public Health Association, Washington DC.

Arizona Instrument Corporation (1997). *Jerome 631-X Hydrogen Sulfide Analyser Opeation Manual*. Part number SS-087 Doc #6J21-0002. Rev B.

Gostelow P. and Parsons S.A. (2000) Sewage treatment works odour measurement. *Wat. Sci.Technol.* **41**(6), 33-40.

Koe, L.C.C. (1985) Hydrogen sulphide odor in sewage atmospheres. *J. Water Air Soil Pollution* **24**, 297-306

Laing, D.G., Eddy, A., Best, D.J. (1994) Perceptual characteristics of binary, trinary and quaternary odor mixtures consisting of unpleasant constituents. *Physiology Behavior* **56,** 81-93.

Laska, M. and Hudson, R. (1991) A comparison of the detection thresholds of odour mixtures and their components. *Chemical Senses* **16,** 651-662.

McIntyre, A. (2000). Odour modelling and monitoring: the use of marker compounds such as hydrogen sulphide. Proc. *CIWEM/Southern Water Approaches to Setting Odour Planning Conditions Workshop*.

Patterson, M.Q., Stevens, J.C., Cain, W.S., and Commeto-Muniz, J.E. (1993) Detection thresholds for an olfactory mixture and its three constituent compounds. *Chemical Senses* **18,** 723-734.

Vincent, A. and Hobson, J. (1998) *Odour Control. CIWEM Monographs on Best Practice*, No. 2, Terence Dalton Publishing, London.

Winegar, E.D. and Schmidt, C.C. (1998). Jerome 631-X portable hydrogen sulphide sensor: laboratory and field evaluation. Report to Arizona Instrument Corporation, 15p.

7

Olfactometry and the CEN standard prEN 17325

Robert W. Sneath

7.1 INTRODUCTION

Olfactometry is the measurement of the response of assessors to olfactory stimuli and is to make comparisons of odours from different sources, for these measurements to be useful they must be made objectively, and reproducibly, the CEN standard prEN 17325 is designed to do that. If objective measurements can be made then the results obtained can be used in settlement of disputes about changes in odour emissions, data can be used with dispersion modelling to predict the impact of the odour on the surroundings. Data collected can then be used to formulate planning conditions on new odorous processes and be used as criteria for design of abatement equipment.

 Odours in Wastewater Treatment: Measurement, Modelling and Control edited by Richard Stuetz and Franz-Bernd Frechen. ISBN: 1 900222 46 9

7.2 THE ESSENCE OF QUANTITATIVE OLFACTOMETRY

7.2.1 The detection threshold

Sensory perception of odorants has four major dimensions: detectability, intensity, quality and hedonic tone.

The first dimension of the sensory perception of odorants is detectability. There is no conscious subjectivity to this dimension: either a person can smell an odour or they can't, but every one will have their own detection threshold and this threshold will vary in each person depending on their situation at the time. The second dimension, intensity, refers to the perceived strengths of the odour sensation. The third dimension of odour is the odour quality, i.e. what the substance smells like. The fourth dimension of odour is hedonic tone, this is a category judgement of the relative pleasantness or unpleasantness of the odour.

Detectability is the only one of those dimensions that can be reduced to an objective perception. The only answers to the question "Can you detect the odour?" are "Yes" or "No" (although the value of the response depends on the honesty of the subject). The threshold of detection is different for each individual and can be affected by where one is, by background odours, by familiarity with that odour etc. Therefore, threshold values are not fixed physiological facts or physical constants, but represent the best statistically estimated value from a group of individual responses.

Odour thresholds are estimated in one of two ways, by getting a 'yes/no' response, as above, or by a 'forced choice' response where the subject is forced to choose which air stream, from two or more, smells. In the former classical evaluation, 'yes/no' answers are, amongst other factors, dependent on the subjects' honesty and motivation. If odours at a range of concentrations, alternating with blanks, are presented a sufficiently large number of times, yes/no answers may be evaluated with the aid of signal detection theory, to eliminate the effects of context.

The forced choice procedure is an attempt to measure a subject's sensitivity, which is not influenced by fluctuations in criterion. Two or more choices are presented to the subject at a range of odorant concentrations and it is the subject's task to choose the one that is odorous from the others that are not odorous. The assumption is made that the observer chooses the one that gives the largest sensory excitation, provided that there is no response bias towards one or more of the options. Provided that the comparison stimuli (blanks) have been carefully defined and controlled, the proportion of correct responses can be used as a measure of sensitivity, because it will always be measured in comparison to blanks.

7.2.2 Transforming the measurement of the subject to the subject's measurement of an odour

The detection threshold value is a measure of the sensitivity of the assessor but what we need to do is to measure, in a reliable way, the odour we are interested in.

In all measurements, two criteria must be satisfied: accuracy and repeatability. This usually means manufacturing a *sensor* that produces the correct answer and will produce the same answer repeatedly. In olfactometry our sensor is the human nose (Figure 7.1). These sensors have been produced in a "manufacturing process" that has no quality control: therefore we must choose from the "production run" sensors that fit our criteria for accuracy and repeatability.

The machine that presents the odour sample to the sensors must equally be constructed and operated to achieve the criteria of accuracy and repeatability. Table 7.1 lists a number of commercial olfactometer manufacturers.

Table 7.1. List of olfactometer manufacturers.

Company/Organisation and location	Model
ECOMA GmbH, Germany (www.ecoma.de/english/ecomae.htm)	TO7 olfactometer
OdourNET, UK (www.odournet.com)	Olfaktomat
St. Croix Sensory Inc, USA (www.fivesenses.com/the_iso.htm)	AC'SCENT
University of New South Wales, Australia (www.odour.civeng.unsw.edu.au)	WANG
Tecnovir International Inc, Canada (www.enviroaccess.ca/fiches_2/F2-02-96a.html)	TECNODOR
McGill University, Canada (www.agrenv.mcgill.ca/AGRENG/STAFF/Barrington/Research/olfactometry.htm)	
University of Singapore, Singapore (www.eng.nus.edu.sg/civil/C_ARG/chai(project).htm)	NUS

7.2.3 Principle of measurement

The odour concentration of a gaseous sample of odorants is determined by presenting a panel of *selected and screened* human subjects with that sample. The concentration of the sample is varied by diluting it with a neutral gas (odour- free air) to determine the dilution factor (Z_{50}) at which there is a 50% probability that the odour can be detected. In practice this means presenting a range of diluted samples to the individual panel members above and below their individual thresholds. That threshold value (the individual threshold estimate, ITE) is the geometric mean of the lowest dilution factor a panel member cannot detect and the next dilution that they can detect. The geometric mean of the ITE panel members is the odour concentration. The odour concentration that the panel experience at point of detection is 1 ou_E/m^3 *by definition*. The odour concentration of the examined sample is then expressed as a multiple (equal to the dilution factor at Z_{50}) of one European Odour Unit per cubic metre [ou_E/m^3] at standard conditions for olfactometry.

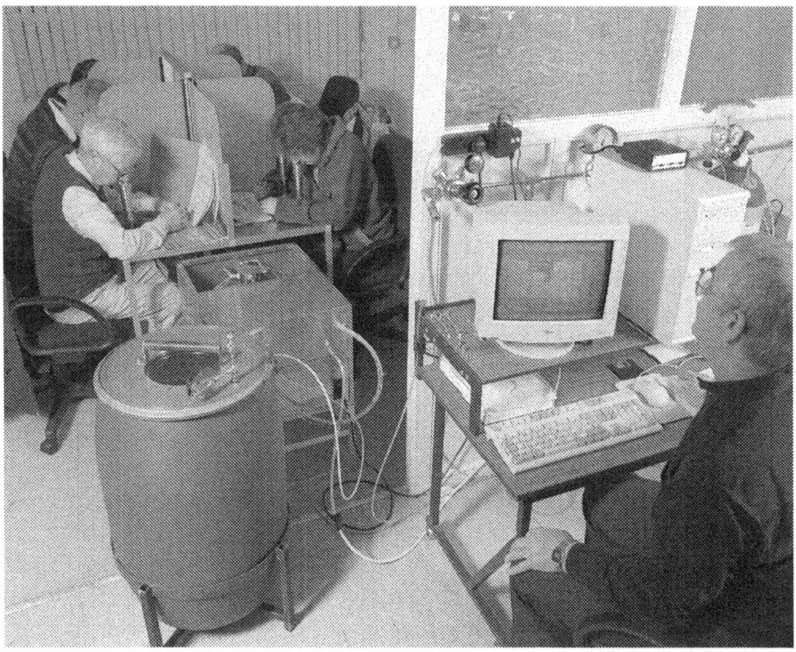

Figure 7.1. A six station forced choice olfactometer in the odour laboratory at Silsoe Research Institute, UK.

7.3 THE DEVELOPMENT OF THE CEN STANDARD

Until about 1998, odour concentration or dilution-to-threshold measurements were made using many different methods in Europe and even more methods around the world but since then many laboratories have adopted the CEN draft standard. A European Union Concerted Action (COST 681) recommended many improvements to the methodologies (Hangartner *et al.* 1989). The Dutch were the first to attempt a statistically-based standard, using selected and calibrated odour panellists. These standards also introduced the concept of measuring inter-laboratory reproducibility and repeatability.

The CEN working group (TC264/WG2), formed in 1992, used their experience of olfactometry, their knowledge of olfaction and incorporated national standards used in Europe (NVN 2820, 1995; AFNOR NF X 43-101, 1986; VDI 3881, 1987). The standard was formulated to be applicable both to yes/no and to forced choice methods; both to single-panellist and to multi-panellist machines. The standard is performance-based, rather than being a prescription for the use of specific equipment. The aim is to ensure that whatever analytical method is chosen, provided the quality criteria are met, the results of odour measurements on the same sample will yield comparable results in any laboratory.

During 1996 the members of the TC264/WG2 organised an inter-laboratory test. Eighteen laboratories in England, The Netherlands, Germany and Denmark participated in it to validate the draft standard (Harreveld and Heeres 1997). The results of this test illustrated that, by implementing the standard in full, laboratories were able to comply with the quality criteria set. Some amendments were nevertheless made to the draft standard in the light of the results of the interlaboratory test before the pre-standard (prEN) was finally issued for consultation in 1999.

The resulting European draft Standard, prEN 17325 (CEN 1999) defines the method for the objective determination of the odour concentration of a gaseous sample using a dynamic olfactometer with human assessors.

The statistical significance of the analysis, as with any other measurement, depends on the precision of the laboratory analysis and on the number of samples analysed. An example will be used to illustrate the importance of calculating the number of samples required for a given purpose.

Previous standards have, in the main, provided a method of measurement of the concentration of the odour. This was previously referred to as the Threshold Odour Number (TON), dilutions to threshold, odour strength, odour threshold or other words to that effect.

The standard now used in many countries in Europe is the CEN/TC264/WG2 (prEN 13725) standard “Air quality – Determination of odour concentration by dynamic olfactometry”. This is in the process of enquiry and is being processed through European national standards organisations in 2000.

Because the standard needs to be understood internationally the European working group agreed a comprehensive glossary of terms and definitions in

English. Where these were already used in other ISO or CEN standards they were adopted. The terms and definitions used here are referred directly from the prEN 17325 and appear in section 7.11. Abbreviations used in this text are explained in section 7.12.

7.3.1 The scope of the draft standard, prEN 13725

The standard carefully defines how and where it can be used. The following statement is quoted directly from it.

"This European Standard (EN) defines a method for the objective determination of the odour concentration of a gaseous sample using dynamic olfactometry with human assessors and the emission rate of odours emanating from point sources, area sources with outward flow and area sources without outward flow. The primary application is to provide a common basis for evaluation of odorous emissions in the member states of the European Union."

"The scope of this standard is the measurement of odour concentration of pure substances, defined mixtures and undefined mixtures of gaseous odorants in air or nitrogen, using dynamic olfactometry with a panel of human assessors being the sensor. The unit of measurement is the European odour unit per cubic metre: ou_E/m^3. The odour concentration is measured by determining the dilution factor required to reach the detection threshold. The odour concentration at the detection threshold is by definition 1 ou_E/m^3. The odour concentration is then expressed in terms of multiples of the detection threshold. The range of measurement is typically from 10^1 to 10^7 ou_E/m^3 (including pre-dilution)."

"The field of application of this EN includes:

- the measurement of the mass concentration at the detection threshold of pure odorous substances in g/m^3.
- the measurement of the odour concentration of mixtures of odorants in ou_E/m^3.
- the measurement of the emission rate of odorous emissions from point sources and surface sources (with and without an outward flow), including pre-dilution during sampling.
- the sampling of odorants from emissions of high humidity and temperature (up to 200 °C).
- the determination of effectiveness of end-of-pipe devices used to reduce odour emissions."

7.3.2 Exclusions

The standard specifically does not cover the measurement of odours potentially released by particles of odorous solids or droplets of odorous fluids suspended in emissions, i.e. dusts and condensates. It assumes that the odour concentration emitted from a source is not variable. The methodologies within the standard are designed to measure the detection threshold and it does not cover the measurement of the relationship between odour stimulus and supra-threshold responses (assessor response above detection threshold), for example recognition thresholds and identification thresholds. Measurements of hedonic tone (or (un)pleasantness) or direct assessment of potential annoyance are also excluded as are field panel methods, used to determine the extent of odour plumes.

7.4 TYPES OF DYNAMIC DILUTION OLFACTOMETRY

7.4.1 Choice Modes

Three different choice modes can be used to obtain an individual threshold estimate. These choice modes and their requirements are described here. They all produce a common result: an individual threshold estimate (ITE). The use of the ITE derived from either of these methods in the calculation of an odour concentration is then identical throughout this standard.

7.4.1.1 Yes/No mode

In the "yes/no" olfactometer (Figure 7.2) either neutral gas or diluted odour passes from the single port. The panel member is asked to evaluate gas presented from the single port and to indicate if an odour is perceived (Yes/No). The panel members are aware that in some cases blanks (only neutral gas) will be presented. (A second port always presenting neutral gas *may* be made available to the assessor to provide a reference.) The samples may be presented to the assessors either randomly or in order of increasing concentration. When using the yes/no mode, 20% of the presentations in a set of dilution series must be blanks to satisfy the operator that the panel members are giving the correct response when there is no odour present. For each panel member the measurement must include a dilution step at which they respond "No" to a diluted odour and for two adjacent dilutions they must respond "Yes".

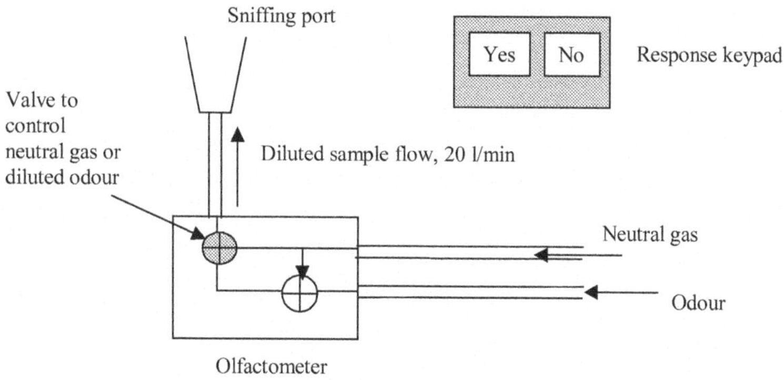

Figure 7.2. Schematic diagram of a "Yes/No" olfactometer.

When the presentations are sorted in order of ascending concentration, the geometric mean of the dilution factors of the last FALSE and the first of at least two TRUE presentations determines the ITE for a panel member. The odour concentration for a sample is calculated from the geometric mean of at least two ITE for each panel member.

7.4.1.2 *The forced choice mode*

A forced choice olfactometer (Figure 7.3) has two or three outlet ports, from one of which the diluted odour flows, while clean odour-free air flows from the other(s).

In this method panel members assess the ports of the olfactometer, from one of which the diluted odour flows, neutral gas flows from the other port(s). The port carrying the odorous flow is chosen randomly by the control sequence on each presentation. The assessors indicate from which of the ports the diluted odour sample is flowing.

The measurement starts with a dilution of the sample large enough to make the odour concentration beyond the panel members' thresholds. The concentration is increased by an equal factor in each successive presentation: this factor may be between 1.4 and 2.4. The port carrying the odorous flow is chosen randomly by the control sequence on each presentation. The assessors indicate from which of the ports the diluted odour sample is flowing, using a personal keyboard. They also indicate whether their choice was a guess, whether they had an inkling or whether they were certain they chose the correct port. Only when the correct port is chosen and the panel member is certain that their choice was correct is it taken as a TRUE response. At least two consecutive TRUE responses must be obtained for each panel member. The geometric mean of the dilution factors of the last FALSE and the first of at least two TRUE presentations determines the ITE for a panel member. The odour concentration for a sample is calculated from the geometric mean of at least two ITE for each panel member.

For measurements on reference odorants, this value can be converted to an individual threshold estimate, expressed as a mass concentration using the known concentration of the reference gas divided by the ITE.

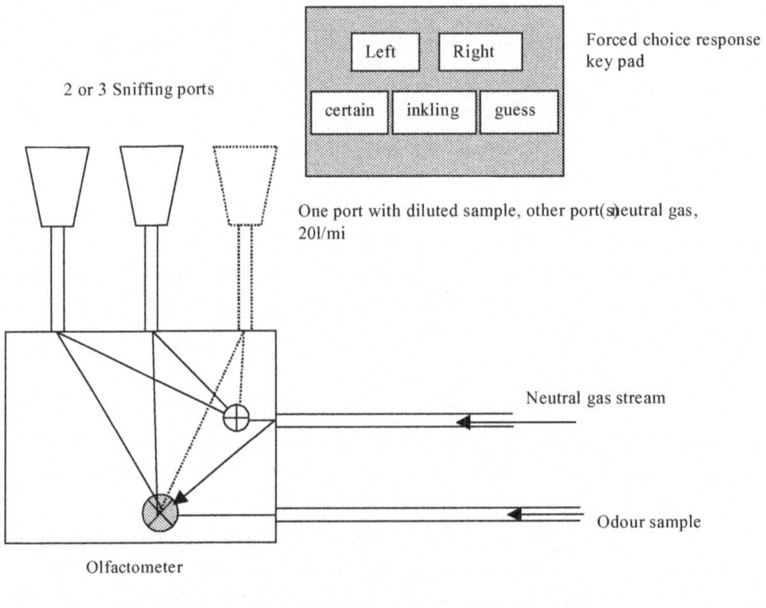

Figure 7.3. Schematic diagram of a forced choice olfactometer.

7.4.1.3 The forced choice/probability mode

In the forced choice/probability mode an olfactometer with three or more ports is used, its construction is similar to Figure 7.3. In this mode the ITE of individual determination of the individual threshold estimate (Z_{ITE}) for each panel member in forced choice/probability mode is done in three stages:

1. Estimation of the approximate value of the dilution factor at the individual perception threshold, Z_d.
2. The value Z_d is used to calculate the presentation series of three steps to be used to determine the Z_{ITE} , the three dilution steps are $Z_1 = Z_d \times 3$, $Z_2 = Z_1/F_s$ and $Z_3 = Z_1/F_s^2$ where F_s, the dilution step factor, $\approx 2^{0.5}$.

3. These three dilutions are each presented at least 10 times randomly at random positions on the olfactometer. Correct choice of horn is TRUE, incorrect is FALSE.

In this mode the panel member is not required to indicate guess, inkling or certain.

7.4.1.4 Calculation of ITE for forced choice/probability

The individual threshold estimate Z_{ITE} is calculated from a set of recorded responses obtained by presenting each of three dilutions, Z_1, Z_2 and Z_3 repeatedly, n times to each assessor, with $n \leq 10$. Because the panel members are not asked to indicate "guess, inkling, or certain" account must be taken of the probability that the assessor produces a random TRUE result when they had not detected the odour, the calculation used below corrects for this.

For each dilution, the observed fraction $f_{observed}$ of TRUE responses in the total of n presentations for that dilution is calculated. This fraction is then corrected for the probability that the assessor produces TRUE results when responding randomly using an olfactometer with p ports:

$$f_{corrected} = \frac{f_{observed} - \frac{1}{p}}{1 - \frac{1}{p}} \qquad (7.1)$$

The dilution factor at the individual threshold estimate, Z_{ITE}, is then calculated by finding the dilution factor that corresponds with $f_{corrected} = 0.5$ from the linear regression formula derived from the three fractions $f_{corrected}$ and the corresponding logarithms of dilution factors Z_1, Z_2 and Z_3.

For measurements on reference odorants, this value can be converted to an individual threshold estimate, expressed as a mass concentration using the known concentration of the reference gas divided by Z_{ITE}.

7.4.1.5 Assessor selection

The key part of accurate odour measurement, according to prEN 17235, is the selection of the odour assessors. In order to select odour assessors, n-butanol (butan-1-ol) has been chosen as the reference material. (While it is recognised that a single component reference gas is not the ideal, no representative odorant mixture has yet been formulated.) Only people with a mean personal threshold for n-butanol in neutral gas of between 20 ppb and 80 ppb and a log standard deviation of less than 2.3, calculated from the last 10 to 20 ITEs, are acceptable. These assessors are continually checked for their detection threshold (at a minimum after every 12 odour measurements) and have to remain within these limits to be a panel member.

This selection criteria used at the Silsoe Research Institute laboratory leads to us having to reject about 43% of those tested because they are not sensitive enough and 12% because they are too sensitive to n-butanol. The complete distribution of sensitivities of all 142 people tested in the Silsoe Research Institute laboratory, to date, is illustrated in Figure 7.4. The butanol thresholds are grouped into 0.3 log intervals, i.e. less than 1.0, 1.0 to 1.3, 1.3 to 1.6, etc. Of those who have a *qualifying* sensitivity, about two thirds have a threshold above the accepted reference value of 40 ppb (log 1.6).

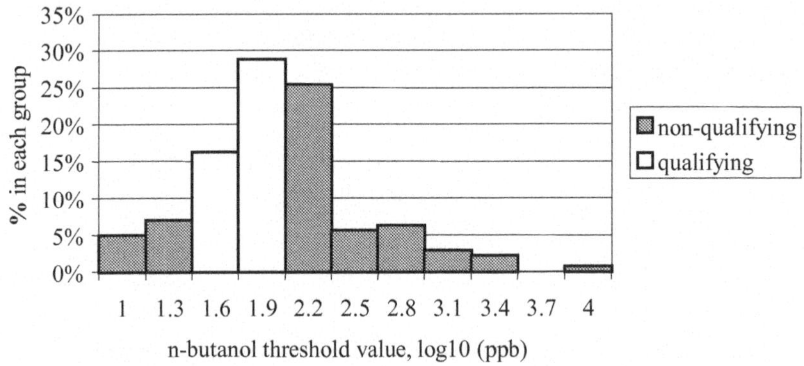

Figure 7.4. Distribution of sensitivities to n-butanol (for 142 subjects tested).

Selection of the panel members using the above method will lead to acceptable accuracy and precision and enable a laboratory to comply with the criteria set in the prEN (section 7.5.1.3).

7.4.1.6 Calculation of the odour concentration

The prEN states that a minimum of two ITEs must be obtained for each of the panel members used to assess each odour, and a minimum of 8 ITEs must be used in the final calculation of the odour concentration. The result is obtained by calculating the geometric mean of the ITEs. It is known that, even if all assessors qualify as panel members using the n-butanol criteria, some will be insensitive (anosmic) and some hypersensitive to environmental odours. In order to eliminate these extremes and improve the repeatability of the measurements, the prEN has adopted a systematic method of excluding the outliers. The ITEs are compared with the geometric mean value of all ITEs. If one ITE varies from the mean by more than a factor of five, above or below, the mean all responses of that panel member are excluded from the calculation and a new mean calculated. This retrospective screening is repeated until all responses are within the +/- factor five variations. The result will be valid, according to the EN, only if at least 8 ITEs remain for calculating the odour concentration.

7.5 COMPLIANCE WITH THE CEN STANDARD

7.5.1 Laboratory practice

7.5.1.1 Laboratory conditions

For laboratories to conform to the required standard, they must be guaranteed to be free from odour. They are usually air-conditioned with activated charcoal filtration. They must also have a source of odour free air, i.e. neutral gas, with which to dilute the odour sample. The olfactometer, which is a dilution device, is made entirely from approved materials, glass, FEP, or stainless steel. Samples are processed within 30 hours of collection.

7.5.1.2 Quality criteria

The Standard is based on the following accepted reference value which shall be used when assessing trueness and precision:

$$1\ ou_E \equiv 1\ \text{EROM} = 123\ \mu g\ \text{n-butanol}$$

When 123 μg n-butanol is evaporated in one m^3 of neutral gas at standard conditions (20 °C) for olfactometry the concentration is 0.040 μmol/mol (40 ppb or a $\log_{10}$ value of 1.6)

Two quality criteria, as below, are specified to measure the performance of the laboratory in terms of the standard accuracy and precision, respectively.

Accuracy reflects the trueness or closeness to the correct value, in this case the true value for the reference material is 40 ppb and the precision is the random error. The standard specifies how these two quality criteria are calculated (CEN, 1999).

The criterion for accuracy A_{od} (closeness to the accepted reference value) is

$$A_{od} \leq 0.217\ .$$

In addition to the overall accuracy criterion, the precision, expressed as repeatability, r, shall comply with

$$r \leq 0.477\ .$$

This criterion for repeatability can also be expressed as:

$$10^{r} \leq 3.0\,.$$

This repeatability requirement implies that the factor that expresses the difference between two consecutive single measurements, performed on the same testing material in one laboratory will not be larger than a factor 3 in 95% of cases.

7.5.1.3 Compliance with the Quality Criteria

The performance of a laboratory is monitored continuously by checking the accuracy and repeatability of the daily measurements of n-butanol. The charts shown below illustrate this over the first three months of 2000 at the Silsoe Research Institute laboratory. Each point on the graphs is the result of the previous 20 panel threshold butanol measurements. The panel thresholds are shown in Figure 7.5. This shows the accuracy to be slightly biased to the high side of the accepted reference value of 1.6. This is explained by reference to Figure 7.4, the distribution of threshold values. Up to the present, panel members are selected randomly from our list of qualified assessors, thus the panel is biased towards the higher n-butanol threshold. Closer agreement with the "accepted reference value" can be achieved by selecting panel members more rigorously.

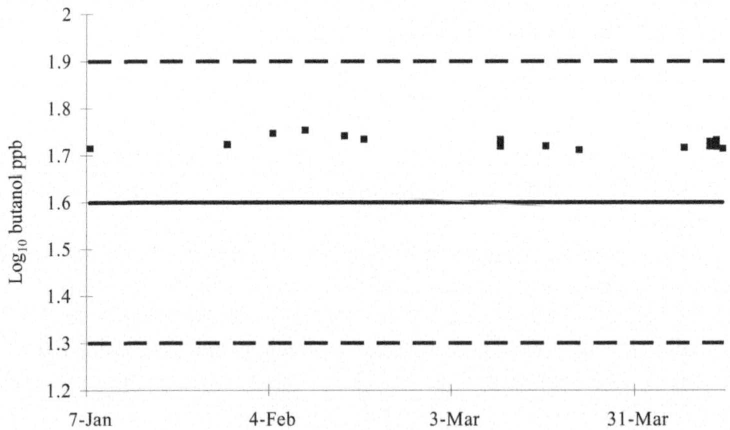

Figure 7.5 Three-month history of average panel threshold at the Silsoe Research Institute laboratory.

In figure 7.6. the record of accuracy and repeatability criteria over the same period shows that the laboratory exceeded the quality criteria of the standard (accuracy criteron shown as — — , and repeatability criteron shown as - - - -).

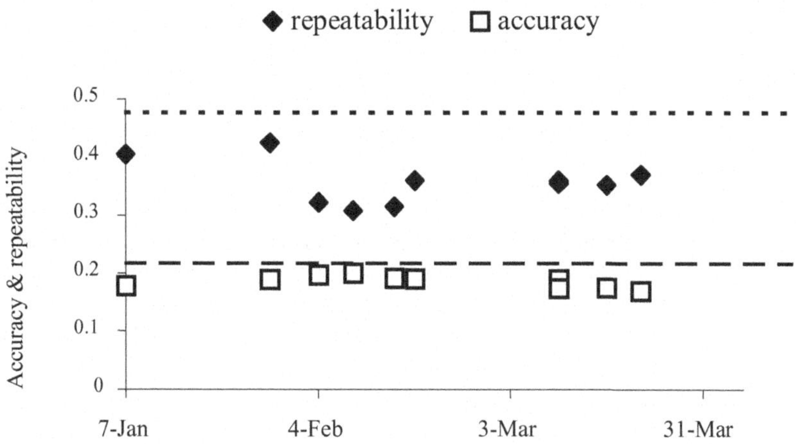

Figure 7.6 Repeatability and accuracy recorded at Silsoe Research Institute laboratory.

7.6 SAMPLING CONSIDERATIONS

7.6.1 Number of samples that should be collected

The number of odour samples required will depend upon the nature of the source and the purpose of the measurement. The number of analyses required to achieve a defined precision is reported in the draft standard.

Table 7.2 shows the 95% confidence limits applicable to the analysis of n identical samples of odour concentration m (=1000 ou_E m^{-3}) where the laboratory's precision *r* is 3.

There is a large improvement in the confidence interval between 1 and 4 samples, but relatively less between 3 and 10 samples. Practically, this usually results in triplicate or quadruplicate samples giving the optimum balance between precision and cost.

Table 7.2. 95% confidence limits with replicate samples.

n	lower limit $ou_E\ m^{-3}$	m	upper limit $ou_E\ m^{-3}$
1	453≤	1000	≤2209
2	571≤	1000	≤1752
3	633≤	1000	≤1580
4	673≤	1000	≤1486
10	778≤	1000	≤1285

Where the performance of an abatement device is to be assessed, the number of paired (inlet and outlet) samples required to measure the removal efficiency will depend upon the assumed removal efficiency and upon the laboratory's precision. Table 7.3 shows the 95% confidence limits applicable to the analysis of n pairs of concurrently collected inlet and outlet samples, where the laboratory's precision is 3 and the calculated removal efficiency, η_{od}, is 90%.

As previously, it can be seen that, under most instances, the optimum balance between precision and cost will be satisfied by having 3 or 4 pairs of samples.

Table 7.3. How the number of replicate samples affects the 95% confidence interval of estimation of the % odour removal efficiency.

		Removal efficiency	
n	η_{od}	Lower limit confidence interval	Upper limit confidence interval
1	90	69.3%	96.7%
2	90	77.9%	95.5%
3	90	80.9%	94.8%
4	90	82.5%	94.3%
10	90	85.7%	93.0%

7.7 QUALITATIVE ASSESSMENTS COMBINED WITH THE CEN STANDARD

7.7.1 Perception of the odour

The prEN 13725 deals with the detection of an odour and no opinion is sought of the assessors other than is there a difference between the neutral gas and the sample.

The second dimension of the sensory perception of odorants, intensity, refers to the perceived strengths of the odour sensation. Intensity increases as a function of concentration. The dependence may be described as a theoretically derived logarithmic function according to Fechner (1860):

$$S = k_W \cdot \log I/I_0 \tag{7.2}$$

Where:
S = perceived intensity of sensation (theoretically determined),
I = physical intensity (odour concentration),
I_0 = threshold concentration,
k_W = Weber-Fechner coefficient.

Stevens (1957) suggests a power relationship should be applied:

$$S = k \cdot I^n \tag{7.3}$$

Where:
S = perceived intensity of sensation (empirically determined),
I = physical intensity (odour concentration),
n = Stevens' exponent,
k = a constant.

Which one of these two descriptions applies depends on the method used. To date, no theory has been able to derive the psychophysical relationship from knowledge about the absolute odour threshold of various substances.

The third dimension of odour is the odour quality, i.e. what the substance smells like.

The fourth dimension of odour is hedonic tone. This is a category judgement of the relative pleasantness or unpleasantness of the odour. Both odour quality and hedonic tone in addition to concentration influence the odour intensity (and potential annoyance).

Although the end use of odour measurement is in reducing odour nuisance, the relation between measured concentrations of odour according to this standard and the occurrence of odour nuisance is highly complex. Atmospheric processes determining the dispersion of the odours, the quality of the odour (hedonic tone) and finally the receptor characteristics of those exposed to the odour affect the level of nuisance the odour may cause. These characteristics not only vary strongly between individuals, but also in time within one individual.

7.7.2 The second dimension of odour: assessment of odour intensity

For all the human senses, including the sense of smell, there are relationships between the magnitude of a sensation and the intensity of stimulus. The form of these relationships depends on the scaling method used. Category estimation derived from Fechner's Law (Fechner 1860) when related to the sense of smell, states that equal ratios of odour concentrations lead to equal differences between perceived intensities (e.g. points on a category scale), thus perceived intensity (I) is a linear function of the logarithm of odour concentration, C:

$$I = k_1 (\log_{10} C) + k_2 \quad (7.4)$$

Where:
k_1 and k_2 are constants.

Odour intensity is measured using this category estimation technique. After determining the odour concentration of the samples, a range of supra-threshold dilutions is presented in random order to panel members. They are required to indicate their perception of intensity at each dilution according to the following scale:

0 No odour; 1 Very faint odour; 2 Faint odour; 3 Distinct odour; 4 Strong odour; 5 Very strong; 6 Extremely strong odour.

Intensity scores are obtained from each panel member at each of 12 presentations of supra-threshold dilutions and the average score for each presentation plotted against $\log_{10}$ concentration. A linear regression is performed on intensity *vs.* $\log_{10}$ concentration and the line of best fit plotted on the graph.

Examples of two such measurements are shown in Figures 7.7 and 7.8. The fresh landfill material has an intensity of 2.5 (faint to distinct odour) at 0.5 log ou/m^3 (3.2 ou/m^3), whereas at the same odour concentration the land fill gas has an intensity of only 1.5 (very faint to faint odour). This means that at the same odour concentration the odour from fresh landfill material will be perceived to be the stronger odour.

If these data had been obtained from an odour source of which an odour abatement plant needs to be designed, then it could be that the intensity of a "faint odour", at a complainant's premises, was considered as the unacceptable limit. In that case the outlet concentration from the abatement equipment would have to be designed to deliver an odour with a concentration of less than 2 ou/m^3 or 6 ou/m^3, respectively, to the nearest complainant.

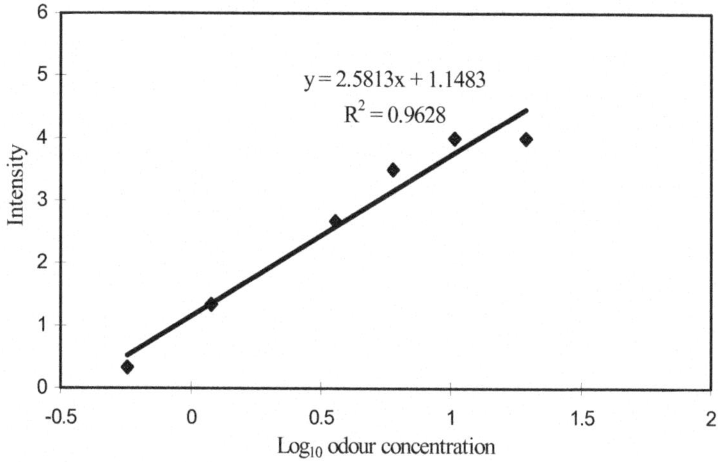

1 Very faint odour | 3 Distinct odour | 5 Very strong
2 Faint odour | 4 Strong | 6 Extremely strong

Figure 7.7. Plot of odour intensity and odour concentration for fresh landfill material.

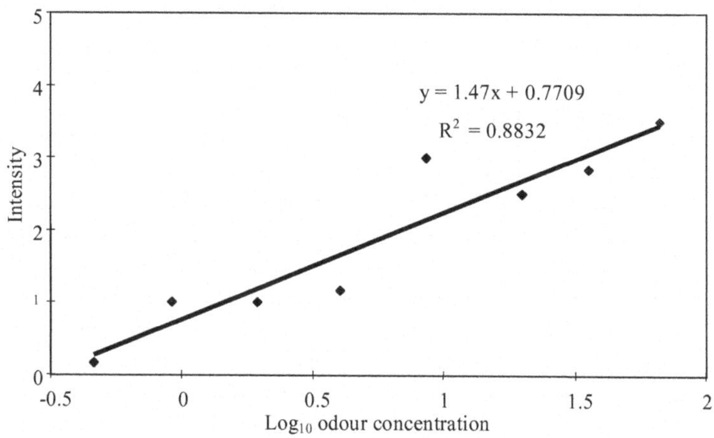

1 Very faint odour | 3 Distinct odour | 5 Very strong
2 Faint odour | 4 Strong | 6 Extremely strong

Figure 7.8. Plot of odour intensity and odour concentration for landfill gas.

7.7.3 The third dimension of odour: odour quality

Some useful information about the characteristics of an odour can be obtained if quality assessments are made at a range of dilution ratios close to the panel detection threshold, although these are not included in the standard.

One assessment we often carry out for customers is a description of the odour. Our odour panel members are asked to smell the odour at a dilution ratio of between 12 and 100 and indicate, from a choice of descriptors, which comes closest to their perception of the odour. Typically we ask if the odour sample smells like these: sewage, fish, rotten cabbage, rotten eggs, bleach, earthy, compost, tarry, smoky, or other. This method is useful for diagnosing if a piece of abatement equipment is changing the odour as well as reducing the concentration.

The data from such an assessment is usually presented as a histogram of the panel's response.

7.7.4 The fourth dimension of odour: hedonic tone:

This assessment is a judgement of the un/pleasantness of the odour. In a similar way to the assessment of the intensity, the panel members are asked to score their perception of the odour on a scale from 1 to 5 at a range of odour concentrations above the odour threshold. A graph similar to the intensity graph can be plotted.

7.8 CONCLUSIONS

(1) Odour measurements no longer need be the arbitrary assessment they have often been perceived as. Olfactometry to the CEN draft standard, prEN 13725, ensures a measurable accuracy criterion for the laboratory, and ensures reproducibility of results between laboratories.
(2) Sampling is equally important as a part of the measurement; the number of samples taken will affect the confidence we can have in the result.
(3) When assessing the efficiency of odour abatement plant, it is important to take into account the measurement precision of the laboratory.
(4) Once an odour concentration measurement has been done on a sample, then the other three dimensions of odour can be investigated systematically. Measurements of odour intensity can give useful indications of the amount of abatement required, especially when combined with an assessment of hedonic tone.
(5) Odour quality assessments are useful diagnostic tools in disputes: they can provide an independent opinion of the possible odour source.

7.9 ACKNOWLEDGEMENTS

In writing this chapter I have used text (reworded) from the CEN draft standard prEN 17325. I am grateful to the convenor and my fellow members of the TC264/WG 2 who contributed to the document.

7.10 REFERENCES

AFNOR NF X 43-101 (1986) Determination of the dilution factor at the perception threshold.

CEN (1999) Air quality - Determination of odour concentration measurement by dynamic olfactometry. Draft prEN 13725, European Committee for Standardisation, Brussels.

Fechner (1860) Elemente der Psychophysik. Leipsig: Breitkopf and Hartel.

Hangartner, M., Hartung, J., Paduch, M., Pain, B.F. and Voorburg, J.H. (1989) Improved recommendations on olfactometric measurements. *Environ. Technol. Lett.* **10**, 231-236.

Harreveld, A.P. and van Heeres, P. (1997) The validation of the draft European CEN standard for dynamic olfactometry by an interlaboratory comparison on n-butanol, STAUB, Gefahrstoffe Reinh. Der Luft, vol. 57.

Stevens, S.S. (1957) On the psychophysical law. *Psychological Review* **64**, 153-181.

VDI 3881 Blatt 2 (1987) Richtlinien, Olfaktometrie Geruchsschwellenbestimmung, Probenahme.

NVN2820 (1995) Air Quality. Sensory odour measurement using an olfactometer.

7.11 TERMS AND DEFINITIONS FROM THE CEN STANDARD

accepted reference value: A value that serves as an agreed-upon reference for comparison, and which is derived as a consensus value, based on collaborative experimental work under the auspices of a scientific or engineering group. [ISO 5725 part 1, abridged]

accuracy: Closeness of agreement between test result and the accepted reference value. [ISO 5725 part 1]. **Note** - The term ‘accuracy’, when applied to a set of observed values, describes a combination of random components and a common systematic error or bias component.

(sensory) adaptation: Temporary modification of the sensitivity of a sense organ due to continued and/or repeated stimulation. [ISO 5492:1992]

anosmia: Lack of sensitivity to olfactory stimuli. [ISO 5492:1992]

assessor: Somebody who participates in odour testing.

bias: The difference between the expectation of the test results and an accepted reference value. [ISO 5725 part 1]. **Note**: Bias is often called 'systematic error'

certified reference material, CRM: A reference material of which one or more property values are certified by a technically valid procedure accompanied by or traceable to a certificate or other documentation which is issued by a certifying body. [ISO 5725 part 4]. **Note**: For the purpose of olfactometry The Netherlands Measuring Institute in Delft certifies reference materials by comparing them with national standard gas mixtures. A European certifying body for gases does not currently exist.

delayed olfactometry: Measurement of an odour with a time-lag between sampling and measurement. The odour sample is preserved in an appropriate container. [AFNOR X43-104E, see bibliography, Annex J.]

detection limit: see Lower Detection Limit.

detection threshold, (for a reference material)**:** The odorant concentration which has a probability of 0.5 of being detected under the conditions of the test.

detection threshold (for an environmental sample)**:** The dilution factor at which the sample has a probability of 0.5 of being detected under the conditions of the test.

diffuse sources: Sources with defined dimensions (mostly surface sources) which do not have a defined waste air flow, such as waste dumps, lagoons, fields after manure spreading, un-aerated compost piles.

dilution factor: The dilution factor is the ratio between flow or volume after dilution and the flow or volume of the odorous gas. [AFNOR X 43-104E, see bibliography, Annex J]

dilution series: The presentation of a sequence of dilutions to one panel member in order to obtain one Individual Threshold Estimate. **Note**: One dilution series can consist of:
- One series of presentations, at odour concentrations in ascending or random order, where, when sorted in order of descending concentrations, a significant change from consistently TRUE responses to a FALSE response occurs.
- A repeated pattern of presentations according to the procedure described for forced choice/probability mode.

direct olfactometry: Measurement of odour concentrations without any time-lag between the sampling (operation) and the measurements; equivalent to dynamic sampling or on-line olfactometry. [AFNOR X 43-104E, see bibliography, Annex J].

dynamic olfactometer: A dynamic olfactometer delivers a flow of mixtures of odorous and neutral gas with known dilution factors in a common outlet. [AFNOR X 43-101E, modified, see bibliography, Annex J]

dynamic olfactometry: Olfactometry using a dynamic olfactometer.

dynamic sampling: Sampling in direct olfactometry.

European Odour unit: That amount of odorant(s) that, when evaporated into 1 cubic metre of neutral gas at standard conditions, elicits a physiological response from a panel (detection threshold) equivalent to that elicited by one European Reference Odour Mass (EROM), evaporated in one cubic metre of neutral gas at standard conditions

European Reference Odour Mass, EROM: The accepted reference value for the European odour unit, equal to a defined mass of a certified reference material. One EROM is equivalent to 123 μg n-butanol (CAS 71-36-3). Evaporated in 1 cubic metre of neutral gas this produces a concentration of 0,040 μmol/mol.

expected value: The value approached by the average value with an increasing number of measurement values.

forced choice method: For this standard the following definition applies: an olfactometric method in which assessors are forced to make a choice out of two or more air flows, one of which is the diluted sample, even if no difference is observed.

fugitive sources: Elusive or difficult to identify sources releasing undefined quantities of odorants e.g. valve and flange leakage, passive ventilation apertures etc.

group threshold: Detection threshold applying to a group of assessors.

identification threshold: See recognition threshold

individual threshold: Detection threshold applying to an individual.

individual threshold estimate, ITE : The detection threshold applying to an individual estimated on the basis of one dilution series.

instability: The change of a characteristic over a stated period of time, consisting of a systematic part (drift) and a random part (dispersion). [ISO 9169, 6.2.2.]

instrumental dilution range: Range between the minimum and maximum dilution factor.

instrumental fall time: Time taken for the reading to pass from (by convention) 10% to (by convention) 90% of the final change in output signal reading. [ISO 6879] For instruments where transient oscillations occur in the approach to the final output signal reading, the fall time is the time taken for the instrument reading to pass from (by convention) 10% of the final change in instrument reading until the oscillations fall to less than 10% (by convention) of the final change in instrument reading.

instrumental lag time: Time taken to reach 10 % (by convention) of the final change in instrument reading. [ISO 6879]

instrumental response time: Time taken for an instrument to respond to an abrupt change in value of the air quality characteristic. It is the sum of the lag time and rise time (rising mode) or lag time and fall time (falling mode). [ISO 6879, modified]

instrumental rise time: Time taken for the reading to pass from (by convention) 10% to (by convention) 90% of the final change in output signal reading. [ISO 6879] For instruments where transient oscillations occur in the approach to the final output signal reading, the rise time is the time taken for the instrument reading to pass from (by convention) 10% of the final change in instrument reading until the oscillations fall to less than (by convention) 10% of the final change in instrument reading.

lower detection limit, LDL: Lowest value of the air quality characteristic which, with 95% probability, can be distinguished from a zero sample. [ISO 6879]

maximum dilution factor: Maximum settable dilution factor of the olfactometer; an instrument property.

measurement: The presentation to all panel members of those dilution series necessary to produce sufficient data to calculate the odour concentration for one sample.

measuring range: The measuring range comprises all odour concentrations which can be measured by a specific olfactometer. It depends on the minimum and maximum dilution factor and the step factor. The numerical values defining the measuring range are the minimum dilution factor multiplied with the step factor to the power three and the maximum dilution factor divided by the step factor to the power three.

minimum dilution factor: Minimum settable dilution factor of the olfactometer; an instrument property.

neutral gas: Air or nitrogen that is treated in such a way that it is as odourless as possible and that does, according to panel members, not interfere with the odour under investigation. **Safety Warning**: Nitrogen is only used to predilute the sample itself. For the olfactometer the neutral gas used to dilute the sample and present a reference shall be air.

objective method: Any method in which the effects of personal opinions are minimised. [ISO 5492]

odorant: A substance which stimulates a human olfactory system so that an odour is perceived. [Hangartner, M, *et al.* 1989, see bibliography, Annex J]

odorant flow rate: The odorant flow rate is the quantity of odorous substances passing through a defined area at each time unit. It is the product of the odour concentration c_{od}, the outlet velocity v and the outlet area A or the product of the odour concentration c_{od} and the pertinent volume flow rate V. Its unit is ou_E/h (or ou_E/min or ou_E/s, respectively). **Note**: The odorant (emission) flow rate is the quantity equivalent to the emission mass or volume flow rate, for example in dispersion models.

odorous gas: Gas that contains odorants.

odour: Organoleptic attribute perceptible by the olfactory organ on sniffing certain volatile substances. [ISO 5492]

odour abatement efficiency: The reduction of the odour concentration or the odorant flow rate due to an abatement technique, expressed as a fraction (or percentage) of the odour concentration in or the odorant flow rate of the untreated gas stream.

odour concentration: The number of European odour units in a cubic metre of gas at standard conditions. **Note**: The odour concentration is not a linear measure for the intensity of an odour. Steven's Law describes the a-linear relation between odour stimulus and its perceived intensity. When using odour concentrations in dispersion modelling, the issue is complicated by the effects of the averaging time of the dispersion model, further complicating the use of the odour concentration as a direct measure for dose. To define a 'no nuisance level', the entire method of dosage evaluation, including the dispersion model, will yield a 'dose'. The relation between this 'dose' and its effect (odour annoyance) should be validated in practical situations to be a useful predictive tool for occurrence of odour nuisance.

odour detection: To become aware of the sensation resulting from adequate stimulation of the olfactory system.

odour panel: See panel.

odour unit: One odour unit is the amount of (a mixture of) odorants present in one cubic metre of odorous gas (under standard conditions) at the panel threshold. **Note**: See also "European odour unit".

odour threshold: See panel threshold.

odourless gas : See neutral gas.

olfactometer: Apparatus in which a sample of odorous gas is diluted with neutral gas in a defined ratio and presented to assessors.

olfactometry: Measurement of the response of assessors to olfactory stimuli. [ISO 5492]

olfactory: Pertaining to the sense of smell. [ISO 5492]

olfactory receptor: Specific part of the olfactory system which responds to an odorant. [after ISO 5492]

olfactory stimulus: That which can excite an olfactory receptor. [ISO 5492, modified]

on-line olfactometry: See direct olfactometry.

operator: Person directly involved in operating the olfactometer and instructing the panel in olfactometry.

panel: A group of panel members.

panel member: An assessor who is qualified to judge samples of odorous gas, using dynamic olfactometry within the scope of this standard.

panel screening: Procedure to determine if the performance of panel members is in compliance with selection criteria. See also panel selection.

panel selection: Procedure to determine which assessors are qualified as panel members.

panel threshold: Detection threshold applying to a panel.

perception: Awareness of the effects of single or multiple sensory stimuli. [ISO 5492]

population (detection) threshold: Detection threshold applying to the general population, if this population is not specified.

precision: Closeness of agreement between independent test results obtained under prescribed conditions. [ISO 5725-part 1]. **Note**: Precision depends only on the distribution of random errors and does not relate to the true value or the accepted reference value. The measure of precision is usually expressed in terms of imprecision and computed as a standard deviation of the test results. Higher imprecision is reflected by a larger standard deviation. 'Independent test results' means results obtained in a manner not influenced by any previous result on the same or similar material.

presentation: One presentation is the presentation of one dilution to one assessor. [NVN 2820]

presentation series: The presentation of one dilution to all panel members in one round.

presented gas flow: The gas flow presented to the assessor. It may be:
-a diluted odour sample
- neutral gas (e.g. as a blank or reference air).

proficiency testing: The system for objectively testing laboratory results by an external agency.

quality: The totality of features and characteristics of a product or service that bear on its ability to satisfy stated or implied needs. [ISO 6879]

quality assurance: All those planned and systematic actions necessary to provide adequate confidence that a product, process or service will satisfy given requirements for quality. [ISO 6879]

random error: Unpredictable errors which average to zero. [ISO 5492]

recognition threshold: The odour concentration which has a probability of 0.5 of being recognised under the conditions of the test (definition not applied in this standard).

reference material: For this International Standard: Substance or mixture of substances, the composition of which is known within specified limits, and one or more of the properties of which is sufficiently well established to be used for the calibration of an apparatus, the assessment of a measuring method, or for assigning values to materials. [ISO 6879]

reference value: See accepted reference value.

repeatability: Precision under repeatability conditions. [ISO 5725-part 1]

repeatability conditions: Conditions where independent test results are obtained with the same method on identical test material in the same laboratory by the same operator using the same equipment within short intervals of time. [ISO 5725-part 1]

repeatability limit: The value less than or equal to which the absolute difference between two test results obtained under repeatability conditions may be expected to be with a probability of 0,95. [ISO 5725-part 1, modified]. **Note**: In this standard the test result is the decimal logarithm of the panel threshold.

reproducibility: Precision under reproducibility conditions. [ISO 5725-part 1]

reproducibility conditions: Conditions where test results are obtained with the same method on identical test material in different laboratories with different operators using different equipment. [ISO 5725-part 1]

reproducibility limit: The value less than or equal to which absolute difference between two test results obtained under reproducibility conditions may be expected to be with a probability of 0,95. [ISO 5725-part 1, modified]. **Note**: In this standard the test result is the decimal logarithm of the panel threshold.

responsible person: Person who is finally responsible for the total of olfactometry in a laboratory.

round: One round is the presentation of one dilution series to all assessors.

sample: In the context of this standard, the sample is the odorous gas sample. It is an amount of gas which is assumed to be representative of the gas mass or gas flow under investigation, and which is examined for odour concentration. [ISO 6879]

sensory fatigue: Form of adaptation in which a decrease in sensitivity occurs. [ISO 5492]

sensory reference: The presented gas flow to which the diluted sample is compared.

single measurement: Identical to Measurement, see also test result.

to smell: To detect or to attempt to detect an odorant.

standard conditions for olfactometry: At room temperature (293 K), normal atmospheric pressure (101.3 kPa) on a wet basis [as in ISO 10780]. **Note**: This applies both to olfactometric measurements and volume flow rates of emissions.

static olfactometer: A static olfactometer dilutes by mixing two known volumes of gas, odorous and odourless, respectively. The rate of dilution is calculated from the volumes. [AFNOR X 43-101E]

static sampling: Sampling in delayed olfactometry.

step factor: The factor by which each dilution factor in a dilution series differs from adjacent dilutions.

subjective method: Any method in which the personal opinions are taken into consideration. [ISO 5492]

substance: Species of matter of definite chemical composition. [Hangartner, 1989]

test result: The value of a characteristic obtained by completely carrying out a specific measurement, once.[ISO 5725-part 1]

traceability: The property of the result of a measurement that can be related through an unbroken chain of comparisons to appropriate reference materials, generally national or international reference materials, using measurement standards of successively increasing accuracy.

trueness: The closeness of agreement between the average value obtained from a large series of test results and an accepted reference value. [ISO 5725-part 1]. **Note**: The measure of trueness usually is expressed in terms of bias.

true value: See accepted reference value.

yes/no method: Olfactometric method in which assessors are asked to judge whether an odour is detected or not.

7.12 ABBREVIATIONS

AFNOR: Association Française de Normalisation (French Standardisation Association)
CEN: Comittée Européen de Normalisation (European Committee for Standardisation)
EN: Euro Norm (European Standard)
ISO: International Standards Organisation
prEN: pre Europäische Norm (preliminary European Standard)
NVN: Nederlands voor-Norm (Netherlands pre-standard)
VDI: Verein Deutscher Ingenieure (Association of German Engineers)

8

Odour analysis by gas chromatography

Phil Hobbs

8.1 INTRODUCTION

Odours are temporal and spatially dimensioned and perhaps one of the more difficult challenges for scientists to investigate. They can act as a trigger to the memory or a feeling in an instant, as physiologically odour recognition is associated with the emotional centre of the brain (Hirsch and Trannel 1996). Of the armoury at the disposal of the analyst gas chromatography (GC) offers the best means of separating odorous components. Sampling and the choice of detector for the GC and its limitations can impose upon chromatographic configurations. There are two major concerns: how does the data obtained from GC analysis relate to any olfactory response; and the low concentrations at which odorants can contribute to malodour

 Odours in Wastewater Treatment: Measurement, Modelling and Control edited by Richard Stuetz and Franz-Bernd Frechen. ISBN: 1 900222 46

make detection difficult. If we consider that the olfactory threshold is the point at which odours are just perceived and for many objectionable odorants this value is as low as about 1 ppb (10^{-9} etc). Quantification at such low concentrations is important as odorants can have a complexity of multiplying and additive interactions (Patterson *et al.* 1993). Therefore it is better we are able to monitor odorants in the part per trillion (ppt; 10^{-12}) range, for ultimately a more complete description. At such low concentrations surface absorption and chemical activity become major concerns and samples can be lost anywhere from preconcentration stages prior to GC to within the tubing to the detector.

The chemical nature of odorous compounds is such that they are generally polar and subject to chemical reactions or adsorption onto surfaces. There are numerous descriptions of odorous compounds; a summary of sewage odour descriptors is shown in Table 1.1. The most offensive of compounds from sewage are the sulphides (Brennan *et al.* 1996) which present difficulties in terms of analysis as they readily adsorb onto surfaces and can oxidise over a few days during storage with air. Sulphides are normally very malodorous by nature, but at low levels dimethyl sulphide and methanethiol have been clearly identified as beneficial to the bouquet of selected cheeses (Kubickova and Grosch 1998) and beers (Scarlata and Ebeler 1999) at ppb concentration range. Sulphides primarily originate from protein decay (Spoelstra 1980) and sulphate whereby sulphur replaces the chemical function of oxygen in anaerobic environments and forms hydrogen and methyl substituted sulphides. These have been identified in anaerobic processes within sewage works, livestock manure (Hobbs *et al.* 1999) and paper mill works (O'Connor *et al.* 2000).

The biggest concern relates to the measurement of the emission rate from a source, which determines the total emissions within a volume probably of a plume and of course its dilution. So surface emission rate determinations are paramount when wanting to understand how to reduce nuisance odours. GC systems can be connected to a system for continuous monitoring and this overcomes some of the major problems associated with sampling. There are few determinations of emissions rates from wastewater sources (Devai and DeLaune 1999), but publications and work will be presented from other relevant sources such as intensive pig units and paper mills.

8.1.1 Sampling

As with most analytical procedures sampling can greatly influence the outcome both in terms of quantitative and qualitative measurements. On the one hand odours are easy to detect by the olfactory senses but for those at concentrations below 10 ppb there are difficulties not only because of the volume of the sample required but because of their chemical character. Accurate knowledge of what compounds are present and the materials by which we sample them are important aspects of identification and determination of the emission rate from a source. Because of the potential loss of volatile compounds from the measuring system good quality control by the use of standards and spiking techniques should be used to validate results. Spiking a sample involves adding known quantities of the compound to be measured, and identifying if all that has been added is present during analysis. If less than that is present after being introduced as a spike then the analytical procedure gives lower concentrations. Often the sample amount lost will be a greater proportion of the initial concentration for samples at an increasingly lower concentration.

At low concentrations, adsorption onto surfaces becomes significant, especially for odorants like the sulphides, which can adsorb onto metal surfaces and oxidise or react chemically. Silicone tubing is hydrophobic by nature and will adsorb sulphides and should not be used for connections where such odorants are being transferred.

In the case of methanethiol polymerisation occurs to produce higher methyl sulphides e.g. dimethyl disulphide (CH_3-S-S-CH_3). Of those odorants that give trouble during analysis sulphides can give considerable problems. At low concentrations adsorption onto surfaces becomes significant. Compounds like the sulphides will adsorb onto metal surfaces and oxidise or react chemically in the case of methanethiol to polymerise and produce higher methyl sulphides such as dimethyl disulphide (CH_3-S-S-CH_3).

8.1.2 Sample odorant profile and its relationship to olfactory description

While GC methods can obtain a profile of odorant concentrations and give useful information additional information in terms of olfactory response should be considered. However, before we can do this we need a means of transforming the odour expressed as odorant composition into one that expresses olfactory response. Development of this possibility would reduce odour measurements by olfactometric methods that can be costly. However mixtures of odorants have been shown to have several effects on perception. Previous workers have attempted to find a relationship between OC and the concentration of a single component in the headspace to simplify odour measurements (Dorling 1977; Schaefer 1977). Establishing an indicative relationship between OC and any component can be difficult because of the variation of odour measurements and emissions. Livermore and Laing (1998) suggest that odours are recognised as an object from a source such as sewage, food or flowers even though the odorants are complex mixtures of chemicals. Some early understanding of odour composition has been developed. Studies using 1-butanol, 2-pentanone and n-butyl acetate have demonstrated that sensitivity to, and stability of the odour was enhanced by composites of the three components rather than the presence of a single component (Patterson *et al.* 1993). They also reported that mixtures can exhibit an additive effect on odour and in some cases hyper and hypo addition where the results are above addition and below addition of their threshold levels respectively. A maximum of four odorants could be identified in a mixture and that this was found to be independent of the type of odorants (Livermore and Laing 1998).

There are few publications on the effects of different malodorous components associated with sewage and sludge cakes that contain mostly sulphides (Leach *et al.* 1999; Winter and Duckham 2000). Such models of malodour mixtures appear to function differently to the perfume mixtures. One looks at a model involving dimethyl disulphide, H_2S and pyridine where the model confirmed the addition of odorants explained the odour intensity of the three components. Laing and Glemarec (1992) also confirmed the addition model for up to four component mixtures with no synergistic effects. They found that H_2S was the least suppressed component in mixtures containing 3-methyl indole, butanethiol and 3-methyl pentanoic acid.

In a recent approach (Hobbs *et al.* 2001) using 4-methyl phenol, hydrogen sulphide, acetic acid and ammonia in concentrations present from decay of pig manure, as expected hydrogen sulphide was found to be the primary odorant. The model showed no effect of the acid–base balance of the odour. Suprisingly 4-methyl phenol gave a negative OC effect with increasing concentration. The model did not follow an additive, geometric or average olfactory prediction.

8.1.3 Emission rates

Whilst odours can be sampled immediately they will only give a profile of the relative concentrations of components present if the appropriate GC configuration is used. It is necessary to take more exacting measurements to determine the emission rate from a source to obtain data for modelling or give realistic estimation of emission rates. This requires planning and normally the use of or access to a vent or experimentally with an open or closed chamber system (Cumby *et al.* 1995). Results can be as mass emitted per unit area or mass emitted per unit area divided by volume of the emitting sample, especially in the case of digestive or decaying processes that are normally biogenic. The latter measurement allows for gas production and the surface area, which will have physical factors that limit the rate of emission. Biogenic sources will also have a history of the age of the waste, the starting material and the degree to which oxygen diffuses into the storage area. The surface to be evaluated should have well characterised airflows and can be non-interfering (laminar) or turbulent. Clean air has to be introduced at one end for the open system of sampling. However, if a closed system is preferred which makes the volatile compounds where the concentrations of odorous components increase to make analysis easier, further complexities of mathematics are required to allow for the increase in concentration that suppresses further emissions. This occurs ideally according to Henry's Law and normally gives an inverse exponential curve limited at the saturated vapour pressure (SVP) for a volatile compound. The curvature is related to the kinetics of the emission process and for compounds with increasing molecular weights the time to reach saturation can be of the order of months e.g. TNT (MWt 261). Generally odorous compounds have a molecular weight of about 30 to about 150.

If simultaneous sensory measurements are required then 6 sample volumes of about 50 litres are used. For a closed system a large chamber volume of about 40 m^3 can be required. Such a large system will leak and differential equations are necessary to calculate the mass lost with respect to time and the mass emitted in the closed chamber.

There is a general consideration that concerns the emission rate from a surface and the wind speed. Laminar flows are present at low air speeds and surface emissions will be diffusion limited. At greater wind speeds turbulence will increase the rate of emission because mixing enhances emission but only to a point where diffusion to the surface limits the rate of emission. Therefore with increasing air speed more turbulence will occur but there will be a speed at which the emission rate will not increase and of course the volatile compound concentration will decrease in the air.

There are some good observations that apply in all situations with regard to wind speed and add to the complexity of total odorant mass emitted and apply to individual odorants and may change the note of the odour. As the wind increases from zero more mass of odorant will be emitted and at some point a maximum transfer of odorant will be emitted from the surface. From this point as the wind increases the odorant will be diluted and hence change odorant profile with increasing wind speed. (Zahn *et al.* 1997) have approached this problem in broad terms in a publication describing emissions from pig lagoons in the USA. After about 3 $m.s^{-1}$ the odour concentration declines.

8.1.4 Transportation of sample

Samples can be transported without being concentrated, but again this is another means by which sample integrity can be affected. Fortunately our sense of smell and more professionally an odour panel have highlighted some of the better means of transporting samples. Stainless steel canisters and poly-fluorinated polymer bags made of Teflon have proved effective for volatile organic compounds. However, odorous compounds are often polar and chemically reactive so the presence of moisture, light or reactive surfaces require that some storage tests should be performed before analytical assurances can be recognised. Teflon bags have proved effective for odours for up to 24 hours, but our laboratory has shown that if sulphides are present the sensory information on odour concentration will decay by 20 %[1]. Therefore the quality of the results can only be assured if tests are performed to identify the stability of sample which may be dependent on a range of factors such as bag material method of collection. They are general rules to conserve different sample types. For hydrophobic gases a film of non-porous layer or bag that has hydrophilic tendencies should be used and vice versa for hydrophilic odorants.

8.2 PRE-CONCENTRATION OF SAMPLE

The chemical composition can be in the ppt range or below the odour threshold in multiplicative or additive effects are present between odorants. Such concentrations are below the detection range of most instruments. Therefore concentration techniques are required if we are to obtain realistic appraisals of the chemical composition. Volatile compounds are normally condensed onto a surface that should be inert to prevent chemical reactions. There are two major means by which to pre-concentrate the sample before introduction to the GC system: one cooling an inert surface and/or employing one that will adsorb the volatile compounds. Either thermal or liquid desorption can be used to displace the sample for presentation into a GC system.

8.2.1 Cryogenic trapping

Volatile compounds are trapped on an inert surface, such as glass, by reducing the temperatures below their boiling point (BP). The temperature should be 20–50 °C below the BP to reduce the vapour pressure to retain the sample quantitatively. Some adsorbents are designed to trap the less volatile compounds such as Tenax TA®. However, they can be placed in the cryogenic trap to extend the range of the length of carbon chained molecules to be trapped from C_{15}-C_6 to C_{15}–C_2. There are two concerns: first, if the cooling temperature becomes too low as with, for example, liquid nitrogen then liquid oxygen will also be trapped. Liquid oxygen will readily oxidise organic compounds. Many instruments with a Peltier-cooling device can be electronically controlled to operate within the required temperature range for primary and secondary re-focusing or concentrating a sample. The peaks are more focussed because part of the column or a surface is cooled so that all the adsorbents are collected on a small surface and often heated with the column oven. Second, if too large a sample volume is collected then moisture will condense to form ice and can block or restrict the flow or spread the profile of the compounds trapped in the sample.

---

[1] Unpublished IGER data.

8.2.2 Adsorption

Numerous adsorbents have a range of specific properties to collect VOCs and their selectivity to certain chemical groups and the mass that they can adsorb are dependent on pore size and surface attraction properties. Adsorbents have a range of different surface areas (per unit mass) and different particle sizes appropriate to different airflow conditions using the same source material. Generally adsorbents with larger surface areas will have a greater capacity for the adsorption of smaller molecules. If too much VOC passes through the adsorbent then a steady state between the mass on the adsorbent and the concentration in the sample will be reached. To ensure this is not reached a breakthrough volume for each adsorbent has been determined for most VOCs and will recommend the largest volume to be sampled without VOC loss. These values are normally at ambient temperatures and some are given for higher temperatures where the breakthrough volume will be less. Such volumes have been determined and are available from manufacturers, the scientific literature and the internet. The following are available on web sites (www.sisweb.com/index/referenc/tenaxtam.htm) including breakthrough volumes for a range of polar compounds showing the dependent on temperature (www.tu-harburg.de/et1/private/gk/break/break.htm). Because of the large number of adsorbents available, we will consider the more commonly researched adsorbents to give an understanding of their general properties. Information on and supply of these adsorbents can be obtained from the major chemical suppliers, but for more detailed information the scientific literature is better. Although adsorbents have proved a successful means of trapping volatile compounds there are some problems. First moisture inhibits and prejudices the VOC profile adsorbed. Second often no one adsorbent can trap the spectrum of compounds present, so different adsorbents are placed adjacent to one another in the tubes. The adsorbents are placed in an order such that the sample migrates through adsorbents of increasing strength. The largest molecules will be trapped first and the more volatile species will be trapped further down the tube. Desorption from the multi-bed tube should be performed when the carrier gas flow is in the opposite direction to adsorption mode. This is because less volatile compounds are more likely to produce broader chromatographic peaks if they pass through the stronger adsorbent for the more volatile compounds. In this event, the limit of detection will be reduced and this is a major problem for analytical methods sampling at such low concentrations.

Adsorption materials have been clearly designated for recognised methods of VOC analysis, whether liquid or thermal desorption is used to transfer the sample from the adsorbent to the GC system. Recognised choices of adsorbents for odour trapping are extracted from the literature. Their use and selection are available on following websites (Table 8.1). Most are legally recognised methods which are designated for groups of compounds generally having similar chemical activity. Some sites require a questionnaire to be filled in first but offer standardisation of approaches has established a range of methods in the National Institute of Official Manual of Methods and the OSHA, which has published fewer methods.

Table 8.1. List of applications notes for methods analysis of VOCs in the atmosphere.

Company and location	Internet address
Marks International, UK	www.markes.com
OI Corporation, USA	www.oico.com/apppvsv
US Geological Survey, USA	water.wr.usgs.gov/pnsp/pest.rep/voc.html
Restek Corporation, USA	www.restekcorp.com/voa/voa.htm
National Institute for Occuptional Safety and Health, USA	www.cdc.gov/niosh/homepage.html
J & W Scientific, USA	www.jandw.com/GCAppnotes.htm
Scientific Instrument Services, USA	www.sisweb.com/index/references/apnoted.htm
Battelle, USA	www.battelle.org/environment/ASAT/canister.html
Gerstel, Germany	www.gerstel.com/solutions/index.htm
SKI Inc, USA	www.skcinc.com/guides.html

8.2.2.1 Porous polymers

Tenax TA® is a porous polymer resin based on a 2,6-diphenylene-oxide polymer resin. It has been specifically designed for the trapping of lesser volatile components from gaseous samples. Tenax TA® has a low affinity for water and is for trapping volatile compounds from high moisture samples. Tenax TA® is therefore suitable for trapping larger organic molecules such as pheromones and large hydrocarbons such as the terpenoids. As with many adsorbents Tenax TA® should be thermally conditioned with a high purity gas containing no oxygen at elevated temperatures to remove any residual components. Tenax TA® has a temperature limit of 350 °C and has been used at lower temperatures in a secondary cryogenic trap. A secondary cryogenic trap is used in several automatic thermal desorption systems to focus the VOCs into a smaller carrier gas volume before entering the GC system to improve detection for small sample masses.

8.2.2.2 Carbon-based adsorbents

These can be identified into activated charcoals, graphitised carbons and carbon molecular sieves. Charcoals were the first adsorbents to be used commercially for adsorbing odours and they still have their uses in analytical science in different forms. Early analytical materials were produced from heating materials such as coconut husks which is an activated charcoal. It is made from material burnt in a super heated, high oxygen atmosphere creating pores about 0.1–0.8 10^{-9}m in diameter throughout the charcoal. One gram of activated charcoal can have about 1000 m^2 of surface. Activated coconut charcoals are effective at adsorbing selected odorants such as carbon disulphide (NIOSH 1600) and hydrogen sulphide (NIOSH 6013), but alkyl halides may also be adsorbed. Desorption for these methods involves using a solvent. Activated charcoals are not considered effective because of memory effects and/or sample broadening problems during desorption.

There are a range of adsorbents within this family that are considered to be chemically non-specific in terms of the surface adsorption characteristics. This is achieved by utilising London attraction forces to perform adsorption. They are graphitised black carbon materials that trap lower MWt compounds down to C_4–C_8. They are often identified as Carbotrap or Carbopack materials that have a range of adsorption capacities from the large to the C_2 chain length molecules. They have been recognised as being more hydrophobic than most adsorbents and minimise the effect of moisture in the sample that may prejudice the VOCs adsorbed. They can operate at temperatures up to 400 °C and have low bleed or background into the GC system.

Carbon molecular sieves have the highest capacities to adsorb and are hydrophobic such that they can operate in humidities of up to 90 % and can be desorbed at temperatures greater than 400 °C. They are used for organic solvents and volatile low molecular weight halides such as CFCs.

8.2.2.3 Other adsorbents

There are several miscellaneous adsorbents that are used for odour analysis. Activated silica is used for trapping amines and amino compounds (NIOSH 2002 and 2010) and compounds that are polar, which covers most odorants. Unfortunately it is very good at adsorbing water (and is used as a desiccant) which can add some uncertainty to the results. Mercuric acetate coated glass-fibre can be used to collect mercaptans or alkyl thiols (NIOSH 2542 and OSHA 26).

8.2.3 Desorption

Once trapped the volatile compounds have to be removed without thermal decay or chemical interactions that may produce false peaks or increase the presence of another volatile component. Solvent desorption has been used extensively but can produce false peaks from impurities even at very low concentrations in the solvent. Thermal desorption is more popular than solvent desorption for this reason and speed and ease of desorption. Each adsorbent has a preferred temperature range for desorption for selected trapped compound. Generally for very low concentrations of VOC desorption should be for several hours at 20-30 °C greater than that used in the analytical method before sample collection. The absence of impurities should be verified by monitoring the chromatogram for those reluctant to leave the adsorbent.

8.3 GAS CHROMATOGRAPHY

Gas Chromatography is ideally suited for the rapid separation of complex volatile components that contribute to odour formation, especially since the development of capillary columns that offer a higher peak resolution. The technique principally involves a carrier gas that passes over a stationary phase for which the volatile components have a differential affinity to effect separation. The efficiency of the chromatography is improved by precision temperature control of the column and constant flow of carrier gas. Generally a range of detectors can be attached to the end of the column with the mass spectrometer proving the most effective for the identification of unknown components of an odour (Figure 8.1). If, and more commonly, a flame ionisation detector is used, volatile components can be recognised by matching the retention time with a known compound on the column. Often the flow to the detector can be split and an odour port used to identify the odour note for a given component. Odour components are introduced into the column through an injector inlet: However, with the introduction of capillary columns small volumes have to be introduced and preconcentration of an odour is required. Good analysis depends upon GC inlet, column and detector configuration, but understanding the principles of chromatography is necessary before we can fully describe cach of these stages.

8.3.1 Principles of chromatography

These have been clearly elucidated in numerous texts and only a brief summary of the main principles appropriate to separation will be given. Packed columns have been mostly superseded by narrower bore columns of the range of about 0.100 to 0.10 mm and a range of porous layer open tubular (PLOT) columns. Manufacturing processes have ensured good quality control in terms of tolerances at dimensions that seem quite astonishing in terms of precision. Consequently, improved separations can be reproduced which are measured in units of theoretical plates N:

$$N{=}16.(\mathrm{RT}/W_b)^2 \qquad (8.1)$$

Where:
RT = retention time,
W_b = the peak width on the baseline.

Measurement of the height equivalent theoretical plate (H) uses the length of the column and N in a simple ratio:

$$H=L/N \tag{8.2}$$

These parameters become important for components that have similar physical and chemical behaviour. For GC systems that have a non-specific detector, recognition of a component can only be by the comparison of the RT. Better peak recognition is possible if an index system is adopted where a relative retention time (RRT) of a compound to a known compound. The more volatile compounds need to be separated and this normally requires a thicker film which lines the capillary column. However, to improve column resolution which gives greater separation between the peaks then we need to consider this as a phase ratio between the film and the internal mobile phase or gas. This is governed as in equation 8.3.

Phase ratio = column radius/2xfilm thickness (8.3)

As the phase ratio decreases then the retention increases and the resolution increases. Practically this means that we can alter the column dimensions and adjust the film thickness to achieve the same retention time. The type of carrier gas will affect the resolution, hydrogen will not only give a better resolution than nitrogen according to the van Deemter Curves, but maintain a more stable separation ability with changing gas flows. There are safety issues with hydrogen, but as oxygen is reduced to water no oxygen trap is necessary for sensitive columns.

8.3.2 Column structure

Columns are normally composed of tubing manufactured from glass, stainless steel or fused silica. They often have a stationary phase that gives additional specificity to separate different groups of chemical compounds. Glass and silica columns can have a protective polyimide coating applied to the tubing to reduce physical damage. The inner surface can be treated with various coatings to assist with the chromatography.

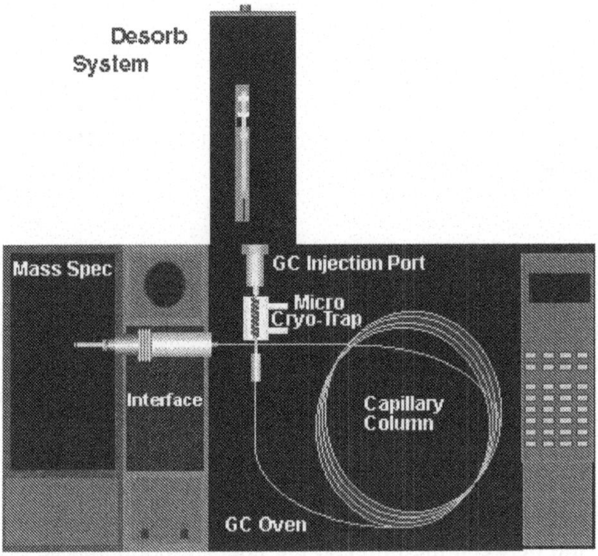

Figure 8.1. Principle components of a gas chromatography.

Most coatings are chemically bonded which reduces bleed from the column and the background of the detector resulting in an enhanced sensitivity or limit of detection to eluting compounds. Coatings normally contain phenyl or methyl-substituted polysiloxanes or a mixture of both. Non-bonded phases such as polyethylene glycol (PEG) can be used to separate fatty acids, however it is less stable and is prone to oxidation and has a lower upper temperature limit of about 200-250 °C. PEG can be chemically modified to reduce acid and base peak tailing. Larger bore columns can have particles that use size exclusion to separate gases that enter into the pores of the particles that are attached to the inner surface of the tubing. PLOT columns are well suited for the separation of sulphide odours as well as hydrocarbons and atmospheric gases. Columns are expensive and protection will enhance their lifetime. Oxygen and water traps are available to reduce concentrations before entry into the column using molecular sieves and metal surfaces respectively. Carbon traps can be used to reduce hydrocarbons which and especially at low concentrations is important for odour measurement.

8.4 CHOICE OF CHROMATOGRAPHY COLUMN

There is a wide selection of column types from a range different manufacturers. A list of the major chromatography column manufacturers is in Table 8.2. These companies also offer technical advice and specialist information, however the choice of column can be difficult as there are several options or combinations necessary to obtain the best information about an odour's components. The choice of column will affect those odorants that are observable and detectable. Often two columns may be used especially during method development and possibly with different detectors to obtain the required sensitivity. Typically some columns will be able to analyse sulphides but may not be able to separate volatile fatty acids and a sample splitter should be used with a dual column system. Choice of column will be limited by the gas volume to be analysed, as well as considering the flow capacity of the detector. Overall a balancing act has to be performed to ensure that sample introduction, gas chromatographic separation and detector specifications are compatible.

Table 8.2. List of column manufactures.

Company and location	Internet address
Alltech Associates Inc., USA	www.alltechweb.com
Anglia Instruments, UK	www.angliainst.co.uk
Environmental Services Associates, USA	www.esainc.com
Enviro Technology Services, UK	www.ssd.rl.ac.uk/news/cassini/huy.html
J & W Scientific, USA	www.jandw.com/gc3.htm
Jones Chromatography, UK	www.jones-chrom.co.uk
Labquip, Ireland	kol.ie/18324e
Perkin-Elmer Analytical Instrument, USA	instruments.perkinelmer.com/index.asp
Scientific Instrument Services, USA	www.sisweb.com/home.htm
Shimadzu	www.shimadzu.com/index.htm
Supelco Inc., USA	www.sigma-aldrich.com/saws.nsf/supproducts?openframeset
ThermoQuest, UK	www.thermoquest.com
Unicam Chromatography, UK	www.unicam.co.uk/Pages/gchome.htm
Varian Chromatography, USA	www.varianinc.com/chrompack/index.html

Capillary columns have dimensions between 0.05 and about 1 mm internal diameter and operate with gas flows lower than 1 ml/min, so a small volume of sample is required for analysis. Here the packed columns have an advantage over the capillary columns in determining low concentrations chiefly because the FID response is a function of the mass of the analyte ionised in the detector. However, peak separation for packed columns becomes a problem with complex mixtures and there are difficulties with materials used. Sulphur-containing compounds will readily disappear into silica or metal or porous polymer surfaces.

8.4.1 Sample components and choice of column

From the analytical point of view the important factors are that the odour sample or its components are not decomposed or lost on the instrument surfaces due to adsorption. This primarily dictates the choice of column, fittings and sampling method. As sulphides are relatively unstable and often present in malodours, then consideration should focus around minimising oxidation, adsorption and chromatographic column choice. Low concentrations of odorants means a high sample volume and column size unless samples can be pre-concentrated onto an adsorbent and adequately retrieved for analysis. Samples obviously have to be volatile and increasing volatility means they are more likely to be eluted quickly from the column, but this is less clear for PLOT columns. Other column types are available for sulphide analysis, however cryogenic cooling is necessary for these less-polar columns designated with a 1 or 5.

8.5 CHOICE OF DETECTOR

The limits of most GC detection systems are in the nanogram (ng) range and each detector has its own characteristics. Only selected detectors that are sensitive to odorants will be discussed. Generally odours and more certainly malodours are mostly polar in behaviour and are flammable so flame ionisation detectors would be a good choice. However odour ports, especially from a split outlet offer options that can give the best hands-on approach preferably with a mass spectrometer for definitive recognition of the odorants.

8.5.1 Odour ports

The olfactory senses are the most intelligent and sensitive detectors and there is still much to understand about the chemical composition of each sample type. Detection is simply by responding to the odour coming from an inverted cone that receives the effluent gas from the GC system. In some instances mixture with an inert gas or preferably clean air increases the information available. This information can be semi-quantitative and identification can be improved by splitting the column effluent into equally dimensional inert tubing to another detector and the odour port so that quantitative data can be achieved with recognition by RRT.

8.5.2 Flame ionisation detector (FID)

Perhaps the most commonly used detector and hence of odorants as well. The FID simply measures the ionisation current of the analyte after combustion.

With sensitivities in the low ng range (lower than the MS systems for most compounds by a factor of 2 or 3) and a dynamic range of 106. FID is a good choice as odorants are mostly composed of hydrogen and carbon, however sensitivity to sulphides can be less for the FID for the same reason. Recognition of odorants is again by RRT for the FID so for complex mixtures of odorants may require confirmation by injecting a synthetic mixture that replicates the odour composition.

8.5.3 Mass spectrometry (MS)

This should be the detector of choice, because although expensive it has the additional capacity to identify unknown compounds. There are limitations to gas flows of often up to 1 ml/min, but the is compatible with capillary columns not packed columns. Normally MS systems operate in the ng range when scanning the full mass scale of 10 to 600 mass units but this can be improved by single ion monitoring. Single ion monitoring with modern quadrupole MS systems means that we can jump to different masses for each peak that elutes. Co-eluting peaks can often be identified because there are mutually exclusive ions present from each compound so that they can be quantified. One of the limitations is that the sensitivity increases with lower gas flows into the MS, because there are less gas molecules to impede travel of ions to the detector. Additional problems include the inflow of oxygen into the ion source from the sample volume especially if it requires desorption from a solid adsorbent after odour sampling in air. Oxygen will react with the hot surfaces of the ion lens reducing the sensitivity by the build up of electrostatic charges on these oxidised surfaces. Chemical ionisation rather than electronic ionisation may be used to increase sensitivity as less fragmentation occurs.

8.5.4 Other detectors

Sulphur chemiluminescence detector (SCD) depends upon the oxidation to sulphur dioxide and the production of light in a dark background to give sensitivities in the picogram range which is ideal for odorant detection.

8.6 REVIEW OF GAS CHROMATOGRAPHY OF ODOURS

The popularity of GC has produced a rapid development of methods in other industrial and research endeavours that are relevant to odour analysis. A review of up to date techniques for odorous components found in the wastewater industry is therefore appropriate.

Odour ports are perhaps one of the easier means of assessing components in an odour and can associate an odour to an unidentified component. However, the odour may have a differing olfactory response even if it has a not too dissimilar composition, as was the case between the GC odour port and the sensory panel responses when assessing odours from drinking water. Sampling was by micro-extraction with hexane followed by GC ion trap analysis which identified geosmin 2-methylisoborneol and assorted aldehydes and ketones (Bao *et al.* 1997). A closed loop trapping system that excludes water was found effective at trapping VOCs. Over 80 compounds were identified from oral odour using an ion trap system (Claus *et al.* 1997). The ion trap mass spectrometer has the capacity to integrate a signal over a longer period to produce very good sensitivities down to about 1 ng/l for each VOC.

The use of GC system with a parallel odour port and FID detector was used to investigate two taste and odour events in the city of Philadelphia's water supply (Khiari *et al.* 1992). The GC analysis was used to detect components that may have an influence on the taste as identified by the sensory panel. The taste sensory panel did not always correlate with the GC sensory analysis. This may have been caused by the antagonistic and synergistic effects of the chemicals present. These case studies illustrated the use of Sensory GC and GC-MS analysis to understand the chemical nature of the odours present.

The difficulties of sulphide analysis are brought to light by an investigation of emissions from cut onion and garlic using different methods of analysis (Ferary and Auger 1996). They compared results from liquid extraction, trapping on adsorption and cold trapping; and using high performance liquid chromatography (HPLC) and GC as well as a range of detector systems. They concluded that there are not so many compounds present as identified in the literature and that there are no disulphides. Such information suggests that different techniques should be compared to ensure that new compounds are not produced from these unstable odours or added to by solvents or unclean instrument surfaces. The source of odorants may also pose problems as with compounds identified from muskmelon which were shown to decay after they left the blended flesh (Wyllie *et al.* 1994). Other unstable sulphides have been identified in garlic odours from the breath as allyl methyl sulphide, diallyl sulphide, diallyl disulphide, p-cymene and d-limonene (Ruiz *et al.* 1994).

Different foods have been sampled using GC with a detector in parallel to an odour port; ranging from boiled potatoes (Petersen *et al.* 1998b), buckwheat (Mazza *et al.* 1999), dairy products (Friedrich and Acree 1998), dried bell peppers (van Ruth and Roozen 1994) and even sampling under-mouth conditions to assess different foods (van Ruth and Roozen 2000). The comparison of odour from beer requires exacting procedures if they are to replicate such samples to sensory panels. Different sensory tests were performed with different adsorbents prior to analysis by GC odour port sampling. XAD resins were found successful in transferring a realistic representation of an odour from beer headspace to the panel (Bao *et al.* 1997) and was able to distinguish taint of the beer odour from a sealing ring (Linssen *et al.* 1998). XAD resin, dichloromethane and ethanol were used to transfer odours from champagne wines for GC odour port analysis (Priser *et al.* 1997). New approaches to sensory responses at the odour port have been identified to measure odour intensities using ethyl butyrate (Etievant *et al.* 1999).

Biofilters have been evaluated for controlling animal rendering odours using a odour port GC system and a GC-MS system and compared with a forced choice olfactometric response from an odour panel. About 300 compounds were identified and 40 were recognised as odorous. Some compounds originated from the biofilter (Luo and van Oostrom 1997). An unusual compound, 3-hydroxy-4,5-dimethyl-2(5H)-furanon, was detected during composting especially when high temperatures were reached (Krauss *et al.* 1992).

8.6.1 Comparison of techniques

There have been a few comparisons performed to demonstrate the differences between analytical methods of odour analysis. Gas chromatography-mass spectrometry (GC-MS) analysis was used to demonstrate differences in the chemical composition of the pig and chicken odour (Hobbs *et al.* 1995). Extraction of odorous compounds from slurries by solvent and purging with air gave different chemical profiles. Some of the major odour compounds were chemically unstable, so rapid, portable devices for odour measurement may have considerable advantages. A photo-ionisation detector and an electronic nose based on polypyrrole sensors, were less sensitive than olfactometry giving responses down to 1,000 and 60,000 ou/m^3 in air, respectively. The electronic nose was able to discriminate between the two odours at different concentrations through the pattern recognition. Electronic nose measurements were compared with analysis from GC-MS to determine if naturally contaminated barley samples were infected (Olsson *et al.* 2000). While ketones and aldehydes were recognised 3 and 6 samples were misclassified from the 40 samples infected for the electronic nose and the GC-MS.

8.6.2 Odours from sewage sources

There are several processes that occur at the sewage works and these can have different odour associated with their function. Bonnin *et al.* (1990) identified that the main source of odours were from the thickeners (H_2S, methanethiol and ammonia) thermal processing (H_2S and acetaldehyde) dewatering (H_2S and ammonia) and storage (ammonia). In some significant research Gostelow and Parsons (2000) look at data collected from odour surveys at 17 different wastewater treatment sites and suggest a power-law relationship between hydrogen sulphide and odour concentration. In this case simultaneous analysis of odorous sewage air by GC and olfactometric means has lead directly to the relationship that reduces the use of the odour panel and enables effort to be directed into preventing or reducing the odour problem. Furthermore, it may be possible to use this information to better regulate the processes that occur at the treatment works. Sludge cakes have been noted as causing odour problems and these have been attributed to the sulphide group, in particular the methyl sulphides up to the dimethyl trisulphides (Winter and Duckham 2000). A number of malodorous compounds were identified in digested sludge and the corresponding sludge cakes with a number of sulphides contributing most to the odour. Analysis by purge and trap sampling of liquefied samples followed by GC-MS identified that the majority of malodorous sludge cakes had higher sulphide levels than less odorous sludge cakes. The most abundant compounds were dimethylsulphide, dimethyldisulphide and dimethyltrisulphide. Dimethyltrisulphide was found to have a proportionally greater impact on olfactory response than any of the other compounds (Winter and Duckham 2000).

Amines have also been identified as contributing to odours from sewage sources (Hwang *et al.* 1995). A purge and cold trap method was used so those compounds at concentrations of ng/l were detectable. Indole, 3-methylindole, trimethylamine, dimethylamine and n-propylamine were detected in wastewater samples, but not by GC analysis. GC-detectable concentrations of dissolved low aliphatic amines and indoles were about 10 μg/g (Abalos *et al.* 1999). Large amounts of nitrogen-containing compounds remained after secondary treatment compared with sulphur-containing compounds However, their determination was in in wastewater, primary and secondary effluents, and sewage-polluted river samples and not the headspace (Abalos *et al.* 1999). A tailor-made PoraPLOT® capillary GC column was used and no cryogenic trap was required.

8.7 EMISSION RATES

Emission rates from the surfaces of odour sources are one of the primary means of measurement as changes are less likely than those found in monitoring an odour plume. There are several fundamental approaches but they all involve covering or containing a sample and using known flow rates across the surface. Individual methods will be discussed and compared plus investigations of the different approaches to measuring fluxes from liquid or sludge surfaces. Emission rates of the components of odours have not received so much attention as organic compounds of anthropogenic origin that are the focus of environmental and human health interest. However odorous compounds such as

H_2S and methanethiol at concentrations in the mg/l range have been shown to be fatal to people and livestock (Donham *et al.* 1982).

There are some examples of data from wastewater treatment plants. Devai and DeLaune (1999) determined the concentrations in the headspace for hydrogen sulphide, methanethiol, dimethyl sulphide, carbon disulphide and carbonyl sulphide. However, no emission rates have been identified and more information on emissions from wastes is available from agricultural sources. These have concentrated mostly on the emissions of ammonia from livestock in terms of animal housing (Hartung 1992), from storage facilities (Petersen *et al.* 1998b) and landspreading (Pain *et al.* 1990) and total emission rates have been used to produce an inventory for the UK of ammonia emissions (Pain *et al.* 1998). However, there are relatively few ammonia emissions from sewage works (Sutton *et al.* 1995) and most studies on sewage works do not include VOC emission rates but concentrations found in the atmosphere. Some VOCs have been performed on livestock wastes, and Zahn *et al.* (1997) have identified 27 VOCs, which decrease the air quality near the facility. The VFAs C_2–C_9 demonstrated the greatest potential for decreased air quality, since these compounds exhibited the highest transport coefficients and highest airborne concentrations. Flux measurements suggested that the total rate of VOC emissions from the deep-basin swine waste storage system was 500- to 5700-fold greater than established VOC fluxes from natural sources. The emission rates were positively correlated with wind velocity between 0.2 and 9.4 m/s and a maximum concentration of VOCs present in the air was observed to occur at a wind velocity of 3.6 m/s. Zahn *et al.* (1997) also identified some bromonated compounds and a phthalate that has a ubiquitous presence in GC-MS traces especially with the use of soft plastic containers which are often used with sampling. This research identifies two important findings regarding the effect of wind velocity and transport coefficients. Wind has a major role to play in emissions, not only in terms of increasing the emission rate, but during certain wind conditions creating a high odour concentration that would be more offensive. Of course, this does not take into account the degree of mixing associated with different weather conditions that may occur, but may help those who model odour plumes.

8.8 CASE STUDY

Although it may be possible to use estimations and models for mass transfer of VOCs from waste surfaces, it is better to measure them because of the differing effects of the surface composition. To be able to understand the complexity of measurement required a case study will include an investigation of the emission rates from pig slurry which contains a range of sulphides, phenolic, indolic and fatty acids (Hobbs *et al.* 1998). However, some of the difficulties and problems should be considered for other approaches to emission rate measurement. The odour concentration will also be determined to give an odour profile to attempt to associate it with the odorous components in the slurry and in the headspace. In order to obtain concentrations of these components with some degree of confidence a closed chamber system (Cumby *et al.* 1995) that initially contained 40 m^3 of air was used to increase the concentration of odorants at constant conditions of temperature and airflow. The slurry was from slatted floor pig housing and was stirred to give a fresh surface rather that achieve a crusted surface. The temperature of the slurry and the air were controlled at 15 and 20 °C respectively to minimise condensation onto the internal chamber surfaces built of stainless steel U-shaped ducting whose ends were connected into a large Tedlar bag. The chamber operated under a slight pressure to prevent external air being drawn in through small leaks to prevent dilution of the sample volume. The emission rates can be calculated by estimating the saturated vapour pressure point and using Henry's Law to estimate the mass transfer. However we determined the emission rate at zero time of the experiment because any suppression of emission by the mass present in the headspace would be minimised and any physical and chemical interactions with other odorants should be reduced. The equation of the total mass emitted was the sum of the mass leaked and that present in the chamber, which was best expressed in a quadratic form. The emission rates were expressed as mass emitted per unit area or area divided by mass to allow for the depth of the sample (Table 8.3). Samples were taken from a spur protruding into the centre of the flow by using the lung principle of sampling so that no sample passes through an air pump. Volatile compounds were pre-concentrated from a 600 ml sample of the headspace volume above the pig slurry by adsorption onto silica (Orbo 52, Supelco Inc., USA) and carbon (Orbo 32, Supelco Inc., USA) such that the sample has to pass through the silica first. The concentrated odorants were then thermally desorbed into the GC-MS system for analysis. A HP-5890 II Series gas chromatograph (Hewlett Packard, USA) and a 5972A mass selective detector (MSD II) were used to analyse all the samples. A 25 m fused silica HP-1 column with an i.d. of 0.2 mm and a 1.00 μm film with a 1 m deactivated fused silica guard column (0.25 mm i.d.) was used to analyse samples directly from the OEC to determine the sulphide components. An increased film thickness of 0.34 μm was used to analyse the headspace for other odorants. The column flow rate was 0.75 ml/min. The optic temperature programmable injector (Ai Cambridge Ltd., UK) was used to thermally desorb headspace samples at 250 °C for 1 minute. The GC oven temperature was initially 27 °C and was increased at 15 °C/min to 220 °C and maintained for one minute. The GC-MS interface was at 280 °C. The mass spectrometer scanned from 32 to 250 mass units every 0.2 seconds to give sensitivities down to 50 pg. Retention time and mass spectral matching were used to confirm the odorant identity.

Table 8.3 Emission rates from agricultural wastes.

	Emission rates ($mg/m^2/min$)	Standard deriation	Minimium	Maximium
Odour concentration ($ou/m^2/min$)	1.36 e^6	1.01 e^6	3.23 e^6	2.65 e^6
Carbon dioxide	1056	352	548	1660
Methane	9.22	4.80	4.8	17.9
Hydrogen sulphide	214.7	83.9	105	337
Ammonia	2.15	1.75	0.35	5.85
Phenol	0.21	0.247	0.0068	0.58
4-methyl phenol	0.44	0.397	0.0125	1.06
4-ethyl phenol	0.07	0.069	0.0001	0.182
Indole	0.20	0.199	0.00001	0.475

There are several factors that will contribute to emission rates from slurries: they are the wind speed (Liu *et al.* 1995), stirring, temperature and bacterial biogenesis of odorants. Indirectly odour emissions may be increased as more carbon dioxide and methane are produced with increasing temperature (Husted 1993) to strip the odorants from the waste. Emission rates also depend on the chemical behaviour of the gas or odorant, for example the biogenesis of methane will generally be greater in a larger store per unit volume because the greater volume to area ratio creates a more stable and necessary anaerobic environment. Methane also has a low solubility and should be emitted quickly after biogenesis. The emission rates of hydrogen sulphide were high when compared with those for ammonia. Hydrogen sulphide is not highly soluble in water and additional stirring may have induced rapid biogenesis of this gas. There is the possibility that by measuring the concentrations in the slurry we can approximate the concentrations in the headspace. Only a good relationship was achieved between the headspace and the slurry concentration for the phenolic components in the pig slurry.

8.9 REFERENCES

Abalos, M., Bayona, J.M. and Ventura, F. (1999) Development of a solid-phase microextraction GC-NPD procedure for the determination of free volatile amines in wastewater and sewage-polluted waters. *Analyt. Chem.* **71**, 3531-3537.

Bao, M.L., Barbieri, K., Burrini, D., Griffini, O and Pantani, F. (1997) Determination of trace levels of taste and odor compounds in water by microextraction and gas chromatography ion trap detection mass spectrometry. *Water Res.* **31**, 1719-1727.

Bonnin, C., Laborie, A. and Paillard, H. (1990) Odor nuisance created by sludge treatment: problems and solutions. *Water Sci. Technol.* **22**, 65-74.

Brennan, B.M., Donlon, M. and Bolton, E. (1996) Peat biofiltration as an odour control technology for sulphur-based odours. *J. Chart. Instit. Water Environ. Manag.* **10**, 190-198.

Claus, D., Geypens, B., Ghoos, Y., Rutgeerts, P., Ghyselen, J., Hoshi, K. and Delanghe G. (1997) Oral malodor, assessed by closed-loop, gas chromatography, and ion- trap technology. *Hrc-J. High Resolution Chromatography* **20**, 94-98.

Cumby T R, Moses B, and Nigro I. (1995) Gases from livestock slurries; Emission kinetics. *Proc. 7th International Conference on Agricultural and Food Wastes.*

Devai, I. and DeLaune. R.D. (1999) Emission of reduced malodorous sulfur gases from wastewater treatment plants. *Water Environ. Res.* **71**, 203-208.

Devos, M., Patte, F., Rouault, J., Lafort, P. and Van Gemert, L.J. (1990) *Standardised Human Olfactory Thresholds*. Oxford University Press, New York.

Donham K L, Knapp L W, Monson R, and Gustafson K. (1982) Acute toxicity exposure to gases from liquid manure. *J. Occupational Medicine* **24**, 142-145.

Dorling, T.A. (1977) Measurement of odour intensity in farming situations. *Agric. Environ.* **3**, 109-120.

Etievant, P.X., Callement, G.,Langlois, D., Issanchou, S. and Coquibus, N. (1999) Odor intensity evaluation in gas chromatography olfactometry by finger span method. *J. Agric. Food Chem.***47**, 1673-1680.

Ferary, S. and Auger. J. (1996) What is the true odour of cut Allium? Complementarity of various hyphenated methods: Gas chromatography mass spectrometry and high-performance liquid chromatography mass spectrometry with particle beam and atmospheric pressure ionization interfaces in sulphuric acids rearrangement components discrimination. *J. Chromatography A* **750**, 63-74.

Friedrich, J.E. and Acree. T.E. (1998) Gas chromatography olfactometry (GC/O) of dairy products. *International Dairy Journal* **8**, 235-241.

Gostelow, P. and Parsons. S.A. (2000) Sewage treatment works odour measurement. *Water Sci. Technol.* **41**(6), 33-40.

Hartung, J. (1992) Emission and control of gases and odorous substances from animal housing and manure stores. *Zentralblatt Fur Hygiene Und Umweltmedizin* **192**, 389-418.

Hirsch, A.R. and Trannel. T.J. (1996) Chemosensory disorders and psychiatric diagnoses. *J. Neurological Orthopaedic Medicine And Surgery* **17**, 25-30.

Hobbs, P.J., Misselbrook, T.H. and Cumby. T.R. (1999) Production and emission of odours and gases from ageing pig waste. *J. Agric. Engin. Res.* **72**, 291-298.

Hobbs, P.J., Misselbrook, T.H. and Pain, B.P. (1995) Assessment of odors from livestock wastes by a photoionization detector, an electronic nose, olfactometry and gas-chromatography mass-spectrometry. *J. Agric. Engin. Res.* **60**, 137-144.

Hobbs, P.J., Misselbrook, T.H. and Pain, P.B. (1998) Emission rates of odorous compounds from pig slurries. *J. Sci. Food Agric.* **77**, 341-348.

Hobbs, P.J., Misselbrook, T.H., Dhanoa, M.S. and Persaud, K.C. (2001) Relationship between the chemical composition and olfaction of decay odours. *Proc. ISOEN2000*, Brighton, pp. 13-14.

Husted, S. (1993) An open chamber technique for determination of methane emission from stored livestock manure. *Atmos. Environ. Part A- General Topics* **27**, 1635-1642.

Hwang, Y., Matsuo, Y Hanaki, K., and Suzuki, N. (1995) Identication and quantification of sulphur and nitrogen odorour compoiunds in wastewater. *Water Res.* **29**, 711-718.

Khiari, D. Brenner, L, Burlingame, G.A., Suffet, I.H. (1992) Sensory gas-chromatography for evaluation of taste and odor events in drinking water. *Water Sci. Technol.* **25**, 97-104.

Krauss, P., Krauss, T., Mayer, J., and Wallenhorst, T. (1992) Examination of odor formation and odor reduction in composting plants. *Staud Reinhaltung Der Luft* **52**, 245-250.

Kubickova, J. and Grosch., W. (1998) Quantification of potent odorants in Camembert cheese and calculation of their odour activity values. *International Dairy Journal* **8**, 17-23.

Laing, D.G and Glemarec, A. (1992) Selective attention and perceptual analysis of odor mixtures. *Physiology Behavior* **52**, 1047-1053.

Linssen, J.P.H., Rijnen, L., Legger-Huiysman, A., and Roozen, A.P. (1998) Combined GC and sniffing port analysis of volatile compounds in rubber rings mounted on beer bottles. *Food Additives Contaminants* **15**, 79-83.

Liu, Q., Bundy, D.B. and Hoff, S.H. (1995) A study on the air flow and odor emission rate from a simplified open manure storage tank. *Trans. ASAE* **38**, 1881-1886.

Livermore, A. and Laing, D.G. (1998) The influence of chemical complexity on the perception of multicomponent odor mixtures. *Perception Psychophysics* **60**, 650-661.

Luo, J.F. and van Oostrom, A. (1997) Biofilters for controlling animal rendering odour - a pilot- scale study. *Pure Appl. Chem.* **69**, 2403-2410.

Mazza, G., Cottrell, T., Malcolmson, L. , Girard, B., Oomah, B.D. and Eskins, M.A.M. (1999) Headspace gas chromatography and sensory analyses of buckwheat stored under controlled atmosphere. *J. Food Quality* **22**, 341-352.

O'Connor, B.I.,. Buchanan, B.E and Kovacs, T.G. (2000) Compounds contributing to odors from pulp and paper mill biosolids - Anaerobic biological activity a contributing cause. *Pulp Paper-Canada* **101**, 57-61.

Olsson, J., Borjesson, T., Lundstedt, T. and Schnurer, J. (2000) Volatiles for mycological quality grading of barley grains: determinations using gas chromatography-mass spectrometry and electronic nose. *Internal. J. Food Microbiol.* **59**, 167-178.

Pain, B.F., van der Weerden, T.J., Chambers B.J, Phillips, V.R. and Jarvis, C. (1998) A new inventory of ammonia emission from UK agriculture. *Atmos. Chem.* **32**, 309-313.

Pain, B.P., Misselbrook, T.H., Clarkson, R. and Rees, Y.J. (1990) Odor and ammonia emissions following the spreading of anaerobically - digested pig slurry on grassland. *Biological Wastes* **34**, 259-267.

Patterson, M.Q., Stevens, J.C., Cain, W.S. and Comettomuniz, J.E.. (1993) Detection thresholds for an olfactory mixture and its 3 constituent compounds. *Chem. Senses* **18**, 723-734.

Petersen, M.A., Poll, L. and Larsen, L.M. (1998a) Comparison of volatiles in raw and boiled potatoes using a mild extraction technique combined with GC odour profiling and GC-MS. *Food Chem.* **61**, 461-466.

Petersen, S.O., A.M. Lind, and Sommer, S.G. (1998b). Nitrogen and organic matter losses during storage of cattle and pig manure. *J. Agric. Sci.* **130**, 69-79.

Priser, C., Etievant, P.X., Nicklaus, S. and Brun, O. (1997) Representative champagne wine extracts for gas chromatography olfactometry analysis. *J. Agric. Food Chem.* **45**, 3511-3514.

Ruiz, R., Hartman, T.G., Karmas, K., Lech, J. and Rosen, R.T. (1994) Breath analysis of garlic phytochemical in human subjects – combined adsorbents trapping and short-path-thermal desorption gas chromatography-mass spectrometry. *Food Phytochemicals For Cancer Prevention I* **546**, 102-119.

Scarlata, C.J. and Ebeler, S.E. (1999) Headspace solid-phase microextraction for the analysis of dimethyl sulfide in beer. *J. Agric. Food Chem.* **47**, 2505-2508.

Schaefer, D.G. (1977) Sampling, characterisation and analysis of malodours. *Agric. Environ.* **3**, 121-127.

Sutton, M.A., Place, C.J., Eager, M., Fowler, D. and Smith, R.I. (1995) Assessment of the magnitude of ammonia emissions in the United Kingdom. *Atmospheric Environ.* **29**, 1393-1411.

van Ruth, S.M. and Roozen, J.P. (1994) Gas-chromatography sniffing port analysis and sensory evaluation of commercial dried bell peppers (*Capsicium annuum*). *Food Chem.* **51**, 165-170.

van Ruth, S.M. and Roozen, J.P. (2000) Gas chromatography/sniffing port analysis of aroma compounds released under mouth conditions. *Talanta* **52**, 253-259.

Winter, P. and Duckham, S.C. (2000) Analysis of volatile odour compounds in digested sewage sludge and aged sewage sludge cake. *Water Sci. Technol.* **41** (6), 73-80.

Wyllie, S.G., Leach, DN., Wang, Y.M. and Shewfelt, R.L. (1994) Sulfur volatiles in cucumis-melo cv makdimon (Muskmelon) Aroma - sensory evaluation by gas-chromatography olfactometry. *Proc. ACS Symposium Series* **564**, 36-48.

Zahn, J.A., Hatfield, J.L., Do, Y.S., DiSpirito, A.A., Laird, D.A. and Pfeiffer, R.L. (1997) Characterization of volatile organic emissions and wastes from a swine production facility. *J. Environ. Quality* **26**, 1687-1696.

9

Odour measurements using sensor arrays

Richard M. Stuetz and Richard A. Fenner

9.1 INTRODUCTION

Public concerns over the release of unpleasant odours from sewage and sludge treatment have increased. The assessment of odours has become highly significant in the control and prevention of odorous emissions and is crucial for new planning applications. The measurement of odours is not straightforward and there are two broad classes of measurement techniques.

Analytical measurements such as GC-MS analysis and H_2S measurements are used to characterise odours in terms of their chemical composition or act as a surrogate for odour strength. These methods can provide an accurate description of each compound in an odour mixture and are suited for use in analytical formation, emission and dispersion models, but unfortunately tell us very little about the perceived effect of the odour. Olfactometry employs a panel of human

edited by Richard Stuetz and Franz-Bernd Frechen. ISBN: 1 900222 46 9

assessors to characterise the odour in terms of their perceived effect and is the usual method for measuring odours. Although this methodology gives the right human sense evaluation and can now be based on a draft European odour standard, prEN 17325, it is strongly influenced by subjectivity (Bliss *et al.* 1996), is time consuming, labour intensive and expensive. Furthermore, olfactometry laboratories are often remote from the odour source and with the increasing need to assess odours on site and provide continuous operation, may be an unsuitable technique for future real-time annoyance odour assessment.

The development of sensor array technology so called "electronic noses" for odour classification may offer an objective and on-line instrument for assessing environment odours. Previous commercial sensor array systems were mainly manufactured for laboratory-based applications, however portable and on-line instruments designed for environmental monitoring have recently become available.

In this chapter we review sensor array technology, data processing techniques, electronic nose instrumentation and discuss the current status of odour assessment using sensor arrays and their potential application for objective measurement of olfactive annoyance.

9.2 SENSOR ARRAY TECHNOLOGY

Sensor array systems are analytical instruments that can characterise an odour without reference to its chemical composition. A typical system is based upon the configuration shown in Figure 9.1. This consists of a sensor array, with appropriate signal conditioning connected to a pattern recognition system for data analysis. Several different system designs are used in today's commercial instruments. However, whatever the design the sample headspace must be presented to the sensor array in a reproducible manner. This can be achieved through the use of automated control systems for sample handling, gas flow and temperature for both the sensor chamber and sample environment.

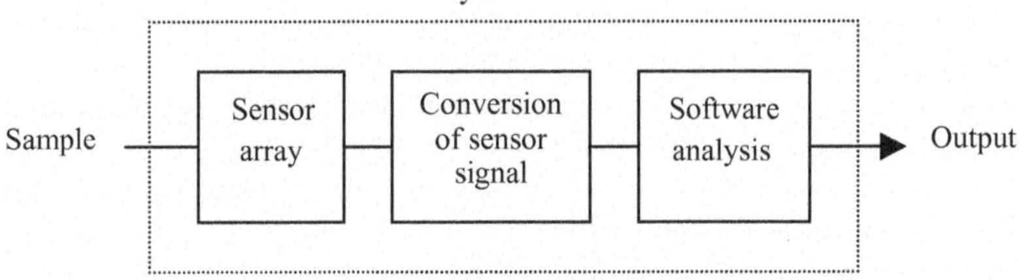

Figure 9.1 Principal components of a sensor array (Hodgins and Simmonds 1995).

9.2.1 Types of sensors

A variety of different sensor technologies are used in sensor array systems (Persaud and Travers 1997). Some of the most common are metal oxide sensors (MOS), conducting polymers (CP), surface acoustic wave sensors (SAW) and quartz crystal microbalances (QCM) and their measurement principles are summarised in Table 9.1. The use of an array of non-specific sensors allows for responses from many thousands of chemical species, due to the broad selectivity of the different sensor surfaces (Persaud *et al.* 1996a). The relative responses between the sensors can be used to produce a unique odour profile that is analogous to the human olfactory system (Gardner and Bartlett 1994). Figure 9.2 shows an example of an odour profile for sewage using a 12-sensor conducting polymer array. The resultant odour-specific response pattern or fingerprint can then be further analysed using pattern recognition techniques (section 9.2.2).

Table 9.1. Most common types of sensors and measurement principles (Fenner and Stuetz 1999)

Sensor types	Mode of action	Comments
Conducting polymers	In the presence of a gas species a change in voltage across polymers such as polyaniline, polypyrrole and polythiopene can be measured.	Selectivity is achieved by controlling surface functional groups or by varying anion chemistry during growth. Reproducibility good. Return to baseline resistances in short times.
Metal oxides	Metal oxide sensor passes an electrical current causing oxidation of gas molecules via electron transfer from the gas to the metal oxide leading to a change of resistance.	Less selective than other sensor types. Can be subject to poisoning Response can be affected by presence of solvents.
Quartz crystal microbalances	Measure change in frequency of oscillation of a quartz crystal when a gaseous species is adsorbed.	Problems of reproducibility in commercial production of sensors.
Surface acoustic wave sensors	Similar to quartz crystal microbalances but operate at much higher frequencies.	Can achieve good sensitivity. Problems of reproducibility in sensor production
Fibre optic sensors	Use fluorescence measurements from photodeposited polymer/fluorescent dyes on bundles of fibre optics.	Provide large quantities of data. Not available in commercial instruments.

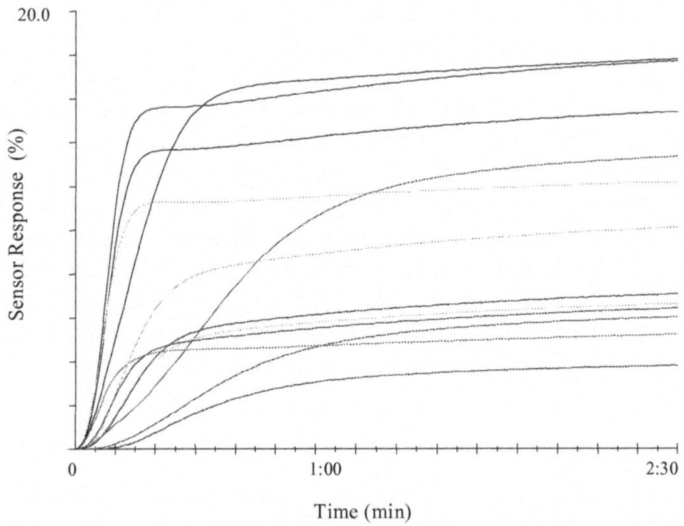

Figure 9.2 Response pattern of a 12-sensor polypyrrole array to sewage odour, showing odour profile over a 2.5-minute acquisition (Stuetz and Fenner 1998).

9.2.2 Data processing techniques

The interpretation of data obtained from sensor arrays usually relies on the use of sophisticated analysis routines. Output can be displayed using a variety of graphical formats that allows comparisons between samples or averaged data over a number of analysers (Hodgins 1995). However, to cope with large number of samples and number of variables (i.e. number of sensors), pattern recognition techniques are often employed to process the sensor array data. These techniques allow underlying relationships between one set of independent variables (i.e. output from an array of *n* sensors) and another set of dependent variables (i.e. odour class or component concentrations) to be determined and can be divided into two basic methods: unsupervised and supervised (Gardner and Bartlett 1999). In an unsupervised technique, one seeks to discriminate between unknown odours by enhancing the differences between the associated input vectors, whereas in a supervised technique unknown odours are analysed using relationships that have been learnt during an earlier calibration procedure, learning or training stage (Gardner and Bartlett 1999). The choice of analysis technique is dependent on the amount and nature of information available and the type of information required from the analysis (i.e. quantitative or qualitative).

9.2.2.1 *Statistical analysis of sensor responses*

Multivariate statistical techniques are employed to reduce the dimensionality of the sensor array data, so that relationships between the observations can be explored using one or two dimensions (Persaud *et al.* 1996b). The most common of these techniques, principal component analysis (PCA), cluster analysis (CA), multiple discriminant analysis (MDA) and canonical correlation analysis (CCA) are used for *qualitative* analysis of multivariate problems (Gardner and Bartlett 1999). Table 9.2 summarises these qualitative statistical methods that have been used for the analysis of sensor array data. A more extensive treatment of these statistical topics can be found in many textbooks on multivariate analysis such as Manly (1986) and Rencher (1998). More specific reviews describing their application to the analyses of sensor array data are Gardner and Bartlett (1991; 1992) and Gardner and Hines (1996).

Table 9.2 Common statistical techniques used for the analysis of sensor array data.

Techniques	Supervised	Linear	Comments
Principal component analysis	No	Yes	Used to reduce a large number of variables to a smaller number of components and in the same time extract maximum variance in the data.
Cluster analysis	No	Yes	Used to find natural grouping or clusters of individual observations within a data-set.
Multiple discriminant analysis	Yes	Yes	Used to understand existing groups and to create new groupings and classification based on known properties.
Canonical correlation analysis	Yes	Yes	Explores linear relationships between dependent sets of variables and independent sets of variables by maximising the correlation in the data.

9.2.2.2 *Artificial neural networks (ANNs)*

Artificial neural networks are a learning architectures that consists of parallel-interconnected processing elements called *neurones*, which is based loosely upon a physical model of the biological nervous systems. Each processing element is weighted mathematically - these weights are determined through a process of training or learning (Gardner and Bartlett 1999). The structure and learning phase of the neural network are therefore used to define ANNs. By far the most commonly used ANNs to analyse sensor array data is a multilayer network called the multilayer perceptron. A multilayer perceptron consists of three types of layer of units, the input layer, a number of hidden layers and the output layer. An illustration of a typical multilayer perceptron is shown in Figure 9.3. Various algorithms can be used to train multilayer networks, the back-propagation technique, which relies on two stages, is the most understood (Gardner and Bartlett 1992). This consists of a learning stage employed to train the network, followed by a recall phase or prediction during which the trained data set is used to classify unknown input data (Gardner and Hines 1996).

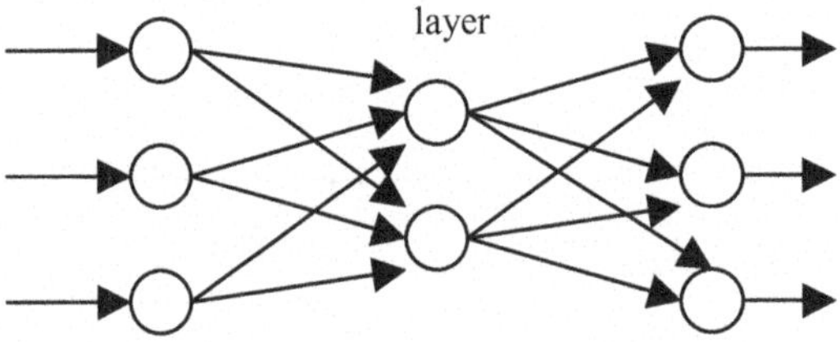

Figure 9.3 Architecture of typical multilayer perceptron.

9.2.3 Electronic nose instrumentation

9.2.3.1 Design and measurement consideration

The basic elements of a sensor array instrument are the sample delivery system, which is designed to transfer the headspace from the sample material to sensor chamber and the sensor array system (Figure 9.1). There are two main ways in which an odour sample can be delivered to the sensor array chamber, either by pumping or injecting the sample headspace from the sample vessel to the sensor array "dynamic sampling" or by generating a headspace above a sample and then introducing this headspace to the sensor's “static sampling”. The different designs of these odour delivery systems has lead to different performance characteristics, which are suited to different fields of application (Gardner and Bartlett 1999).

The simplest odour delivery system is that of a manual headspace sampling, whereby a sample material is stored at a constant temperature and allowed to reach equilibrium (Mielle and Marquis 1999). After which a small volume of odorant headspace is removed from the sample vessel using a glass syringe and injected into a sensor chamber (Gardner and Bartlett 1999). Many researchers have employed this method as the apparatus for this technique is simple and inexpensive, however, the method is labour intensive, time-consuming and can have poor repeatability due to the manual operation of the syringe. In order to improve the repeatable of this type of static odour delivery system, a sensor array can be mounted above the sample vessel with the facility to raise and lower the sensor head into the sample vessel, thereby minimising any system measurement errors (Hodgins and Simmonds 1995). An illustration of such a static sampling system is shown in Figure 9.4. Alternatively, the use of a robotic headspace injector for automated transfer of the headspace from the sample vessel to the sensor chamber, can reduce the variation in the sample temperature, injection rate and headspace concentration (Gardner and Bartlett 1999). These delivery systems also allow for the use of automated purging of the sensor chambers and sample vessels, however, the overall sampling rate is still dependent on the slow reaction kinetics during sample equilibration.

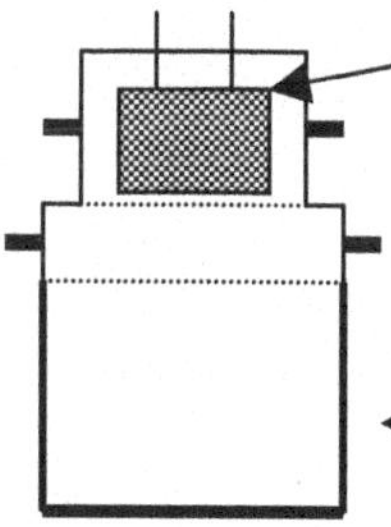

Figure 9.4 Schematic of static sampling system.

The second type of odour delivery system is based on pumping or purging the sample headspace into the sensor chamber (Mielle and Marquis 1999). A carrier gas or gases (i.e. zero grade air or N_2) is used to strip the volatiles from the sample material, which is then transferred to the sensor chamber, usually involving a cleaning, purging and sensor acquisition phase. It should be noted that a dynamic headspace may not be the same as a static headspace–either in composition or concentration due to the mechanism of headspace generation (Gardner and Bartlett 1999). Dynamic sampling systems have a number of advantages over static sampling systems: there is no dilution effect and it is possible to use different levels of humidity and different temperatures to deliver the headspace to the sensor chamber with a high level of accuracy (Gardner and Bartlett 1999). However, this increased complexity in the operation of the system results in higher capital costs.

9.2.3.2 Commercial instruments

A range of bench-top sensor array instruments are available from a number of commercial manufacturers (Table 9.3). These instruments comprise of arrays of different sensors over which the headspace gases is passed. The arrays are usually modular plug-in devices, which makes the instruments very flexible (Mills *et al.* 1996). Severals systems are also able to incorporate a number of different sensor types (i.e. MOS and CP) in the same device (Gardner and Bartlett 1999).

Table 9.3 List of sensor array manufactures (Gibson *et al.* 2000).

Company and location	Models
Alpha MOS, France (www.alpha-mos.com)	Fox 2000, 3000, 4000, 5000, AlphaKronos, AlphaPrometheus, AlphaCent
Bloodhound Sensors, UK (www.bloodhound.co.uk/bloodhound)	Bloodhound BH114
Cyrano Sciences, USA (www.cyranosciences.com)	Cyranose 320
Etherdata, Iceland (www.etherdata.is)	FreshSense
HKR Sensorsysteme, Germany (www.home.t-online.de)	QMB6/HS40XL
Hewlett Packard, USA (www.hp.com)	HP4440A
Lennartz Eectronic, Germany (www.lennartz-electronic.de)	MOSES II
Marconi Applied Technologies, UK (www.marconitech.com)	eNOSE 5000, ProSAT
MoTech Sensoric, Germany (www.motech.de)	VOCmeter, VOCcheck
Nordic Sensor Technologies, Sweden (www.nordicsensor.se)	NST 3210, NST 3220, NST 3220A
Osmetech, UK (www.osmetech.co.uk)	MultiSampler-SP
RST Rostock, Germany (www.rst-rostock.de)	Sam
Smart Nose, Switzerland (www.smartnose.com)	Smart NOSE-300
WMA Airsense, Germany (www.airsense.com)	PEN

Most commercial systems have precise temperature control of the sample delivery system and sensor chamber, as these variables are know to affect sensor responses. The sample to be analysed is usually heated or mixed in an enclosed vessel or vial and the headspace sampled directly (Mills *et al.* 1996). The analysis of a sample headspace usually takes about 2 minutes and another 4–10 minutes is required for the sensor to recover to its baseline (during sensor cleaning and sensor pre-purge). Most commercial sensor array system require a host computer for instrument control, data analysis and manufacturers offer specific software for data acquisition and display and have direct links to spreadsheet packages (such as Microsoft Excel) and statistical packages (such as Unistat or Statistica) to enable more detailed data analysis. An example of a commercial sensor array system is shown in Figure 9.5.

Figure 9.5 Photograph of BH114 sensor array system (Courtesy of Bloodhound Sensors Ltd, UK).

9.2.3.3 Future prospects

Sensor arrays as a commercial device have not reached their full potential. Almost all the instruments sold to date are being used in research and development applications (Gibson *et al.* 2000). Future products are likely to be classified into three application groups: (i) complex laboratory-based instruments; (ii) on-line systems and (iii) portable devices for field measurements. To date only laboratory-based instruments have been commercially available. These systems are mostly second-generation instruments, that offer sensitivity and reliability over a wide range of QC/QA applications and with further development could play a part in effective product development by allowing rapid, accurate assessment of production line samples (Gibson *et al.* 2000). An example of such a system with an autosampler incorporated into the instrument design is shown in Figure 9.6. The recent development of on-line sensor array systems for process monitoring (Figure 9.7)

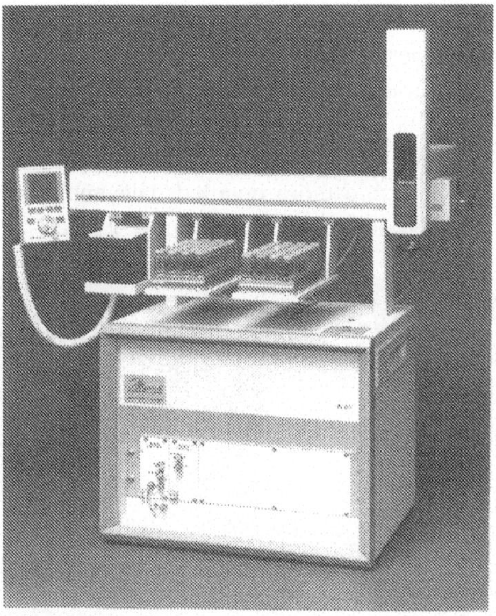

Figure 9.6 Photograph of eNOSE 5000 instrument with autosampler (Courtesy of Marconi Applied Technologies, UK).

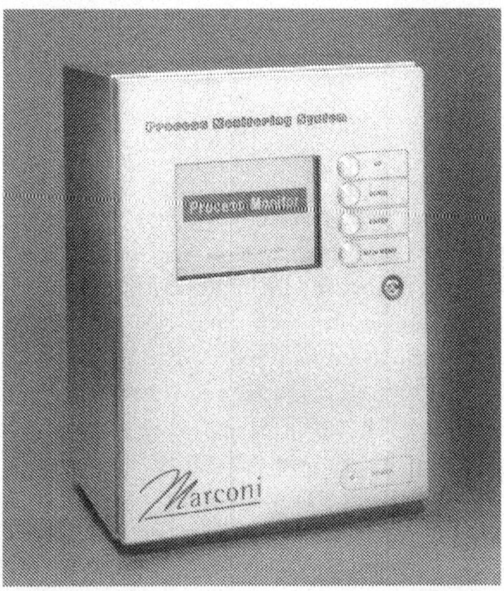

Figure 9.7 Photograph of ProSAT on-line process monitoring system (Courtesy of Marconi Applied Technologies, UK).

and portable devices for environmental monitoring (Figure 9.8) has resulted from the reduced need for flexibility in system design and could offer a more user-friendly operation. These application-specific devices could be used for a wide range of tasks, which may include continuous on-line monitoring of odour abatement units and field odour intensity measurements.

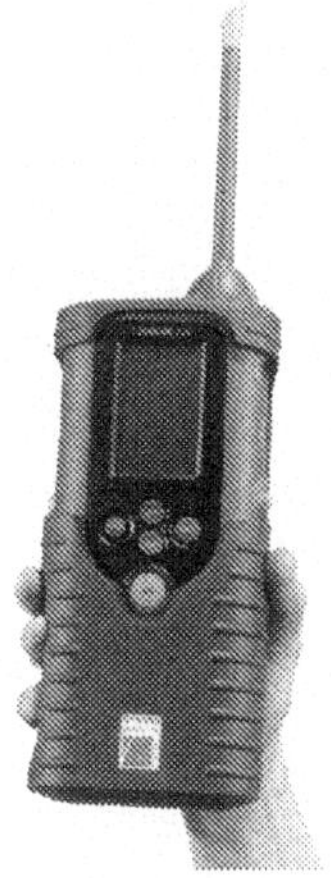

Figure 9.8 Photograph of Cyranose 320 portable sensor array monitor (Courtesy of Cyrano Sciences, USA).

In addition to the development of application-specific devices, new sensor techniques are being incorporated into sensor array systems. These include the use of mass spectrometry (coupled with pattern recognition) and solid-state spectrometry, which involves bypassing the traditional sample preparation stages and introducing whole samples into the mass spectrometer or the solid-state sensor to give a mass spectrometry fingerprint or a spectroscopic trace (Gibson *et al.* 2000). Alternatively, new sensor types are being developed that have either a very low response to water vapour or are sensors that are described as water-insensitive chemoresistors (Gibson *et al.* 2000).

Traditional processing of sensor array data (section 9.2.2) has used classical algorithms (such as PCA, MDA) which are usually integrated into commercial sensor array for off-line analyses. With the development of application-specific devices the need for dynamic data processing is increasing. Specially designed adaptive ANN and fuzzy logic algorithms are being developed for these on-line applications. However, more sophisticated pre-processing and data analysis protocols are required for predictive classification of unknown odours in the more challenging applications such as detecting sub-ppm concentrations and the use of hand-held systems which must work without reference gases or known odour profiles (Gardner and Bartlett 1999).

9.3 APPLICATION OF SENSOR ARRAYS TO ODOUR MONITORING

Applications of sensor array technology have evolved from the food and beverage industries (Hodgins and Simmonds 1995). However, recently studies have investigated its application to medical, industrial and environmental problems. For example, they have been used in predicting different bacterial types and culture growth phases (Gibson *et al.* 1997; Gardner *et al.* 1998; Holmberg *et al.* 1998), monitoring bioprocess for detecting microbial contamination (Namdev *et al.* 1998), detecting chemical pollutants in water supplies (Stuetz *et al.* 1998a) and monitoring wastewater effluents (Stuetz *et al.* 1999a,b). Studies have also reported the use of sensor arrays for the assessment of odour pollution; however this environmental application has been limited (Romain *et al.* 2000).

9.3.1 Odour measurements

The assessment of environmental odours by sensor array systems has until recently been based on the use of prototype or commercial laboratory-based instruments. This has involved the collection of odorous samples (from agricultural and wastewater sources) in collection bags (made of Teflon or Tedlar) by either directly sampling from the environment or by purging a slurry or wastewater sample with odourless air. These odour samples were subsequently used to evaluate the application of using sensor array systems for measuring odours by either (i) comparing the sensor responses for different sample types or (ii) by correlating the sensor responses to known parameters such as threshold odour concentrations (using olfactometry), specific analytical components (using GC-MS) or surrogates for odour strength (using H_2S and NH_3 measurements).

Sensor arrays have been used to assess a wide range of environmental samples including sewage, agricultural and landfill odours. Hobbs *et al.* (1995) initially showed that an electronic nose (consisting of an array of 20 conducting polymer sensors) could discriminate between the different odours from livestock wastes (pig and chicken slurry); however, this early instrument was reported to have a low sensitivity compared with olfactometry measurements. Persaud *et al.* (1996a) showed that with further sensor development, conducting polymers were able to differentiate between the individual volatile components in a pig slurry and that the intensity of the signal response is proportional to the concentration of the volatiles being presented to the sensors (Figure 9.9). Additionally, the same sensor array system has also shown that the odour components being detected from an artificial alkaline pig slurry appear to the associated with patterns obtained from indole, skatole and ammonia (Figure 9.10), which suggests that the indole, skatole and ammonia content of the slurry may dominate the chemical species to which the conducting polymers show more sensitivity and thus may serve as odour markers (Persaud *et al.* 1996a). Persaud *et al.* (1996b) also reported that a single conducting polymer sensor could be correlated to odour intensity (using olfactometry measurements) and could distinguish between odour emissions from pig slurry with different diets.

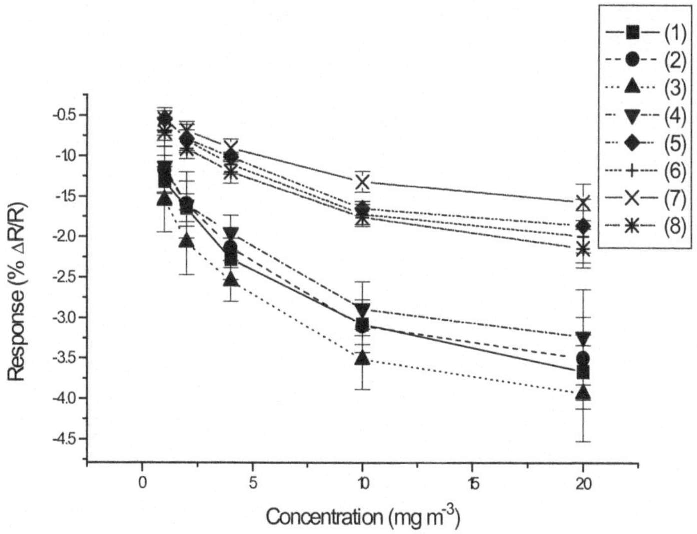

Figure 9.9 Plot of concentration response curves for acetic acid showing 8 sensors from an array of 20 conducting polymer sensors (Persaud *et al.* 1996b).

Misselbrook *et al.* (1997) also showed using cattle slurry that when the output of a 32 sensor array was averaged and compared with odour concentrations (using olfactometry) a reasonable fit could be obtained (Figure 9.11). Another significant feature of this study was the concentrations of the odours being considered (100-1000 ou/m^3), which was considerably lower than had previously been reported. However, when sensor measurements were compared between the 3 field experiments, the difference in response to background samples (in experiments 2 and 3) was entirely explained by the difference in the relative humidity; other laboratory based studies have reported on the sensitivity of conducting polymers to changes in relative humidity (Gardner and Bartlett 1999).

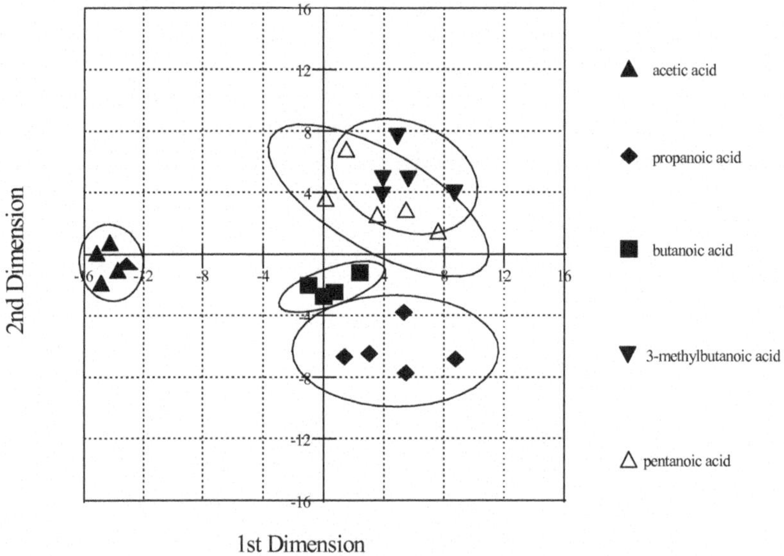

Figure 9.10 Sammon map showing distribution of 4-methylphenol, acetic acid, ammonia, butanoic acid, basic pig slurry, acidic pig slurry, indole, 3-methylbutanoic acid, p-cresol, phenol, propanoic acid, sktole and pentanoic acid for a range of concentrations using a sensor array consisting of 20 pyrrole polymers (Persaud *et al.* 1996a).

Previous studies (using sensor arrays) have used odour samples that are either from the same source or are relatively low in odour intensity. Stuetz *et al.* (1998b) reported on the sensor array analysis of odour samples from 10 sewage treatment works, the odour bag samples were collected in the field and consisted of a range of odour concentrations (125–781,066 ou/m^3). The canonical correlation analysis of the 12 sensor responses with odour concentrations showed that when all the odour samples were analysed no correlation could be found, however when samples with odour concentrations less than 4000 ou/m^3 were used, the correlation was improved (Figure 9.12). When odour samples from a single sewage works were considered a strong linear correlation was evident (Stuetz *et al.* 1999c). Similar relationships were also found when the odour potential (Hobson 1995) of sewage liquors was compared with the sensor responses (Stuetz *et al.* 1999c). These observations suggest that the sewage odour profiles are site specific to individual treatment works, which is most likely dependent on the composition of the wastewater (Stuetz *et al.* 1999c). A study by Romain *et al.* (2000) using odour samples from five different odour sources (paint shop, composting facility, wastewater treatment works, rendering plant and printing houses) also showed different odour sources produced different sensor response profiles (Figure 9.13). These findings suggest that a sensor array could be used differentiate between different types of odour annoyance.

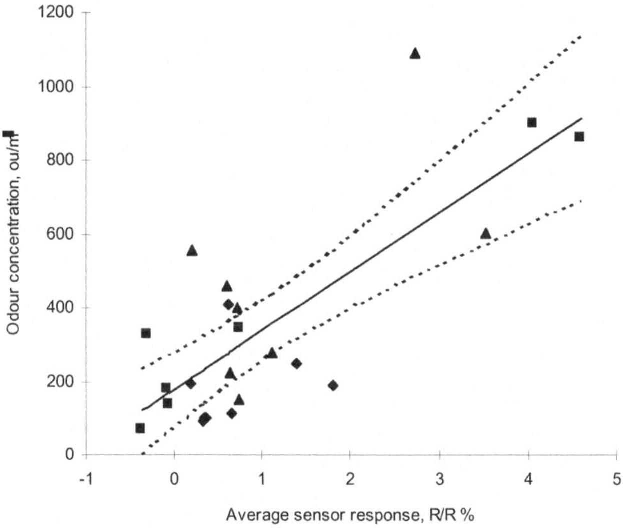

Figure 9.11 Plot of average sensor responses against odour concentrations (ou/m^3) using an Aromoscan electronic nose, consisting of 32 polypyrrole sensors. Samples are from 3 experiments showing a fitted line with upper and lower 95% confidence intervals (Misselbrook *et al.* 1997).

9.3.2 Potential applications in odour assessment

The use of laboratory-based sensor arrays systems for measuring environmental odours have shown that odours emitted from sewage, agricultural and landfill practices can be correlated to the assessment of odour annoyance (using olfactometry) in controlled environments. However, in order to understand the effects of localised odour pollution, it is necessary to translate these laboratory-based experiences into formats that can be applied to making measurements under variable conditions (Flint *et al.* 2000a).

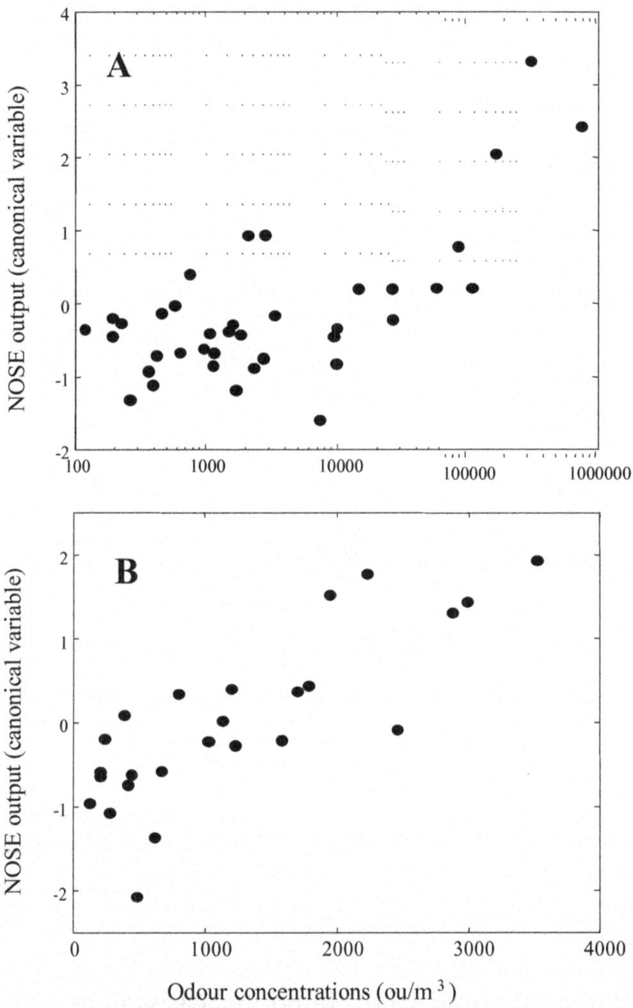

Figure 9.12 Canonical correlation analysis showing the relationship between sensor responses (canonical variable) and odour concentrations (ou/m^3) for 46 duplicate odour samples from 10 sewage works (A) and 25 duplicate odour samples with odour concentration < 4000 ou/m^3 (B) using a Neotronics eNOSE (model D) sensor array, consisting of 12 polypyrrole sensors (Stuetz *et al.* 1999c).

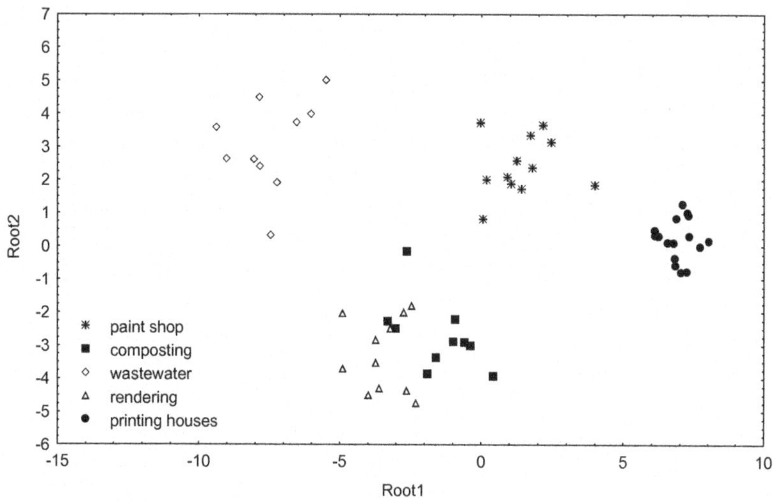

Figure 9.13 Discriminant analysis of 59 environmental samples from 5 odour sources using a sensor array consisting of 12 MOS sensors (Figaro Engineering). Samples were collected from 5 sources, four times during a 7-month period using at least two odour bags (Romain *et al.* 2000).

Recent studies (Nicolas *et al.* 1999; 2000a) have shown that a portable instrument is able to predict an unknown odour in the environment and is able to monitor it continuously, on the basis of a previously calibrated classification model (Figure 9.14). These results also demonstrated that in spite of the influence of environmental parameters (such as climate, source characteristics, sampling location, sampling time and operational staff) on sample humidity and temperature, a simple sensor array system with suitable data processing can detect and identify typical olfactive annoyance (Romain *et al.* 2000). Nicolas *et al.* (2000a) suggest that the air humidity will not influence the odour recognition dramatically, as long as many different humidity conditions are included in the learning phase for a given odour.

Several obstacles still remain before either the direct measurement of odours in the field or the continuous monitoring of odour abatement systems using sensor arrays becomes a reality. Further work will need to (Stuetz *et al.* 1999c; Nicolas *et al.* 2000b; Flint *et al.* 2000b):

(1) understand and control the influence of ambient parameters (such as temperature and humidity) on sensor response baselines;

(2) develop sensor array calibration procedure against olfactometry measurements with a suitable reference gas;
(3) improve sensor sensitivity and noise reduction in order to be able to detect local changes in concentration at a resolution that will permit meaningful measurements to be made and reflect actual site conditions.

However, results to date have shown that although the continuous monitoring of environment odours in the field looks like a challenge, the actual findings in the field are promising and the potential of applications is enormous (Nicolas *et al.* 2000b).

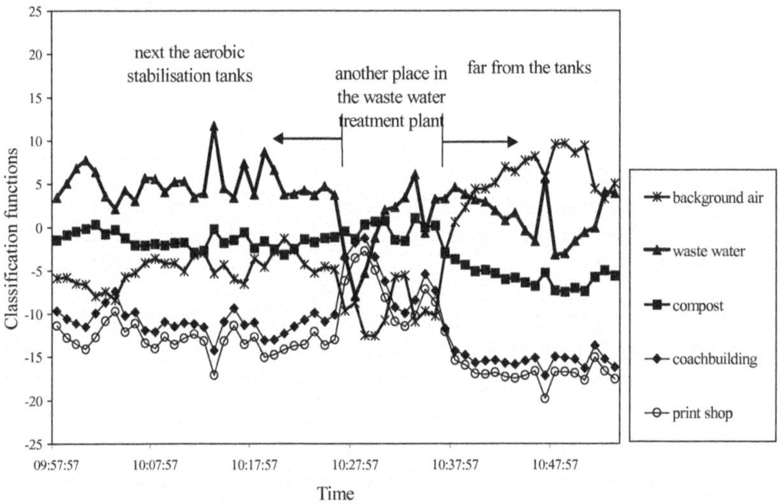

Figure 9.14 The classification functions when a portable sensor array using 12 MOS sensors (Figaro Engineering) is moved around a wastewater treatment works. The discriminant analysis is based on a learning phase with 5 odour sources (Nicolas *et al.* 2000a)

9.4 REFERENCES

Bliss, P. J., Schulz, T. J., Senger, T. and Kaye, R. B. (1996) Odour measurement - factors affecting olfactometry panel performance. *Water Sci. Technol.* **34**(3-4), 549-556.

Fenner, R.A. and Stuetz, R.M. (1999) The application of electronic nose technology to environmental monitoring in the water industry. *Water Enviro. Res.* **31**(3), 282-289.

Flint, T.A., Persuad K.C. and Sneath R.W. (2000a) Automated indirect method of ammonia flux measurement for agriculture: effect of incident wind angle on airflow measurements. *Sensors and Actuators B* **69**, 389-396.

Flint, T.A., Persuad K.C. and Sneath R.W. (2000b) Development of a practical distributed ammonia flux measurement system for the outdoor environment. *Proc. ISOEN2000*, Brigton, pp. 155-156.

Gardner, J.W and Bartlett, P.N. (1991) Pattern recognition in gas sensing. In: *Techniques and mechanisms in gas sensing* (P. Moselt, J. Norris and D. Williams, ed.), Chapter 14, Adam-Hilger, Bristol.

Gardner, J.W and Bartlett, P.N. (1992) Pattern recognition in odour sensing. In: *Sensors and sensory systems for an electronic nose* (J.W. Gardner and P.N. Bartlett, eds.), pp. 161-180, Kluwer, Dordrecht.

Gardner, J. W. and Barlett, P. N. (1994) A brief history of electronic noses. *Sensors and Actuators B* **18**, 211-220.

Gardner, J. W. and Barlett, P. N. (1999) *Electronic nose: principles and applications.* Oxford University Press, New York.

Gardner, J.W. and Hines, E.L. (1996) Pattern analysis techniques. In: *Handbook of Biosensors and Electronic Noses.* (E. Kress-Rogers, ed.), pp. 633-652, CRC Press, Boca Raton.

Gardner, J.W., Graven, M., Dow, C. and Hines, E.L. (1998) The prediction of bacterial type and culture growth phase by an electronic nose with a multi-layer perception network. *Meas. Sci. Tech.* **9**, 120-127.

Gibson, T. D., Prosser, O., Hulbert, J. N., Marshall, R. W., Corcoran, P., Lowery, P., Ruck-Keene, E. A. and Heron, S. (1997) Detection and simultaneous identification of microorganisms from headspace samples using an electronic nose. *Sensors and Actuators B* **44**, 413-422.

Gibson, T., Prosser, O. and Hulbert, J. (2000) Electronic noses: an inspired idea? *Chemistry and Industry* (April), pp. 287-289.

Hobbs, P. J., Misselbrock, T. M. and Pain, B. F. (1995) Assessment of odours from livestock wastes by a photoionization detector, an electronic nose, olfactometry and gas chromatography-mass spectrometry. *J. Agric. Engng. Res.* **60**, 137-144.

Hobson, J. (1995) The odour potential: a new tool for odour management. *J. Chart. Inst. Wat. Enviro. Manag.,* **9**: 458-463.

Hodgins, D. (1995) The development of an electronic nose for industrial and environmental applications. *Sensors and Actuators B* **26-27**, 255-258.

Hodgins, D. and Simmonds, D. (1995) The electronic nose and its application to the manufacture of food products. *J. Auto. Chem.*, **17** (5), 179-185.

Holmberg, M., Gustafsson, F., Hornsten, E. G., Winquist, F., Nilsson, L. E., Ljung, L and Lundstrom, I. (1998) Bacteria classification based on feature extraction from sensor data. *Biotech. Tech.* **12**, 319-324.

Manly, B.F.J. (1986) *Multivariate statistical analysis.* Chapman and Hall, London.

Mills G., Walsh F and Whyte I. (1996) A sense of (electronic) smell. *Chem. Technol. Europe* (July/ August), pp.26-30.

Mielle, P. and Marquis, F. (1999) An alternative way to improve sensitivity of electronic olfactometers. *Sensors and Actuators B* **58**, 526-535

Misselbrook, T. M., Hobbs, P. J. and Persaud, K. C. (1997) Use of an electronic nose to measure odour concentration following application of cattle slurry to glassland. *J. Agric. Engng. Res.* **66**, 213-220.

Namdev, P.K., Alroy, Y and Singh, V. (1998) Sniffing out trouble: use of an electronic nose in bioprocesses. *Biotech. Prog.* **14**, 75-78.

Nicolas, J., Romain, A.C., Wiertz, V., Maternova, J. and Andre, Ph. (1999) First trends towards a field odour detector for environmental applications. *Proc. ISOEN99*, Tubingen, pp. 368-371.

Nicolas, J., Romain, A.C., Wiertz, V., Maternova, J. and Andre, Ph. (2000a) Using a classification model of an electronic nose to assign unknown malodours to environmental sources and to monitor them continuously. *Sensors and Actuators B* **69**, 366-371.

Nicolas, J., Romain, A.C., Monticelli, D., Maternova, J. and Andre, Ph. (2000b) Choice of a suitable E-nose output variable for the continuous monitoring of odours in the environment. *Proc. ISOEN2000*, Brighton, pp.127-128.

Persaud, K. C., Khaffaf, S. M., Hobbs, P. J. and Sneath, R. W. (1996a) Assessment of conducting polymer odour sensors for agricultural malodours measurements. *Chem. Senses* **21**, 495-505.

Persaud, K. C., Khaffaf, S. M., Hobbs, P. J., Misselbrook, T.M. and Sneath, R. W. (1996b) Application of conducting polymer odour sensing arrays to agricultural malodour monitoring. *Proc. Air Pollution from Agricultural Operations*, Ames, pp.249-253.

Persaud, K. C. and Travers, P. J. (1997) Arrays of broad specificity films for sensing volatile chemicals. In: *Handbook of Biosensors and Electronic Noses*. (E. Kress-Rogers, ed.), pp. 563-592, CRC Press, Boca Raton.

Rencher, A.C. (1998) *Multivariate statistical inference and applications*. John Wiley and Sons, New York.

Romain, A.C., Nicolas, J., Wiertz, V., Maternova, J. and Andre, Ph. (2000) Use of a simple tin oxide sensor array to identify five malodours collected in the field. *Sensors and Actuators B* **62**, 73-79.

Stuetz, R. M. and Fenner, R. A. (1998) Electronic nose technology: a new tool for odour management. *Water Quality Internat.* (July/August), pp. 15-17.

Stuetz, R.M., White, M. and Fenner, R.A. (1998a) Use of an electronic nose to detect tainting compounds in raw and treated potable water. *J. Water Supply Res. Tech. - Aqua.* **47**(5), 223-228

Stuetz, R. M., Engin, G. and Fenner, R. A. (1998b) Sewage odour measurements using a sensory panel and an electronic nose. *Water. Sci. Technol.* **38**(3), 330-335.

Stuetz, R. M., Fenner, R. A. and Engin, G. (1999a) Characterisation of wastewater using an electronic nose. *Water Res.* **33**, 442-452.

Stuetz, R.M., George, S. Fenner, R.A. and Hall, S.J. (1999b) Monitoring wastewater BOD using a sensor array. *J. Chem. Tech. Biotech.* **74**, 1069-1074.

Stuetz, R. M., Fenner, R. A. and Engin, G. (1999c) Assessment of odours from sewage treatment works by an electronic nose, H_2S analysis and olfactometry. *Water Res.* **33**, 452-461.

Part IV

ASSESSMENT AND PREDICTION OF ODOURS

10

Prediction of odorous emissions

Franz-Bernd Frechen

10.1 INTRODUCTION

Why should one attempt to predict odorous emissions? The only reason that makes sense is that, prior to construction of a new or expansion of an existing plant, information is required concerning future odour impact and thus odour nuisance in an adjacent sensible site. This information is important or even essential, as a nuisance-free plant operation is of high interest for the owner and operator of the plant as well as the neighbour.

In the case of odour, the prediction of future extent of nuisance would be the desired content of the forecast. However, the respective facilities do not emit nuisance. The facilities emit odorous substances which then are dispersed into the atmosphere and cause nuisance when arriving at the nose of a human being located in an adjacent sensitive area. Therefore, a correlation is needed between the actual emission as a technical process and the actual nuisance impact it causes, which is a sensory, psychologically interpreted process.

edited by Richard Stuetz and Franz-Bernd Frechen. ISBN: 1 900222 46 9

Thus, it is necessary to find the link between future nuisance and future odour emissions in order to assess whether this impact (nuisance) is tolerable or not. This is not just one simple link, but a chain of links, generally described as follows:

- It must be possible to distinguish between "tolerable nuisance" and "no more tolerable nuisance". It is quite unlikely that any government would demand a zero nuisance situation, thus setting this limit to zero. In addition, it must be possible to express this limit in values obtainable from measurements. Such a limit will always have three components: time, impact and location of impact. The most common approach is to limit the percentage of time during which a set limit odour concentration will be exceeded, depending upon the type of location where the impact is present (residential area, industrial area, etc.). Setting a maximum odour concentration which must not be exceeded at any time increases the time component to 100%. Limits referring to the percentage of people who may be more or less affected by malodours must be transferable into this "technical" type of limit in order to become valuable for derivation of measures at the emitting sites. The link between nuisance and "technical" impact limits may be created by assessment of the stakeholders or of the judge who has to decide on a case, or it may be derived by a more general investigation of the connection between impact, expressed in percentage of time during which a specific impact odour concentration is exceeded, and percentage of annoyed people. It may be stated for the time being that this kind of research work was done with local residents beneath several types of industry in Germany to established the new "Directive on Odour Impact" (1998) and the technical limits it stipulates. Extent of impact was connected with the nuisance response of the local population via questionnaires to derive the limit for acceptability of nuisance, and by doing so linking "technical" impact and "psychological" response.
- Having a limit in the form of a fixed impact odour concentration, which must not be exceeded during more than a limited percentage of time for the location in question, it is then necessary to establish a link between this issue and the emission of the odour source. This is done by atmospheric dispersion calculations. As mentioned in Chapter 2 already, there is no specific dispersion calculation for odours, as odours behave like ordinary gases do. However, some atmospheric dispersion

calculations, specially applied for odours, include some kind of impact assessment, which actually should be part of the first link mentioned.

- The last link needed now is the emission of the odour source in question (and, of course, the emission of all other odour sources in the surroundings that do exist already and that may have an impact on the particular site). With new facilities, an emission prognosis is required, and this prediction of the odorous emissions is what this chapter deals with.

With the prediction of odorous emissions, the saying applies that "every prognosis is difficult, especially when dealing with the future". There are multiple reasons for this: low precision of measurements, inclusion of that are not totally understood physiological processes, inclusion of psychological processes which cannot be predicted, variations in operation and process conditions of the odour emitting plants and more.

However, as it is necessary to examine the environmental impact of future projects, it is necessary to predict the odour emissions from a facility as precisely as possible, although it is clear that accuracy in the case of odour (emission prediction as well as nuisance forecast) is humble compared with every other environmental impact, such as noise or others.

As the whole process consists of chain links, no single link can be examined without considering the others. Thus, as the prediction of odorous emissions has to provide the necessary data for the atmospheric dispersion calculation, it is necessary to know which data the dispersion model needs. All relevant models need emission data in the form of emitted mass per time. Thus, with odour it is necessary to provide the emitted mass flow as the total emitted *odorant flow rate* given in ou/h (or ou/s, ou/min or Mou/h, respectively). It is the product of the odour concentration c_{od} and the pertinent volume flow rate $\dot{V}$ – if these two parameters are directly measurable. If not, two special conditions must be regarded:

- a special sampling method is necessary, which is not a major problem, as described in Chapter 5, and
- there is no knowledge concerning the true odour concentration at the emitting source. This may cause problems as will be explained later.

10.2 WHAT CAN WE PREDICT?

10.2.1 Nuisance?

Nuisance can be caused by an odour stimulus which can be characterised multidimensionally. Besides the most common aspect:

- strength of the odour perceived, which may be given in terms of odour concentration or an odour intensity number,

further aspects that are commonly denominated as "should be taken into account" are:

- kind of odour,
- hedonic odour tone,
- several time-dependent characteristics as for example
  - total duration of impact,
  - rhythm of impact,
  - frequency of impact,
  - time of the day / of the week / of the year of impact.

Unfortunately, no one can give equations that would allow an exact dose-effect-calculation for most of the above mentioned "should-be-taken-into-accounts". Even a clear "dose-description" for many of the aspects is not available. How would "rhythm", "time of the week", "kind of odour" or "frequency" be covered by numbers that make them available for comparison with each other or even calculations or limits? Concerning some of the aspects, only presumptions are merchandised, such as "this phenomenon will give more nuisance than that".

It is necessary to return to more solid ground.

10.2.2 Can dispersion calculation help?

Thus, a simplification is necessary that must be based on respective research to give a system of interrelationships which is of any use in practice for prediction purposes.

This is done pragmatically due to the second link listed above – dispersion calculation. Commonly, dispersion calculation programs need information concerning the emitted mass flow of all relevant sources as an input (and, of course, some more information such as meteorological data and geographical information, which are not an issue in this chapter), and then they will output a two-dimensional distribution of impact concentration versus the time (duration) during which the respective concentration is exceeded (or not exceeded).

As mentioned earlier, the emitted odorant flow rate is needed. Either the odour concentration and the associated volume flow rate or the total emitted odorant flow rate must be predicted. In the end, all sources are predicted in the form of total emitted odorant flow rate, given in ou/h. This is what this chapter deals with. Nevertheless, the ability of the common models to predict impact *concentrations* with emissions given only in emissions *mass flow rates* gives some chance to under- or overestimate the "real" (annoying) impact effect.

Example: Downwind a source emitting 1 Mou/h, a dispersion model calculates an impact concentration of 30 ou/m^3 for a specific meteorological situation. The emitted odorant flow rate may be formed by a small volume flow rate of e.g. 2,000 m^3/h with an odour concentration of 500 ou/m^3 or by a large volume flow rate of 50,000 m^3/h with an odour concentration of 20 ou/m^3 – both emission constellations produce the same result on the impact side. But it is obvious that in the second constellation the numerical result is not correct, as a dilution and not any increase of concentration of the air occurs during transport to the location of the impact.

This example gives evidence that the simplifications included in most dispersion models sometimes lead to ridiculous results which will not happen in reality. In other words: if the emission source is of the type "low concentration, high volume", as for example a large aeration basin, then its impact is regularly overestimated. If the emission source is of the type "high concentration, small volume", e.g. an open thickener, then its impact is regularly underestimated.

It is the art of predicting odour emissions to compensate for this effect.

10.3 HOW CAN WE PREDICT?

10.3.1 Totally new plant

There may be the necessity to give a prognosis for a totally new plant. Then, no measurement of the actual situation is possible, and the prognosis must be based on the experience of the expert and his results from measurements at plants with similar constellations at different locations.

Also, of course, the prognosis must be based on the design of the plant. Thus, the first action is to obtain as much information as possible on the engineers thoughts when designing the plant. A careful revision of the design work is one of the two most extensive activities in this case.

The second important and extensive activity then is the prognosis of the emitted odorant flow rates of all identifiable odour sources.

These two steps are what the expert's job consists of. He is paid for this, but even more for his *experience with odour emissions* and his *knowledge of wastewater treatment technology*. An expert on odours in conjunction with wastewater must combine both skills.

The outcome of the experts work is the emission prognosis, which is one of the two decisive input data sets for the dispersion calculation that will possibly allow one to judge over the expected impact and its possible legality.

10.3.2 Expansion of an existing plant

10.3.2.1 Reasons for a measurement program

If the prognosis is needed in conjunction with upgrading or expansion of an existing plant, then it is possible, highly recommended and in Germany usually demanded by the local authorities that a measurement program is performed at the existing site. In addition to the above mentioned basis, consisting of the experience and measurement results from similar plants in different locations, this adds experience concerning the actual site and its special circumstances, if such are present – and it is common experience that these are always present.

Besides this, a general aim of measurement programs is that experience is gathered. If it would be usual to predict without measurement, no measurement results would exist which provide information and experience. This would lead to a deterioration of the quality of the prognoses.

Another reason for a measurement program at an existing plant is that the emission prognosis must include the emissions of the plant as far as it exists already – changes in design, operation and process regarded.

Furthermore, a measurement program is very useful as it will disclose the weak points of the existing plant under the current operation and process. Existing sources or processes emitting excessive odours can be detected, and countermeasures can be designed.

In this context a hint may be of some importance when considering possible success control measurements after implementing countermeasures. Success control can be done by questioning local residents, as for example is explained in VDI guideline 3883 (1993, 1997), by field inspections, as for example are explained in the VDI guideline 3940 (1993), both inspecting the impact. Success control can also consist of another measurement program with subsequent atmospheric dispersion calculation, thus inspecting the emission. Both strategies do have their specific advantages and disadvantages, but when working with an impact method, it will not easily show the weak points of the emitting facility.

10.3.2.2 Detecting important liquid streams with the help of the Odour Emission Capacity (OEC)

When undertaking a measurement program, as a matter of course the state of the art concerning sampling and olfactometry, as described in this book, must be applied. All these techniques deal with gaseous sampling and measurement, and they all characterise gases.

In addition, however, it is of high value and interest, especially when striving for the detection of weak points in an existing plant or system, to use the method of measuring the Odour Emission Capacity (OEC) as introduced by Frechen and Köster (1998). The OEC is presented in the cited paper in detail, so just a short explanation is given here.

The OEC, although characterising liquids, is based on olfactometric measurements. The respective liquid is aerated with odourless air, thus the odorants are stripped off the liquid. The off-gas of the test reactor is periodically measured by analytical methods as well as by olfactometry. Knowing the volume of liquid aerated (30 litre), the volume of air passed through the liquid and the respective concentrations (odorant concentration and any analytically measured compounds concentrations), one can integrate the amount of odour units or of compounds analysed, resulting in a value which characterises the liquid concerning its possible emission (Emission Capacity) of odours or other compounds. These capacities are given in the unit of "ou/m^3_{Liquid}" in the case of the OEC or in the units of for example "$H_2S/\ m^3_{Liquid}$" in the case of the H_2S Emission Capacity and so on.

This method has two major advantages that make it very advisable to use it. On the one hand, it makes it possible to detect small, but highly contaminated liquid flows, and at least an approach to a balance can be made. As a recent example, measurement of the filtrate from a raw sludge dewatering facility in Germany made it evident that this flow, although only about 5% of the influent by volume, carried more than 90% in terms of the odorant mass flow. The small flow of filtrate contaminates the whole aerobic wastewater treatment process, and its "noseprints" were easily detectable even at the final sedimentation tank.

On the other hand, this method is able to give hints concerning the nature of the problem which may exist and may be discovered by the method. The examples following will illustrate this.

In Figure 10.1, the test run is shown with a septic domestic wastewater. As can be seen, odorant concentration and H_2S concentration decrease in a nearly identical manner from the liquid. This indicates a septicity, H_2S related problem.

Figure 10.2 shows the test of a wastewater that surely was septic (and thus had a H_2S problem), but with additional odorants that are not septic-borne and behave different from H_2S concerning stripping from the liquid.

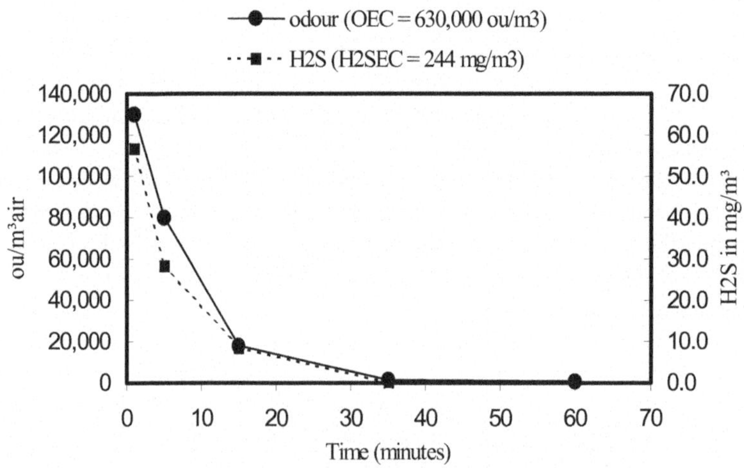

Figure 10.1. OEC and H_2S-EC of a septic domestic wastewater.

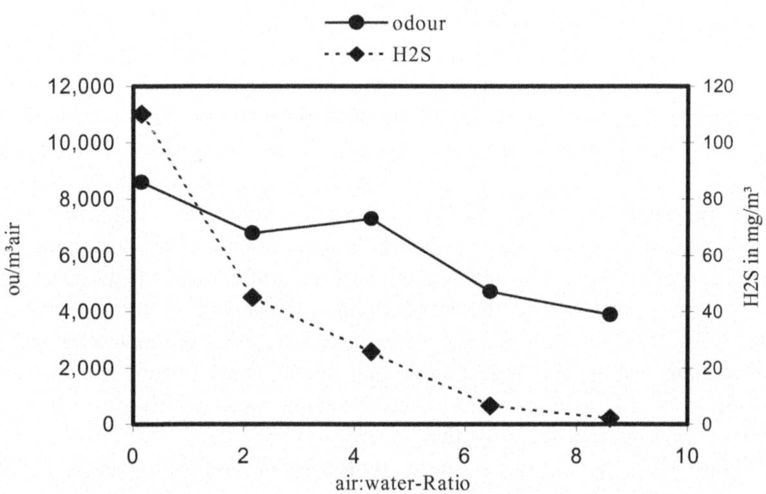

Figure 10.2. OEC and H_2S-EC of a mixture of domestic and industrial wastewater.

These odorants were discharged into the sewer by a large food factory, and their typical *kind of odour* was perceivable in all samples that were taken from the test reactors off-gas.

Thus, the Odour Emission Capacity method allows the detection of the sensible areas and helps to design suitable countermeasures at the sewerage system as well as at the wastewater treatment plant itself.

10.3.2.3 Regarding changes in existing processes

It is surely unnecessary to mention that the newly built parts of the extended plant must be assigned appropriate specific emitted odorant flow rates. But, when working on a plant that will be redesigned and expanded, it also is important to regard a possibly new and different use of existing facilities.

An example: usually, when extending a WWTP, the load of the biological parts, e.g. aeration basins, is lowered. As known, low loaded tanks produce less contaminated air, as the biological process is more complete than it would be in the (actual) high loaded tank, which will decrease the emitted odorants flow rate.

10.4 WHAT WILL WE PREDICT?

10.4.1 Preliminary remark

As noticed already, the data needed for prognosis are the emitted odorant flow rates from the respective odour sources of a plant.

Such data – including those that follow in the next table – must not be used for prognoses without a careful study of the respective circumstances. At least, an inspection of the plant, if it exists already, is necessary; however, with existing plants, measurements should always be done.

Also it has to be kept in mind that the expert has to introduce his experience into his prognosis. This includes an assessment of the effect of extraordinary low and exceedingly high actual emitted odorant concentrations (see above) as well as a careful assessment of the kind of odour with regard to its provenance as well as to its hedonic odour tone.

10.4.2 Overview over measurement results

Keeping all the above– mentioned hints in mind, at least an overview over results from a number of measurements made during recent years at several

WWTPs should be given here. Table 10.1 shows the results of some hundred measurement results from different parts of different WWTPs (Frechen 2000).

Table 10.1. Overview over specific odorant flow rates from wastewater treatment plant sources.

	Specific odorant flow rate	
	from	to
Part of plant	ou/(m^{2}h)	ou/(m^{2}h)
Water surface influent channel	200	1,200
Aerated grit chamber	500	20,000
Grease trap	2,000	40,000
Screenings	1,000	5,000
Grease from grease trap	1,000	15,000
Sand from grit chamber	1,000	6,500
Primary sedimentation tank: surface	500	4,000
Primary sedimentation tank: weir area	500	5,000
Aeration tank: anaerobic part	850	3,000
Aeration tank: anoxic part	600	2,000
Aeration tank: aerobic part	300	1,700
Final sedimentation tank	150	500
Filtration	100	200
Primary sludge thickener	12,000	35,000
Stabilised sludge thickener	500	5,000
Stabilised sludge, dewatered	600	16,000

Some additional remarks are necessary. The range given in this table should be understood as a “usual range” under normal conditions at a well operated plant without major industrial influent and other specific, odour-relevant circumstances. In specific cases, this range may be exceeded, which means that at specific sites values may be measured that are below the given lower value or above the given higher value.

Some of the most important remarks concerning the area sources mentioned in the table as well as concerning some other, non-area sources, are given below.

- Influent area (including pumping station, screen, grit chamber and grease trap): emissions here depend strongly on the plant and process design, i.e. whether or not filtrate, sludge liquor, liquids from night soil/ septic tanks and others are discharged into this part of the plant – values up to 75,000 ou/(m^{2}h) were measured already.
- Grease trap and grease container: even higher values are likely depending upon plant operation and maintenance.
- Air inside building for screen and influent pumping machines: measurement results ranging from 50 ou/m^3 to 400 ou/m^3 with natural ventilation do indicate that air treatment presumably will not make

sense, as no remarkable reduction of total plant emissions will be possible. Note: values up to 6,700 ou/m^3 were measured!

- Stormwater tanks: under regular (usually observable) conditions, during filling phase emissions may be slightly below those of primary sedimentation tanks. During the rest of the year, one may assess the emission to be zero or up to 1,000 $ou/(m^2h)$, depending upon installation and operation.
- Anaerobic operated basins for biological phosphorus removal: values between 15,000 $ou/(m^2h)$ and 50,000 $ou/(m^2h)$ were measured.
- Primary sedimentation tank: high emissions from the liquid surface can often be avoided by correct operation of the tanks, see chapter 4.
- Final sedimentation tanks: usually no countermeasures will make sense except in very extreme cases. Final sedimentation odours belong to the type of "high volume flow, low odorant concentration", and in many cases it would be most appropriate to assign to them a zero emission or a reduced emission in the range of at maximum 50% of the measurable value. The same applies to filtration units following the final sedimentation.
- Deodorization installations: as a result from the recommendations given in the new "Directive on Odours" (1998), in Germany biofilters are assigned a zero emission if the biofilter "is working properly" and some other weak restrictions are met. No comments are to be found concerning other deodorization techniques. As properly working deodorization units are sources of the kind "high volume, low concentration", a reduced emission assignment is pertinent.

All devices of sludge treatment show emissions which strongly depend upon the process used and the operation of the plant. Thermal conditioning always causes strong emissions at several steps of the plant (owing to the high odour emission capacity of the sludge liquid). The same applies with night soil and sludge from septic tanks. Also, some techniques of sludge conditioning, e.g. dosing of lime, may give specific problems. All treatment of primary sludge tends to be very critical concerning odour emissions.

Air inside sludge dewatering building showed measurement results in the range of 20 ou/m^3 to 400 ou/m^3 with natural ventilation. Immediately beside the dewatering machines, concentrations of up to 1,000 ou/m^3 (stabilised sludge) or up to 16,000 ou/m^3 (thermal conditioned sludge) were observed. Thus, exhausting systems must take this (source inside building) into account.

With the conditioning of primary sludge and not sufficiently stabilised sludge with lime, in a specific case emissions of up to 740,000 $ou/(m^2h)$ and NH_3-emissions of up to 9,000 mg $NH_3/(m^2h)$ were measured immediately after the deposition of this sludge. Afterwards, emissions rapidly decreased but increased again when the sludge layer was moved mechanically. After 1 week of storage and subsequent mechanical movement, for example, emissions were as high as 280,000 ou/(m^2h) and 1,200 mg NH_3/(m^2h), respectively.

If sludge or liquid is drawn off digesters conventionally via open sludge chambers at the top of the digester, emissions are to be expected which may vary in a broad rang. They also might not even be noticed by the plant operator due to the emission height.

10.5 QUALITY CONTROL

Of course it is necessary to undertake a control of the quality of prognoses as well as of the success of measures that possibly were taken to minimise odour emissions.

Success control is possible after finishing the respective project and implementing all countermeasures that were assessed to be necessary and thus recommended by the expert working on the case for prognosis purposes.

As mentioned above, quality control can be done by questioning local residents, as for example is explained in VDI guideline 3883 (1993, 1997), by field inspections, as for example are explained in the VDI guideline 3940 (1993), both inspecting the impact. Successful control can also consist of another measurement program with subsequent atmospheric dispersion calculation, thus inspecting the emission. Both strategies do have their specific advantages.

When performing field inspections or questionnaires, the advantage is that complex orographic and topographic constraints or other special circumstances that cannot be taken into appropriate consideration by a dispersion model are included automatically in the results. However, it may be a disadvantage that one is bound to the actual meteorological conditions that exist during the duration of the inspection (usually ½ or one year) or during the questionnaire.

When inspecting the emission by means of another measurement program, the chance exists that weak points can be detected which are still present or which possibly did escape from the expert's notice when creating the forecast.

The control of the quality of prognoses, of course, must not be done by the expert who made the prognosis!

10.6 REFERENCES

Determination and Assessment of Odour Impact – Directive on Odour (1998). *The States Commission on Environmental Impact Control, Germany.*

Frechen, F.-B. and W. Köster (1998) Odour emission capacity of wastewaters – standardization of measurement method and application. *Water Sci. Technol.* **38** (3), 61-69.

Frechen, F.-B. (2000) Overview over olfactometric emission measurements at wastewater treatment plants. *IWA Specialist group on Odours and Volatile Emissions, Newsletter No. 3, September 2000.*

VDI-guideline 3883 part 1 (1997) Effects and assessment of odours – Psychometric assessment of odour annoyance – questionnaires, *VDI-handbook on Air Pollution Prevention,* Vol. 1.

VDI-guideline 3883 part 2 (1993) Effects and assessment of odours - Determination of annoyance parameters by questioning – repeated brief questioning of neighbour panellists, *VDI-handbook on Air Pollution Prevention*, Vol. 1.

VDI-guideline 3940 (1993) Determination of odorants in ambient air by field inspections, *VDI-handbook on Air Pollution Prevention,* Vol. 1.

11

Odour mapping using H_2S measurements

John Hobson and Gong Yang

11.1 INTRODUCTION

Hydrogen sulphide (H_2S) mapping, put at its simplest, is the technique of taking a large number of H_2S measurements within and around a sewage treatment works and using a surface contouring technique to produce a contour map of H_2S concentrations.

The technique was developed and first used at the UK's Water Research Centre (WRc) in the late 1980s. It was around then that the sensitivity of portable H_2S monitors rose sharply from around 1 ppm (vapour part per million) using methods based on electrochemical cells to around 1 ppb (vapour part per billion) using gold film sensor technology. While the former was not sensitive enough to detect H_2S levels in unconfined spaces in the open environment within and around sewage treatment works, it soon became apparent that the latter was.

 Odours in Wastewater Treatment: Measurement, Modelling and Control edited by Richard Stuetz and Franz-Bernd Frechen. ISBN: 1 900222 46 9

The production of a map is simply an aid to interpretation of data. The resulting H_2S maps were the first instance of the visualisation of an odour problem associated with wastewater treatment. Not only were they, and still are, very helpful in determining adverse impact due to odour, and in helping to diagnose its cause, such maps were also very valuable in moving the field of study of odour associated with wastewater treatment from one of almost total subjectivity to one in which quantitative measurement could play an important role.

Since that time other techniques have developed for the quantification of odour - most notably olfactometry, the quantification of source emission rates has become very important and the use of dispersion models allows the visualisation of impact due to odour under a wide range of conditions. Currently there is considerable argument in the field of wastewater treatment as to the merits of basing odour quantification almost exclusively on H_2S measurement or on olfactometry. In this chapter we will demonstrate that H_2S mapping is still a valuable tool with an important role to play in the quantification of odour impact and the diagnosis of its cause.

11.1.1 Why H_2S?

H_2S is an odorous gas produced by the degradation of organic matter in the absence of oxygen. It is produced both from the sulphur contained within organic matter and from inorganic sulphate generally present in water. The conditions that generate H_2S also generate a range of other odorous compounds (see Table 1.1), most notably organic sulphur molecules such as methyl mercaptans and dimethyl disulphide as well as amines, volatile fatty acids and protein breakdown products such as indole and skatole. H_2S is therefore both responsible for odour annoyance and acts as a marker associated with a range of other odorous compounds.

The most suitable of this range of odorous compounds for mapping is that with the highest ratio of its concentration of interest to its limit of detection using a convenient portable measuring device. H_2S has a threshold odour concentration in the region of 0.5 ppb and can be measured at levels close to 1 ppb using a portable H_2S analyser. It also transpires that the natural background level of H_2S in the atmosphere is frequently in the range of 1–2 ppb and that concentrations are frequently raised in conjunction with sewage-related odour.

11.1.2 The purpose of H_2S maps

As stated above, a map is a very valuable method of data visualisation. In the case of odour and H_2S, however, a map has additional value. Isolated measurements of H_2S cannot be related to odour. This is because of both the existence of measurable background levels and H_2S derived from road traffic. It is only when raised levels of H_2S can be clearly traced to a sewage-related source that its full value as a marker for sewage odour is revealed. This requires a map.

11.2 THE MECHANICS OF PREPARING AN H_2S MAP

An exercise to prepare an H_2S map comprises:

(1) Planning,
(2) Data collection,
(3) Use of a surface contouring package,
(4) Interpretation.

11.2.1 Planning

Ideally an H_2S map would be a snapshot of H_2S concentrations within and around a sewage treatment works in as much detail as possible. However, it takes time to take the measurements, typically 20–30 seconds per reading using the portable H_2S analyser. There is then a trade-off between the time taken to complete a survey and the number of measurements that can be made.

To make an H_2S map requires the following:

- An H_2S monitor sensitive to concentrations down to 1 ppb or as low as possible,
- Anemometer,
- Thermometer,
- A good quality map or plan of the site,
- A preliminary visit to the site is advisable,
- Planning the selection of sampling points,
- A surface contouring package, Geographical Information System (GIS) or similar.

Prepare a rectangular grid of sampling points on a photocopy of the works plan. The grid should be oriented along the major axes of the treatment plant processes, which are usually roughly rectilinear. There should be 6 to 10 points along each side of the grid, the number being selected to fit in with the layout of the treatment processes. Ideally the grid should be evenly spaced but some flexibility can be allowed to fit in with the process layout. As few points as possible should be over water and where this is unavoidable, extra points can be added at the nearest accessible locations. Overall the number of pre-planned sampling points will generally lie between 50 and 80 in number.

If possible, while preparing an H_2S map, off-site plumes of raised H_2S concentrations should be followed downwind and laterally until they merge with the background concentration. Some thought should be given at planning as to how off-site measurements might be taken. Frequently, 100% access around a sewage treatment works is not easy, which coupled with the difficulty of predicting wind directions in advance, means that it is often not possible to follow off-site plumes as frequently as desired.

11.2.2 Data collection

At the beginning and end of a monitoring exercise, the wind-direction should be noted and its velocity measured together with the air-temperature. This data could be obtained from an on-site weather station if present, or by using one of a number of proprietary monitors based on either vane or hot-wire anemometers. It is sufficient to orient the anemometer until the maximum wind-speed is noted and record the direction with reference to a site plan with north delineated (a compass could be used). The percentage of cloud cover should be estimated. This is good practice and in addition will allow an estimate to be made of atmospheric stability class should it be desired to use the results alongside the output from dispersion models.

To conduct the survey, at least two and preferably three H_2S readings should be taken and recorded at each sampling point. The survey should be performed as rapidly as possible. It is permissible however to take additional readings during a survey. This might be done for a number of reasons:

- To give a better definition of those locations with raised levels of H_2S.
- To take readings where sewage type odours are noticed.

It might be felt that this introduces an element of subjectivity but the technique of preparing a contour map together with the correct approach to interpretation should overcome this. On the other hand, the same considerations mean there may be less value in taking extra readings than might appear at the time of the survey. Some works, many during certain periods of the year, will produce very low levels of H_2S (this is particularly true in the UK). The areas of raised levels of H_2S may be quite small and be missed if the original grid is adhered to. Since levels may rise during other seasons these low-level sources may turn into high level sources and it is valuable to identify them.

There is another reason for taking extra readings. Surface contouring packages (section 11.2.3) interpolate between the actual measurements made. Generally they do this in a symmetrical fashion so odour may appear to disperse upwind from a source. During interpretation this can be discounted, but for the sake of appearance there is value in taking extra readings to define where raised levels fall off to background levels, upwind of sources.

If this procedure is followed, readings should be taken at several locations upwind of all processes, which will allow a representative background level to be estimated. This level will change from site to site and over time even at one site. If for any reason it is felt a representative set of background values has not been obtained, additional readings should be taken upwind of the site.

Resist the temptation to record very high values extremely close to, or even within, sources. Some sources such as for example semi-enclosed sludge wells, may not have a significant H_2S emission rate. Within the well itself however, high levels of H_2S can build-up (Box 11.1) and lead to still high readings within 1 or 2 metres of the entrance. The section on interpretation explains that isolated high values not associated with a plume have little significance.

Box 11.1 Safety note

H_2S is a highly toxic gas, being fatal at concentrations exceeding 600 ppm, rapidly so above 1000 ppm (Irving 1984). It has an occupational exposure limit (OEL) of 10 ppm (Health and Safety Executive 2000) at which concentration it is extremely odorous. It is extremely rare for concentrations in the open to exceed 1 ppm, even at odorous sewage works. Where sewage from a rising main, or certain types of sludge or sludge liquor, pass through highly turbulent structures, open air concentrations can briefly approach the OEL, but are most unlikely to pose a hazard. Concentrations of H_2S in confined spaces on the other hand can rise to several hundred ppm or higher and be rapidly fatal. Anyone undertaking an H_2S survey should be aware of the hazard of confined spaces and should never enter a confined space without both appropriate training and equipment. Entry into confined spaces is never a requirement when performing an H_2S survey.

11.2.3 Surface contouring

A number of commercial software packages are available to produce a contour map of H_2S concentrations (Table 11.1). The details of this will vary depending on the package being used.

Table 11.1 List of commercial surface contour packages.

Software package and location	Notes
NAG Library Fortran Routines	First method used. Contouring routines for linking to a Fortran program.
ARCINFO™ (www.esri.com/software/arcinfo/index.html)	GIS package originally for workstations before PCs developed sufficient power.
MAPINFO™ (www.mapinfo.com)	GIS package for PC.
SURFER™ (www.goldensoftware.com)	Surface contouring package with some GIS features.

ARCINFO™, MAPINFO™ and SURFER™ can handle a wide range of data formats, which is very valuable. They can accept site plans or maps in most formats. The quality of an H_2S contour map is greatly enhanced if sited on a good quality plan or map. If a map is used, it should be ascertained that permission is obtained from the copyright holder.

For several years the authors have used SURFER™ on a standard personal computer. This package together with most others will accept data in the form of triplets - x-distance, y-distance and z-H_2S concentration. It is however rather tedious to have to measure the x and y values for all points and then input them into the package. Most packages will also accept data in grid form, i.e. a matrix of z-values (H_2S concentrations) but this requires a uniform rectangular grid, which for the reasons described above will not be the norm.

The SURFER™ package together with most GIS packages will allow data entry in the following manner provided an electronic version of the site plan is available. Open the electronic site plan. Use the mouse to position a data point, the precise procedure depending on the package. Enter the H_2S concentration and repeat. In this way a file of triplets can be automatically created. This file can be saved. If repeat surveys are carried out which use essentially the same sampling points - and it is recommended that they should - the new H_2S concentrations can be very quickly entered simply by editing the saved data file using a standard editing or word-processing package.

If this method is used a single file is obtained containing the contours of H_2S concentration overlying a site plan. It is usually necessary to adjust the number and spacing of the contours to produce the most informative presentation. It is suggested that the lowest contour should be 1 ppb above the background concentration. Inclusion of a contour level at the background level can produce an extremely confusing result. A suitable set of contours, in ppb, might be 2, 3, 4, 6, 8, 10, 15, 20, 30, 40 etc. but this may need to be adjusted depending on the levels detected.

SURFER™ contains an option for setting an asymmetry or directional bias to how levels decay from the data points. It should be possible to use this to limit the tendency for H_2S levels to apparently disperse upwind and so reduce the requirement to take additional readings upwind of sources. The authors have not yet however investigated this option in practice.

11.3 H_2S MONITORS AND INTERFERENCES

The production of H_2S maps are usually prepared using the portable H_2S detectors such as the Jerome 631-X H_2S analyser (Arizona Instruments, USA) a gold film resistance monitor. For more details on H_2S measurements and a list of commercial manufacturers of H_2S instruments see Chapter 6.

Portable gold film detectors are operatored by drawing an air sample (usually ambient air) across the gold film, causing an increase in its resistance. As a result gold film resistance monitors also detect alkyl thiols (mercaptans) and di-alkyl di- and poly-sulphides at a sensitivity about one third that of H_2S and other sulphur-containing organic molecules at varying levels of sensitivity (see Table 6.3). In some ways this is no disadvantage at all and could even be seen as an advantage. The main purpose of preparing an H_2S map is to provide a visual representation of odour generated and dispersing from a sewage treatment works. These other sulphur compounds are also generated from sewage and sludge under septic conditions and are also highly odorous. An 'H_2S' map still provides a good representation of odour even if there is uncertainty as to how much of the H_2S response might be due to other sulphur containing organic species.

In most cases H_2S is the dominant species emitted from septic sewage and sludges. Treatments specifically aimed at reducing the level of un-ionised H_2S in solution - the addition of iron salts or the raising of the pH value, can lead to organic sulphur species predominating over H_2S, as can also be found in the air emitted by certain biological and chemical odour control technologies.

Currently regulators and planners in some countries are considering setting standards for H_2S from sewage treatment works. If H_2S measurements are to carry legal weight, the uncertainty caused by a response to other species can be viewed as disadvantageous.

Not an interferant in the chemical sense, there is always a background level of H_2S. This can vary from below 1 ppb in rural areas (Arizona Instruments once produced an instrument for WRc nominally sensitive to 0.1 ppb, but this never became generally available) to commonly in the range 1–2 ppb. It is believed most of this background is due to emissions from road traffic. In Rome during one January, after a prolonged period of high pressure and light winds, the background level of H_2S was 6 ppb. Interestingly there was no odour of H_2S

associated with this background level. This contrasts with the authors' experience that whenever H_2S levels are raised by as little as 1 ppb above the background and clearly associated with a sewage related source, then a sewage odour is generally detectable. The main value of H_2S is as a marker for odour, not as being the component chiefly responsible for odour. Only an H_2S map, however, can clearly demonstrate the association between a level of H_2S and a sewage-related source.

Figure 11.1 shows the results of continuous H_2S monitoring at a single point close to a sewage treatment works. This shows an interesting diurnal variation, probably reflecting the general level of road traffic activity. Wind direction was also recorded. Very few of the short-lived peaks of H_2S could have been due to the treatment works, as the wind was in the wrong direction. These peaks were probably due to the close passage of a road vehicle. Analysis suggests that detection of odour may be associated with wind blowing from the works but that H_2S concentration correlates neither with detection of odour nor with wind direction. Results such as this cast doubt on the value of continuous monitoring for H_2S as is now being proposed for new sewage treatment schemes in the UK and elsewhere. An H_2S map provides a much better estimate of the impact due to odour from a sewage treatment works. To counter this is the argument that the levels of septicity that are the fundamental cause of odour can vary. Since an H_2S map is prepared over a few hours, periods of high septicity in the sewage works can easily be missed. The only way out of this conundrum is to use a large array of permanent monitors that would allow the generation of frequent H_2S maps, clearly impractical at the current time.

It is this conundrum that highlights the value of source quantification (either as an emission of H_2S in mg/s or odour in ou/s) and the use of a dispersion model to estimate impact under a wide range of conditions. Even so an H_2S map remains the best direct estimate of odour impact.

11.4 INTERPRETATION OF H_2S MAPS

At its simplest, an H_2S map requires little interpretation. During preparation of the map, the background level of H_2S will have been noted. The map therefore shows all areas with raised levels of H_2S. Generally these raised levels will all be traceable to particular sources on the treatment works. Occasionally isolated off-site raised levels may not clearly be linked to the treatment works and it should not be assumed that the raised levels at such sites are due to the treatment works.

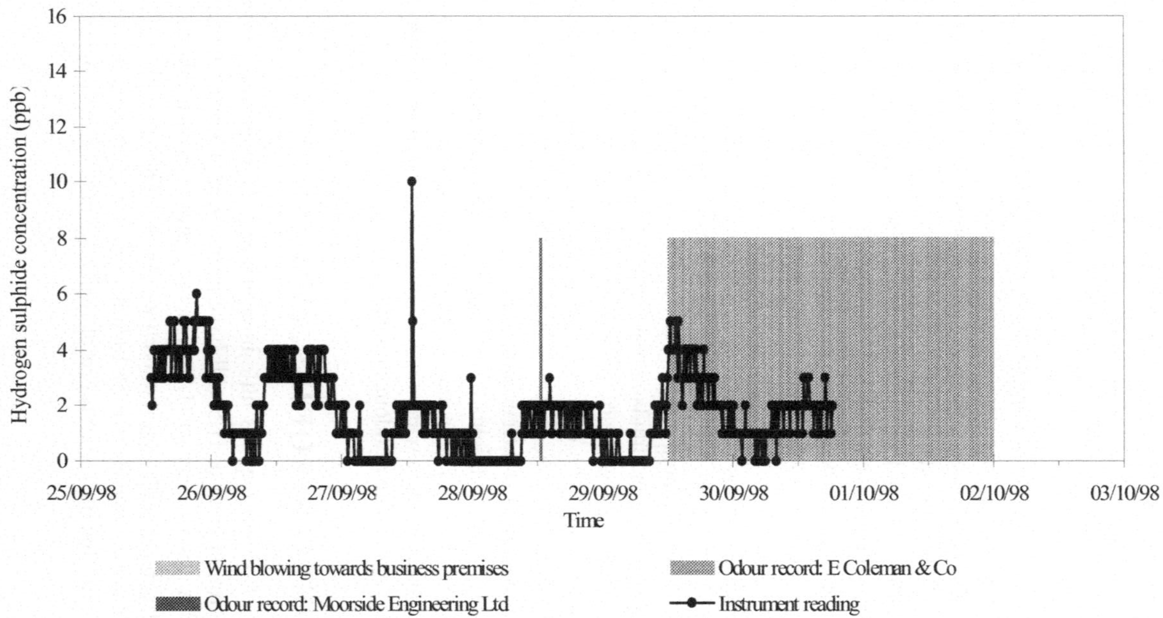

Figure 11.1. Long term H_2S concentrations at a single point outside a sewage treatment works

Because an H_2S map is effectively a snapshot, albeit with a long exposure, it says very little about the long-term odour impact from a site. Because the wind will generally remain from a constant direction during a survey, most off-site locations will show no impact because they are not down wind of the works. The impact illustrated by an odour map is indicative only, though it is reasonable to assume that the impact in other directions would be similar whenever the wind was blowing to that direction.

An H_2S map can be used to attempt to rank the importance of odour sources at a sewage treatment works. Generally an H_2S map will show a number of separate plumes originating from different areas of the works, commonly as follows:

- Works inlet;
- Primary tanks;
- Sludge tanks;
- Sludge treatment.

And less frequently from secondary treatment, particularly the inlet zones of an activated sludge plant if the feed has turned septic or from the whole treatment area if it is very highly loaded.

When assessing the importance of a source, the eye is at first drawn to the peak level of H_2S at the source of a plume. This is an error. The peak value depends on many factors, the size of the source, the precise manner in which odour is transferred to the atmosphere and the degree of approach to the source particularly into areas where little mixing with the atmosphere has occurred (see Box 11.1). The actual magnitude of the source is a lesser factor. What correlates most with the magnitude of the source is the area of the plume over which H_2S levels are raised above the background.

Figure 11.2 shows an example of an on-site H_2S map. This shows no H_2S generation associated with secondary treatment. The apparently raised levels over the downwind end of the aeration tanks are due to the contouring program allowing for dispersion upwind. The levels of H_2S clearly rise at the primary tanks. There is a small further increase at the works inlet and a slightly bigger increase at the sludge tanks and a still bigger increase at the CHP house, presumably due to leaking digester gas. Even this latter source does not appear as big as the primary tanks. It might have been better at this site to conduct an H_2S survey with the wind blowing across the processes from either the east or the west. This would have given better resolution for sources, which are downwind of each other in this figure. Significant levels of H_2S still exist at the site boundary.

Figure 11.3 illustrates a map prepared from off-site readings taken at around the same time. Though the H_2S plume does not extend in a perfectly straight line from the works, it is clear that there is a significant plume of H_2S extending a good 500m to the south of the works and clearly associated with the works. Adverse impact would be expected wherever H_2S levels are above around 1–2 ppb above the background. To the south-west corner can be seen a smaller region of raised H_2S levels. This does not appear to be associated with the works and is most likely associated with road traffic at the crossroads there.

Figures 11.2 and 11.3 illustrate a works with a significant odour problem, which is expected to generate significant off-site impact – always assuming there is a population to be impacted around the works. A set of odour potentials (see section 11.5) and associated H_2S levels would demonstrate whether the problem at the primary tanks was:

- due to a septic feed;
- due to septic return liquors;
- due to septicity generated within the primary tank as a result of an aged sludge blanket.

11.5 OTHER USES OF H_2S MAPS

It should be possible to use an H_2S map as a method of calibrating a dispersion model. This seems to have been done rather rarely and when it has the results are not always that encouraging though sometimes a rather general agreement is seen. There are many reasons why the agreement is less than might be hoped for:

- Though it may take an hour or two to complete the readings for an H_2S map, the individual readings are made over a few seconds. These short-term measurements will show much more variation than the output of a dispersion model, which is usually based on hourly average values. An H_2S map generally shows much finer detail than appears in the output of dispersion models.
- The plumes shown in H_2S maps often die-off much faster than those shown by dispersion models. Dispersion models can allow for the extra turbulence and hence mixing due to particular large buildings but they may fail to adequately allow for the general raising of turbulence, and hence mixing and decay, caused by the sum of smaller buildings and structures in urban areas and within a treatment works itself.
- It is very difficult to determine accurate or even realistic source strengths in terms of mg/s H_2S for sewage treatment processes. These source strengths are fundamental inputs into any dispersion model

whose outputs can be no more reliable than the estimates of those source strengths.

Figure 11.4 illustrates an H_2S map and corresponding output from a dispersion model. The H_2S emission rates were mostly estimated using the WRc STOP model, a set of mass transfer relationships established using a wind tunnel (Yang and Hobson 1998). This H_2S survey shows that the H_2S plume from the primary tanks dies off somewhat quicker than the dispersion model suggests but is of the same general magnitude. Traces of H_2S were observed around the secondary treatment area, rather more than predicted by the dispersion model. It appears as if the edge of the major plume at the south west of the site was just detected by the survey. The unlabelled contour at the bottom of the map represents 5 ppb. This plume was generated from an enclosed point source and the associated emission rate was obtained by direct measurement.

H_2S mapping can be used to calibrate a dispersion model, provided good estimates of the source strengths are available. A variant of H_2S mapping can also be used to estimate source strengths. This is demonstrated in Figure 11.5. A set of H_2S measurements was made downwind of a point source. This could have been widened to collect enough data for a full map, but in this case this was not necessary. A dispersion model was then set up to model this source and the magnitude of the source term adjusted by trial and error until the best fit was obtained. The best fit was as shown in Figure 11.4. The value of the source term used to get this fit was then taken as the best estimate. A problem with this procedure is that the actual plume centre line rarely follows a perfectly straight line. It seems appropriate, when taking the measurements, to allow lateral movement to obtain the maximum concentration for that downwind distance, as this should represent the centre line. Most dispersion models however are based on long term average concentrations, while in practice atmospheric concentrations fluctuate significantly in the short term. This fluctuation is believed to be significantly greater than would be expected simply from lateral movement of the plume centre line. When making the measurements it is difficult to know whether one is following a plume centre line as it shifts, or simply following random short-term fluctuations. Nevertheless, even allowing for this difficulty, the estimated emission rate using this procedure in Figure 11.5 was 59 mg/s as against a directly measured value of 95 mg/s. This level of agreement is considered good given the current state of quantification in the field of odour. It is for instance within the level of uncertainty of the dispersion models themselves.

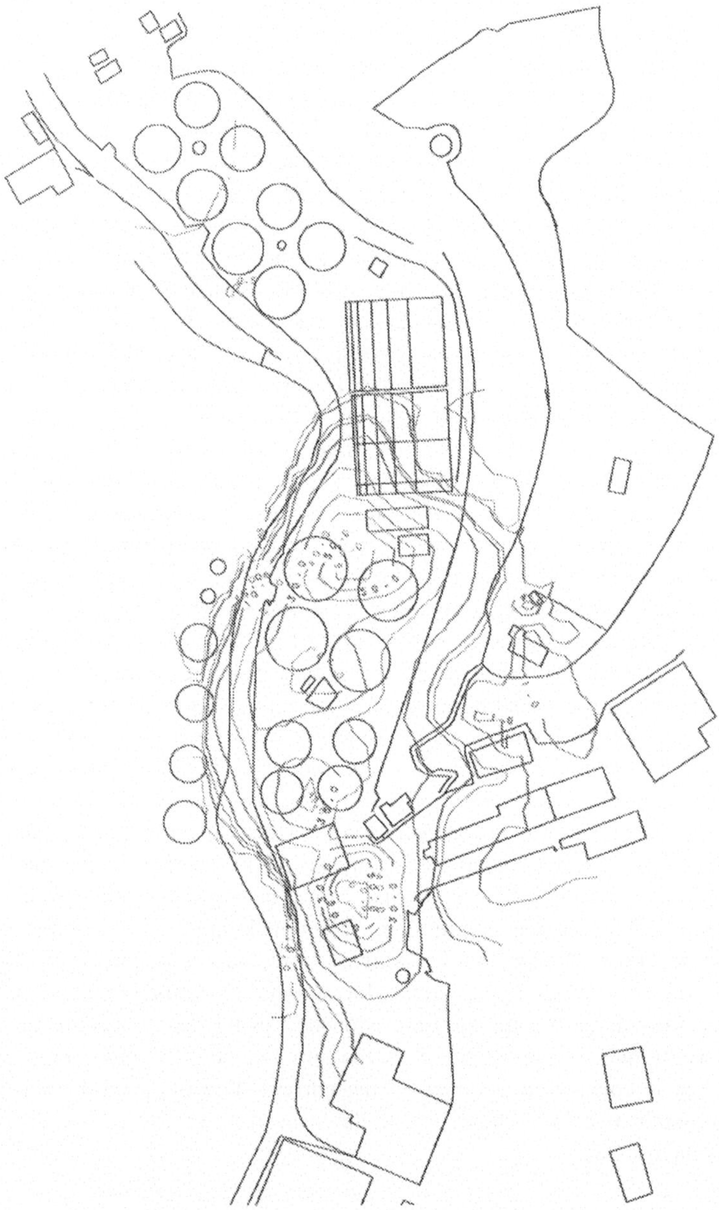

Figure 11.2. H_2S map prepared using on-site H_2S measurements.

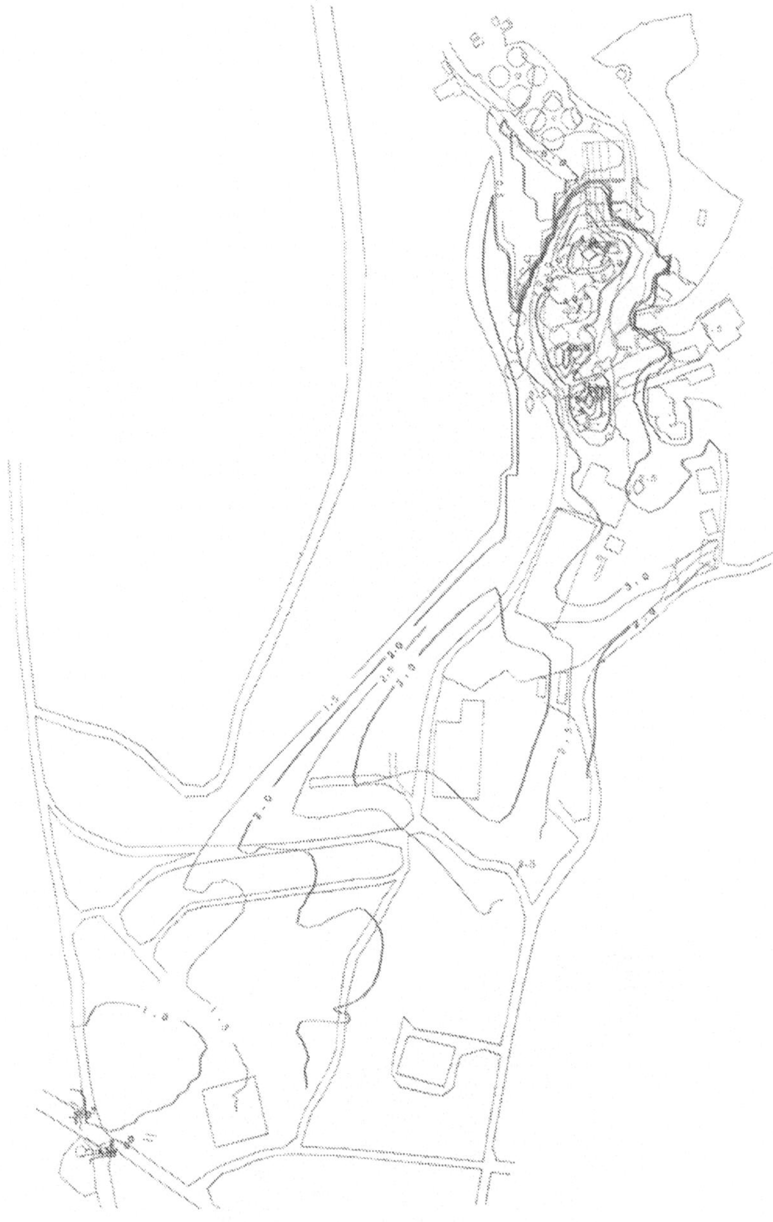

Figure 11.3. H_2S map prepared using off-site H_2S measurements.

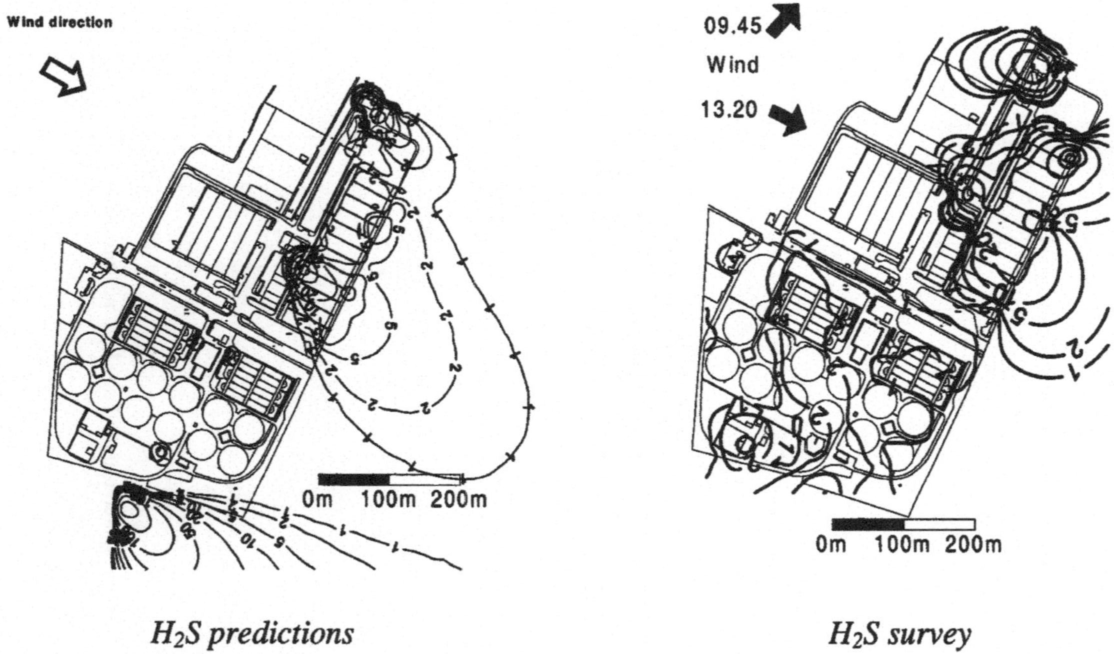

Figure 11.4. Comparison between an H_2S map and the output from a dispersion model (Yang and Hobson 1998).

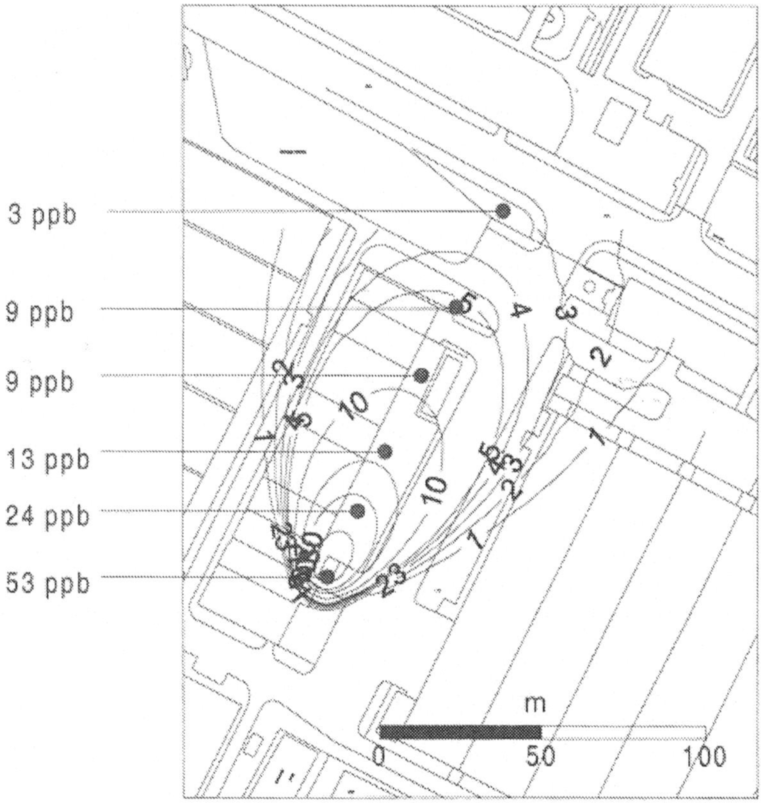

Figure 11.5. Use of an H_2S map to quantify a source.

A second method of increasing the value of an H_2S map, describes a measurement known as the odour potential (Hobson 1995). This is the odour strength, in ou/m^3 (CEN 1999) of air blown through a sample of wastewater. Measurements of odour potential prove very valuable when carrying out diagnostic investigations into the problems of odour at a sewage treatment works. Yang and Hobson (1998) also describe using a set of mass transfer relationships together with the odour potential to make a direct estimate of odour emission rates from sewage treatment processes. The concentration of

H_2S should always be measured in the air collected for an odour potential measurement for any sample of wastewater. Similar mass transfer relationships allow this to be converted into an H_2S emission rate for given processes units carrying that wastewater. The two measurements, the odour potential and its associated H_2S value, allow a ratio to be calculated between odour, in ou/m^3, and H_2S in ppb. To a reasonable approximation this ratio will hold for all emissions associated with that particular wastewater. The ratio can be used to make an estimate of odour strength for any H_2S level, which is clearly associated with a particular source. In this way an H_2S map can be converted into a map of odour concentration. In practice it is not recommended that this be done formally. Odour potentials and more significantly in this context, the ratios between odour concentration and H_2S concentration, will change as sewage flows through a treatment works. Whenever H_2S plumes from different sources coalesce, as they frequently do, there will be uncertainty about the appropriate ratio to use. Nevertheless this technique can be used to make a reasonable estimate of atmospheric odour levels from an H_2S map. (In theory it would be possible to estimate the H_2S:odour concentration ratio simply in a grab sample of air. It is however rare for odour in open air samples as measured by olfactometry to be significantly above the background level which makes interpretation very difficult.)

This technique can also be used when interpreting an H_2S map containing a plume from a direct atmospheric emission such as from an odour treatment plant. In this case the odour concentration:H_2S concentration ratio is simply measured in the air as it is emitted from the process. Also of interest in this context (when investigating emissions from anaerobic digestion), is to measure H_2S:methane ratios and odour concentration:methane ratios.

11.6 CONCLUSIONS

The preparation of an H_2S map was one of the first procedures for visualising and quantifying the problem of odour associated with sewage treatment. Nevertheless it remains an important tool to use when quantifying an odour problem and in diagnosing its causes. The strengths of H_2S mapping are its simplicity, particularly with the recent development of easily available software for producing contours, and its visual impact. Its chief weakness is that, even when allowing for a response to organic sulphur molecules, it is not responding to all odours that typically arise from sewage treatment. In particular an H_2S map will give little impression of odours from secondary treatment. When odour issues started to become major source of public complaints during the 1980s, most problems of odour nuisance related to problems caused by septic sewage (including the development of septicity in primary tanks) and sludge treatment.

H_2S is a good marker for these odours. If however septicity is controlled, and sludge odours are satisfactorily contained and treated, odour from secondary treatment can be the biggest source at a treatment works. Secondary treatment is after all deliberately designed for mass transfer. Odour standards are now being considered which do not allow for uncovered secondary treatment processes. In such an environment, H_2S maps may be of less value. Others argue that secondary treatment odours are intrinsically less offensive and should not be considered in the same way as septic and sludge odour which contains H_2S. The jury would seem to be still out on this issue. H_2S maps have an important role to play provided these limitations are appreciated.

11.7 REFERENCES

CEN (1999) Air quality - Determination of odour concentration measurement by dynamic olfactometry. Draft prEN 13725, European Committee for Standardisation, Brussels.

Health and Safety Executive (2000) *Occupational Exposure Limits* EH40.

Hobson, J. (1995) The odour potential: a new tool for odour management. *J. Chart. Inst. Wat. Enviro. Manag.* **9** (5), 458-463.

Yang, G., and Hobson, J. (1998) Validation of the wastewater treatment odour production (STOP) model. *Proc. 2nd CIWEM National Conference on Odour Control in Wastewater Treatment*, London.

Irving Sax N. (1984) *Dangerous properties of Industrial Materials*. Van Nostrand Reinhold Company.

12

Dispersion modelling

Peter Gostelow, Simon A. Parsons and Alun McIntyre

12.1 INTRODUCTION

So far in this book, the compounds likely to cause odour nuisance and the transfer of these compounds from the liquid to gaseous phases have been considered. A final parameter is required to determine the degree of odour nuisance, which is a description of the transport of the gaseous odorants from the source to the receptor. The receptor is a point at some distance from the odour source and is usually considered to be at or close to ground level. A height of around 1.5 m is often used, this being close to nose height for most adults.

Following an emission into the atmosphere, two factors are important in determining the extent of the subsequent dispersion: wind speed and direction and atmospheric stability. Firstly, the wind direction determines the direction of

edited by Richard Stuetz and Franz-Bernd Frechen. ISBN: 1 900222 46 9

odour transport. Wind speed and atmospheric stability both influence turbulence, with increased turbulence leading to better mixing, and subsequently improved dispersion. Turbulence increases with increasing wind speed, and so in general it can be said that lower odour concentrations are likely when wind speeds are high.

12.1.1 Atmospheric stability

Atmospheric stability refers to the amount of convective (vertical) mixing in the atmosphere. It is affected by the environmental and adiabatic lapse rates.

The environmental lapse rate is the vertical temperature profile of the atmosphere. Typically, the air temperature decreases by about 0.65 °C per 100 m rise.

Deviations from the average are common, however. One example is the inversion condition as shown in Figure 12.1. Where this occurs, there is an 'inverted' layer where air temperature increases with height, due to the ground temperature being very low. Above the inverted layer, the temperature decreases with height as normal. This is a common condition in the winter and is often associated with poor air quality as pollutants released at ground level are trapped within the inversion layer.

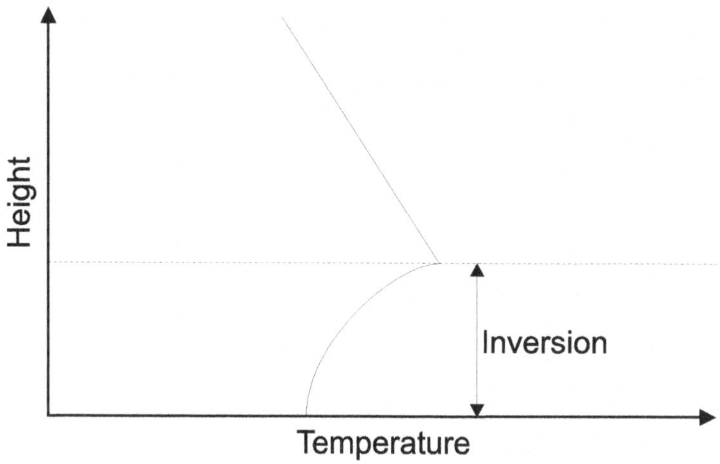

Figure 12.1. Inversion condition.

Owing to poor heat exchange in air, a rising 'parcel' of air cools at a different rate than its surroundings (the environmental lapse rate). The rate of cooling for the rising air parcel is called the adiabatic lapse rate. Adiabatic processes refer to changes in state variables (temperature, pressure, density) of a body of air where no heat is added or withdrawn. If a body of air is compressed, its temperature increases. Conversely, if it expands, its temperature drops. Atmospheric pressure decreases with height. Thus if a body of air is lifted, it expands due to the reduced pressure and its temperature drops. For dry air, the rate of adiabatic cooling is about 1 °C per 100 m rise.

The adiabatic lapse rate varies with atmospheric moisture. Condensation of moisture releases latent heat which ameliorates the temperature decline. Hence the rate of adiabatic cooling is less for moist air than for dry air.

Whether a body of air rises or not is dependent on its density. If a 'parcel' of air is less dense than the surrounding air, the parcel will rise. As it rises, it expands and cools. It will continue rising as long as its temperature is greater than the surrounding air. For this to be the case, the adiabatic lapse rate must be less than the environmental lapse rate. It is the difference between these two lapse rates that determines atmospheric stability.

12.1.2 Pasquill-Gifford stability classes

The degree of atmospheric stability has been formalised using Pasquill-Gifford stability classes (Table 12.1). These are divided into six categories (A–F) that describe the dispersive capacity of the atmosphere. In general, dispersion of pollutants in the atmosphere decreases from A to F. The Pasquill-Gifford stability class can be estimated from field observations using Tables 12.2 and 12.3. Appropriate insolation classes can be determined from Table 12.3.

The Pasquil-Gifford stability classes are a simple, albeit fairly crude representation of atmospheric stability. Modern dispersion theory has moved away from discrete stability classes towards a more fundamental description of stability, typically employing a parameter termed the Mohin-Obukhov length, which is defined as the height at which buoyancy and wind generated turbulence are equal. The Mohin-Obukhov length is itself derived from friction velocity and heat flux.

Table 12.1. Pasquill-Gifford stability classes.

Stability class	Condition	Time of day
A	Extremely unstable	Day
B	Moderately unstable	Day
C	Slightly unstable	Day
D	Neutral	Day or night
E	Stable	Night
F	Very stable	Night

Table 12.2. Estimation of stability class from field observations (USEPA 1992)[a].

Wind speed at 10 m (m/s)	Day Incoming solar radiation (Insolation)			Night[b] Cloud cover	
	Strong	Moderate	Slight	Thinly overcast or ≥ 4/8 low cloud cover	≤ 3/8 cloud cover
< 2	A	A - B	B	F	F
2 – 3	A – B	B	C	E	F
3 – 5	B	B - C	C	D	E
5 – 6	C	C - D	D	D	D
> 6	C	D	D	D	D

[a] The neutral class (D) should be assumed for all overcast conditions during day or night.
[b] Night is defined as 1 hour before sunset to 1 hour after sunrise.

Table 12.3. Insolation categories (USEPA 1992).

Sky cover (opaque or total)	Solar elevation angle > 60°	Solar elevation angle ≤ 60° but > 35°	Solar elevation angle ≤ 35° but > 15°
4/8 or less or any amount of high thin clouds	Strong	Moderate	Slight
5/8 to 7/8 Middle clouds (7000 to 16,000 ft base)	Moderate	Slight	Slight
5/8 to 7/8 Low clouds (less than 7000 ft)	Slight	Slight	Slight

12.1.3 Gaussian dispersion models

To predict the concentration downwind of an odour source, a model is required that includes the effects of wind direction, wind speed and atmospheric stability class. The most commonly applied dispersion model is the Gaussian dispersion model, which predicts Gaussian concentration profiles in the y (crosswind) and z (height) directions. Recent theory suggests a skewed Gaussian distribution is more applicable in the z direction for convective (unstable) conditions, which is reflected in several modern dispersion models. In the x (downwind) direction, mass transport dominates over diffusive mixing, and the diffusion term in the x direction is usually ignored. Figure 12.2 shows the co-ordinate system used in Gaussian dispersion models.

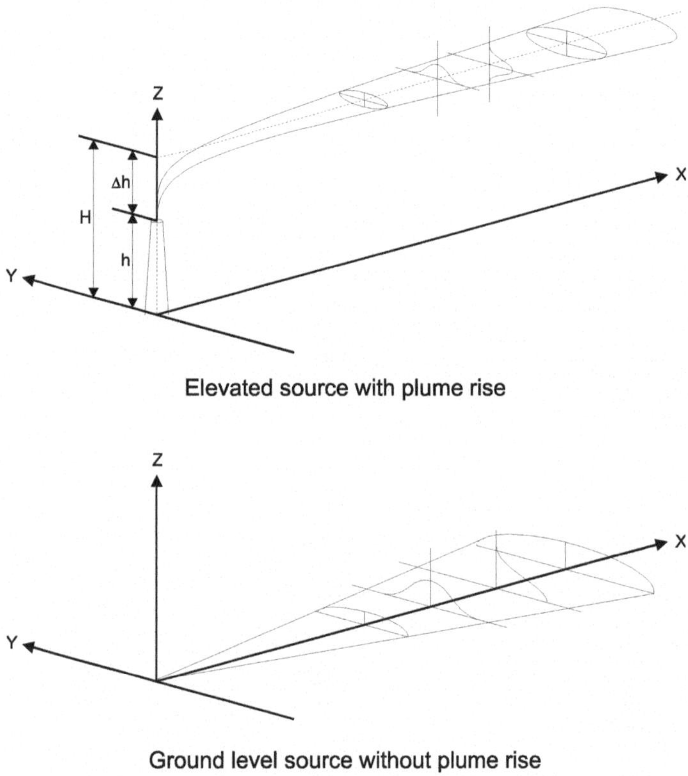

Figure 12.2. Co-ordinate system used in Gaussian dispersion models.

The Gaussian dispersion equation is:

$$C_{(x,y,z)} = \frac{E}{2\pi\sigma_y\sigma_z u}\exp\left[-\frac{1}{2}\left(\frac{y}{\sigma_y}\right)^2\right]\left\{\exp\left[-\frac{1}{2}\left(\frac{z-H}{\sigma_z}\right)^2\right]+\exp\left[-\frac{1}{2}\left(\frac{z+H}{\sigma_z}\right)^2\right]\right\} \quad (12.1)$$

Where:
$C_{(x,y,z)}$ = concentration at point located at co-ordinate *x,y,z,*
E = emission rate,
σ_y = horizontal dispersion parameter,
σ_z = vertical dispersion parameter,
H = Emission height.

The dispersion parameters σ_y, σ_z are effectively the standard deviations of horizontal and vertical dispersion. The dispersion parameters increase with increasing distance away from the source and are usually functions of stability class and surface roughness. Their derivation is empirical.

The term H in the Gaussian dispersion equation is the emission height, which is usually greater than the stack height when stack emissions are considered. This is because the plume initially rises as it exits the stack, due to momentum, and also due to buoyancy if the density of the stack gas is less than the surrounding air. The emission height is given by the stack height plus the plume rise. The wind speed required is the wind speed at the emission height. For emissions from tall stacks, this may be significantly higher than the near ground-level wind speed. Wind speed profiles are usually described by equations of the form:

$$\frac{u_s}{u_{\mathrm{ref}}} = \left[\frac{h_s}{h_{\mathrm{ref}}}\right]^P \qquad (12.2)$$

Where:
u_s = wind speed at the emission height (h_s),
u_{ref} = wind speed at reference (measurement) height (h_{ref}),
P = power law exponent, dependant on meteorological stability and surface roughness.

Dispersion in the y direction can effectively occur for an infinite distance either side of the centreline. For the vertical direction, this is not the case as the ground forms a boundary below the emission height, and atmospheric conditions may form a boundary above the emission height. Where the plume reaches these boundaries, it is usually assumed to be "reflected" downwards or upwards. The reflected plumes are usually modelled using "imaginary sources".

The vertical space in which dispersion can arise is termed the mixing depth. This is in turn determined by atmospheric stability. For strongly stable or inversion conditions, the mixing depth is very low and gases are trapped close to ground level. For strongly unstable conditions, there is effectively no upper layer to the mixing depth and hence dispersion is very good.

Older dispersion models describe atmospheric stability in terms of Pasquill-Gifford stability classes. Modern models are employing more recent stability theory, encompassed in the use of parameters such as the Monin-Obukhov length. The improved stability theory allows for better accuracy in dispersion models, albeit at the cost of increased meteorological data requirements.

12.1.4 Area sources

The Gaussian equation shown above assumes a point emission source. Unfortunately, the majority of emission sources at a sewage treatment works are likely to be area sources. Emissions from area sources may be modelled by a Gaussian dispersion model by dividing the area into *n* strips of width δX perpendicular to the wind direction (Figure 12.3). These strips are then considered as line sources. The downwind concentration is determined by numerical integration of the contribution from each strip.

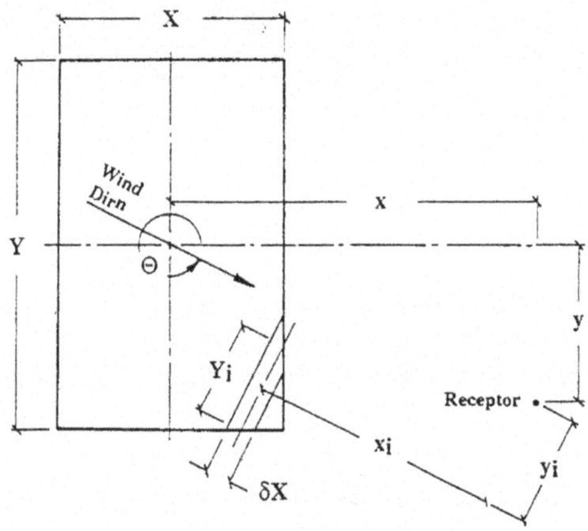

Figure 12.3. Definition sketch for an area source (Smith 1995).

It should be noted that the differences in ground level concentration from area or point sources of equal emission rate diminish with distance. Figure 12.4 shows the predicted ground level concentration - distance characteristics from a 100 m^2 area source and a point source of equal emission rates. As can be seen, at distances greater than about 55 m from the source, the difference is less than 5%. At 100 m, the difference is less than 1.5%.

Emissions from area sources are usually assumed to have negligible upward velocity and little buoyancy. As a result, there is effectively no plume rise.

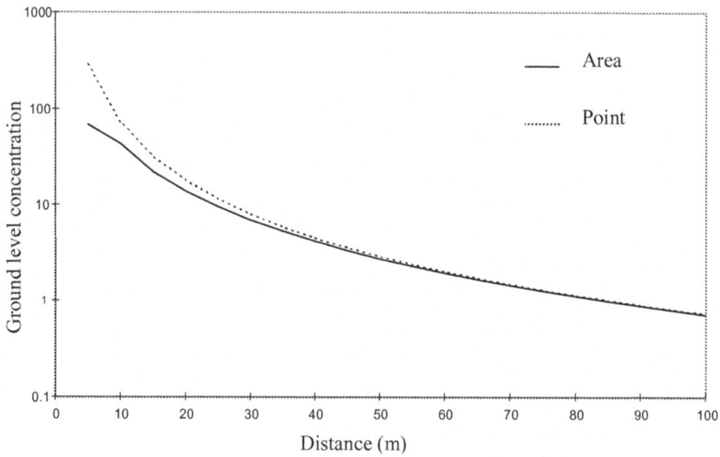

Figure 12.4. Ground level concentrations from area and point sources with equal emission rates.

12.2 ODOUR DISPERSION MODELLING IN PRACTICE

The basic Gaussian dispersion equation has been incorporated into many computer packages, enabling practical modelling of multiple odour sources and receptors. The more complex models will allow for both point and area sources, and will also consider factors such as variable topography and turbulent effects caused by buildings on the site.

Windows-based modelling packages are designed to be user friendly, and will employ a graphical user interface for both model input and output. Input is usually via a 'drawing board', enabling easy input of the location, size and emission parameters of odour sources. Model output typically consists of an odour contour plot, overlaid on a plan of the site and/or the surrounding area. Figure 12.5 shows an example of a typical odour contour plot.

There are a number of commercial models available for modelling odours, which are widely used for regulatory and planning purposes. Detailed descriptions of the range and type of models available for dispersion modelling in general can be found at www.epa.gov/scram001/. Table 12.4 summarises the details of the four most commonly applied models to the dispersion of odours from wastewater treatment. Comparisons between AERMOD, ADMS and

ISCST3 have also been reported by a number of authors including Hobbs *et al.* (1997) and Petts and Eduljee (1999).

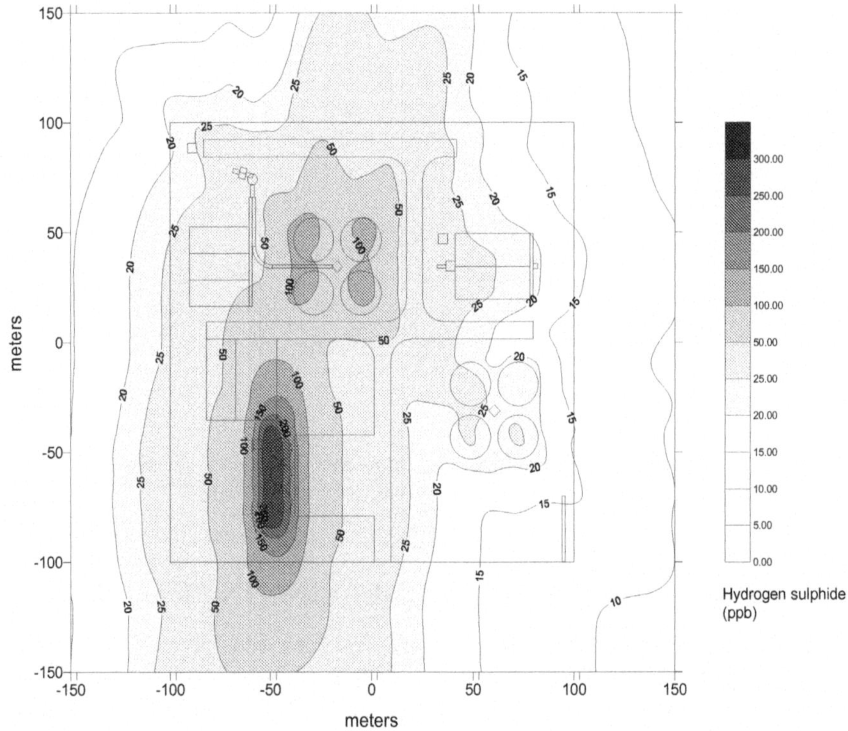

Figure 12.5. Typical odour contour plot derived by a dispersion model.

Table 12.4. List of the four most common dispersion models used for modelling odours from wastewater treatment.

Models	Developer	Type
ISC (www.epa.gov.scram001)	USEPA	Gaussian plume model
AERMOD (www. epa.gov.scram001	Amercian Meterological Society and USEPA	Steady-state gaussian plume model
SCREEN (www. epa.gov.scram001) (home.pec.com/screen3.htm)	USEPA	Steady-state gaussian plume model
ADMS (www.cerc.co.uk)	Cambridge Environmental Research Consultants	Advanced gaussian type

The basic forms of these models have only recently changed with the development of the AERMOD and ADMS models. Both of these models use boundary layer physics concepts and a number of comparative studies between ADMS, ISCST and AERMOD (Department of the Environment 1996) show significantly improved results under convective conditions.

The Industrial Source Complex Short Term (ISCST3) model is the US EPA's current regulatory model but is proposing AERMOD as a refined model for regulatory applications in simple and complex terrain. The ISCST3 model was developed in the 1970s and is based on a steady-state Gaussian plume algorithm, and is applicable for estimating ambient impacts from point, area, and volume sources out to a distance of about 50 km.

AERMOD is a steady-state plume model that is designed to estimate near-field (less than 50 km) concentrations from most types of industrial sources. The AERMOD modelling system consists of three programs, the model itself (AERMOD), a meteorological preprocessor (AERMET), and a terrain preprocessor (AERMAP).

The ADMS (Atmospheric Dispersion Modelling System) model, like AERMOD is based on planetary boundary layer turbulence structure, scaling and concepts.

12.2.1 Data required for odour modelling

The following data are required as inputs to odour dispersion models (Table 12.5).

Table 12.5. Data Requirements for dispersion modelling.

Point Sources	Area sources	Point or area sources
Emission rate	Specific emission rate	Atmospheric stability class or related parameters
Gas flow rate	Area dimensions	Wind speed and direction
Gas exit velocity	Release height	Location of receptor
Stack height	Wind direction relative to area	Type of area (urban or rural) or surface roughness
Gas temperature		Ground level heights
Air temperature		Location of adjacent buildings which may affect dispersion

Emission rate data are the most problematic. Point source data may be relatively easy to collect or derive, but area sources are likely to form the majority of odour sources at a sewage treatment works. Emission rate estimations are detailed in Chapters 4, 5 and 10.

Minimum meteorological data required are the wind speed, direction, boundary layer height and the atmospheric stability class. Modern models may also require additional parameters relating to heat flux. These data are usually available in hourly-averaged form from airports or meteorological agencies.

Model output is usually in the form of the highest concentrations predicted for each receptor or some percentile of that predicted for the range of conditions contained within the meteorological data file. The results are usually expressed in the form of a contour plot.

12.2.2 Case studies

This section offers a commentary on two case studies, where dispersion modelling was successfully employed in problem analysis, complaint comparison and problem resolution.

12.2.2.1 Case 1

The first case deals with a large wastewater treatment works treating a combined domestic/industrial flow of approximately 90 ML/d. Large numbers of complaints had been received from local residents, some living up to 1.5 kilometres distant from the plant boundary, and a detailed odour emission measurement study was implemented to identify the major sources and provide input data for a modelling study. Given the size of the site and amount of infrastructure present, it was likely that any abatement solution would be very costly. It was considered essential, therefore, that the mechanism of production as well as the source of odour should be investigated.

Initial modelling studies could not achieve a match between the 5 ou/m^3 98th percentile odour contour and the complaints profiles (Figure 12.6), although monitoring of hydrogen sulphide concentrations and wind direction had confirmed that elevated odour levels were emanating from the treatment works. This monitoring also confirmed that odour emissions from the treatment plant appeared to increase in the mid-late evening time during the summer. The scale and source of the problem was eventually resolved with a combination of measurement and modelling and inspection of climatic episodes of low mixing heights and wind speeds during the evenings. The nature of the problem was septic settled sewage being sprayed over 4.5 hectares of trickling filters, the wastewater having travelled first along an 11-km trunk sewer.

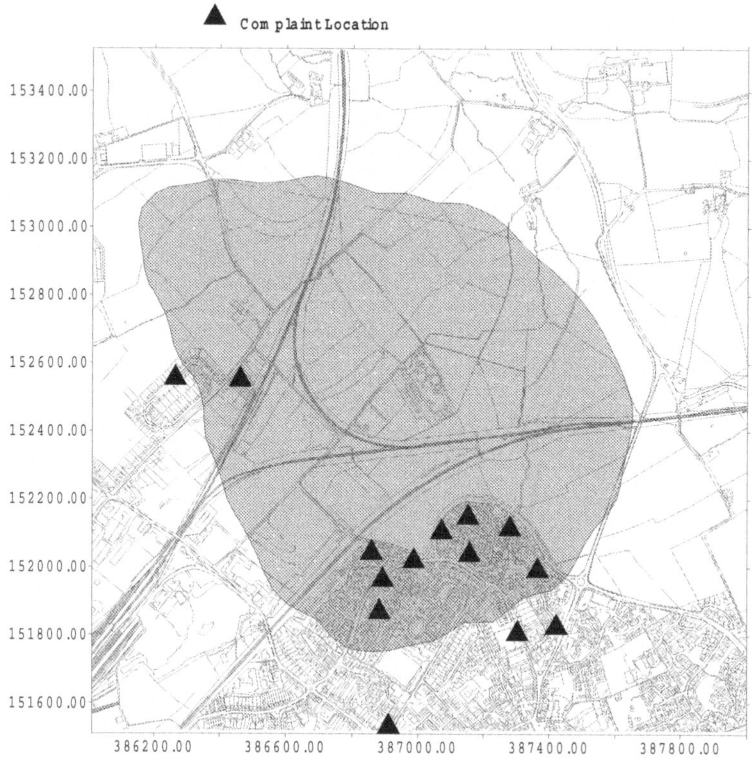

Figure 12.6. Odour annoyance contour plot and recorded complaints.

It was discovered that the highest odour emissions occurred during the mid–late evenings following hot days when the trickling filter stone media began to release absorbed heat as the atmosphere cooled rapidly after sunset. This phenomenon was further exacerbated by a combination of low wind speeds and minor temperature inversions, which resulted in poor dispersion of emitted odours. A plot of wind speed and atmospheric mixing height for a typical evening/night time is given in Figure 12.7.

12.2.2.2 Case 2

The second case presents the results of a model comparison study of odour emissions from a point source located close to a large building that interferes with dispersion patterns from the stack. Identical model input files were compiled for the ISCST3 and AERMOD dispersion models, in relation to source terms, meteorological data and receptor grids. A full year's worth (1988) of sequential hourly meteorological data from the same site were processed by RAMMET and AERMET for the ISC and AERMOD models respectively. Output files containing the sequential hourly average odour concentrations at each of 441 receptors (3.87 million values) were then post-processed using the PERCENTVIEW software package to yield the 98th percentile odour concentrations at each receptor.

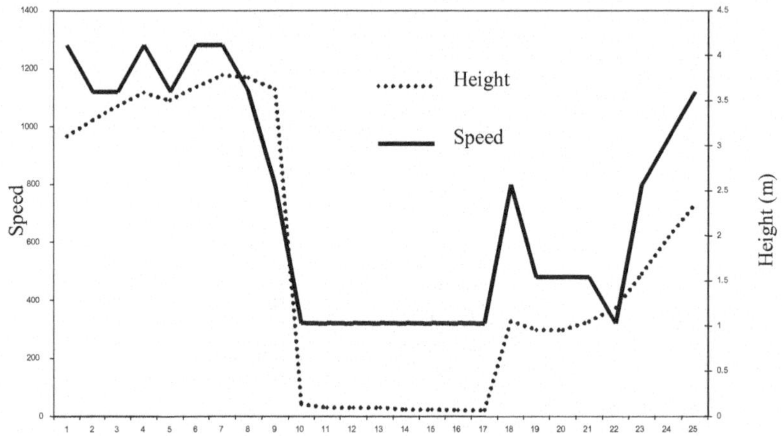

Figure 12.7. Plot of Wind Speed and Mixing Height on Typical Summer Evening/Night

The results of the modelling portrayed as contour plots constructed using the SURFER software package, ensuring that the gridding (Kriging) and contouring algorithms were identical in each case. Figure 12.8 shows the results for the AERMOD run whereas Figure 12.9 shows the results for the ISCST3 run.

The main differences between the two outputs are, evidently, that ISC predicts both a larger area over which emissions from the source have an influence (viz. the 1 ou/m^3 contour) and higher concentrations closer to the source (the peak concentration predicted by AERMOD was 6.5 ou/m^3, whilst ISC predicts a peak of 13 ou/m^3). These results clearly have implications for the way in which odour assessments are conducted in the future and for the promoters of the different models. It is perhaps sufficient to state at this point in the process that the "precautionary principle" will, in all probability, be adopted by practitioners and regulators. This means that, in simple terms, the model that produces the highest results will be taken to be the benchmark assessment case. It is likely that such a stance will remain until additional research and experimental data (model comparison and measurement studies) are available.

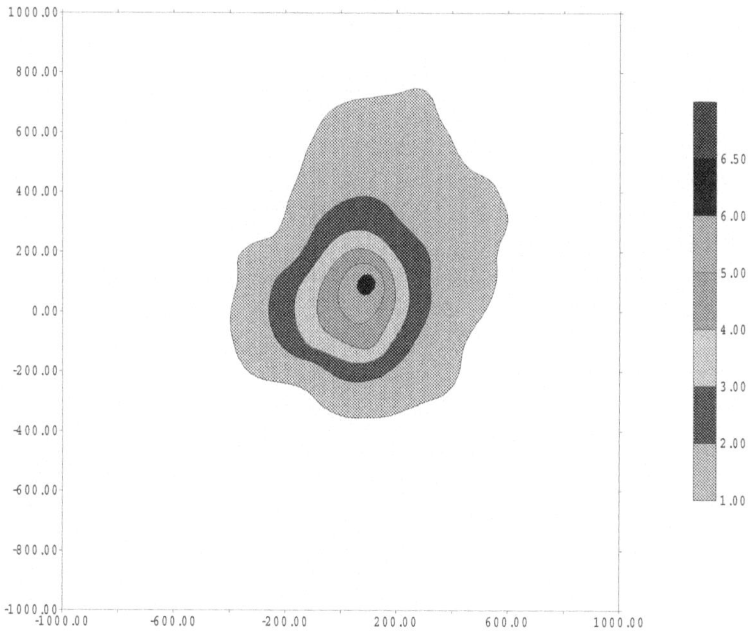

Figure 12.8. Odour analysis by AERMOD dispersion model.

12.3 LIMITATIONS OF DISPERSION MODELLING

Any modelling exercise is an approximation of the true behaviour of odours in the environment. It is impossible to account for every subtle variation in atmospheric conditions and yet keep the model within the bounds of practicability. In addition, there is the problem of determining emission rates, particularly from area sources, which is likely to be the area of most uncertainty.

It is often the case that practitioners of odour modelling claim greater accuracy for model predictions than can be achieved in practice. There is a tendency for model predictions to be incorporated in odour nuisance standards, whereby the dispersion model is used to determine the maximum allowable concentration that can be discharged from an odour control unit stack. Compliance is then demonstrated by measuring concentrations at the stack, rather than at the site boundary. This approach is advantageous to sewage works operators as it means that relatively simple measurements can be made at a point where the concentration is likely to be higher, giving increased confidence in the accuracy of the concentration measurement. Unfortunately, this approach is simplistic, for the following reasons:

- Significant numbers of odour sources may not be odour controlled (i.e. primary tanks).
- The odour concentration is not the only parameter determining concentrations at remote receptors: air flow and stack exit velocity are equally important.
- There is too much reliance on model predictions and the assumptions made when running the model.

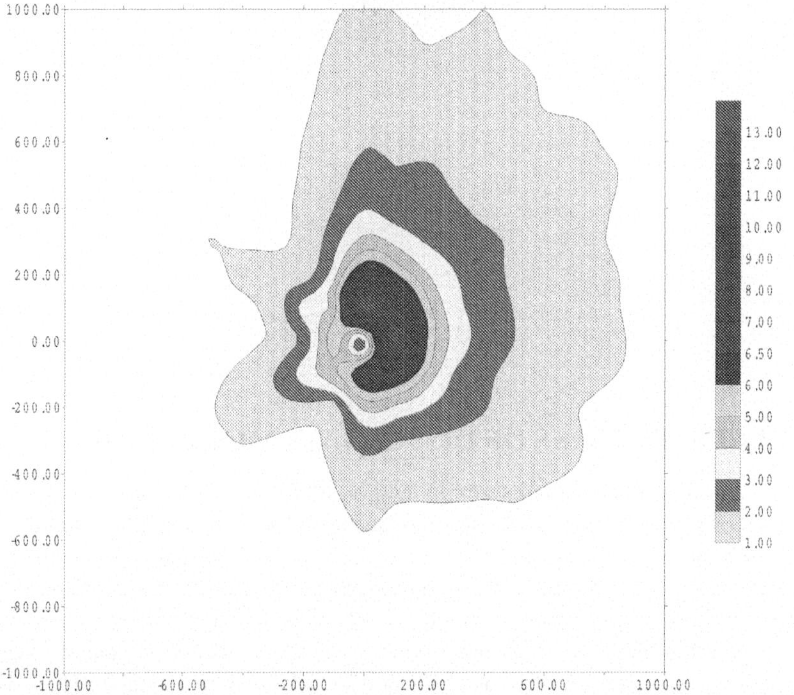

Figure 12.9. Odour analysis by ISCST3 dispersion model.

This approach may be valid for sites where all or the majority of odour sources are included in the scope of the odour control system, hence the majority of odours are vented through the odour control unit stack. An example would be a totally enclosed works. For stack conditions to be used to demonstrate compliance, it would be necessary to specify air flows and stack exit velocities in addition to the maximum allowable concentration. For reliable use as a demonstration of compliance, considerable margins of safety would need to be included in the derivation of maximum allowable concentrations. In addition, a follow-up calibration exercise would be desirable on completion of the odour control scheme whereby model predicted concentrations are compared with concentration measurements made in the field.

There is an increasing tendency for odour nuisance standards to be quoted as percentiles: for example, 5 ou/m^3 as the 98th %. As a result there is a demand for model output to be quoted as percentiles. Where this is done, the model is usually run using a year or more of hourly meteorological data, and the 98th percentile value predicted at each receptor point for this range of meteorological conditions quoted. The value of model predicted percentile values must be questioned for two reasons:

(1) The only variation considered is in the meteorological conditions – a fixed emission rate is used when modelling. The emission rate is likely to be variable, however, due to changes in the quantity and quality of the influent sewage and also numerous operational parameters. In addition, the emission rate will be a function of the meteorology itself, particularly the wind speed. This variation is ignored, however.
(2) Models are not applicable to all meteorological conditions likely to be encountered. The Gaussian dispersion model assumes that dispersion is a function only of turbulence and is therefore dominated by wind speed and stability class. Many models will have a minimum applicable wind speed, typically 1 m/s. Unfortunately, for the majority of odour sources, the highest concentrations are likely to occur for stable conditions where the wind speed is low and dispersion poor. Hence higher concentrations are likely when the wind-speed is less than 1 m/s, and outside the range of model applicability. Hourly average wind speeds of less than 1 m/s may be expected to occur for greater than 2% of the time in many locations.

Quoting model output as percentiles implies a far greater confidence in the model predictions than is reasonable. It is more honest to report the highest concentrations predicted by the model at each receptor for the range of meteorological data used.

It is important to appreciate the limitations of the meteorological data used. Although the UK coverage of meteorological stations providing hourly-averaged meteorological data is good, the nearest station will usually be some distance from the site being modelled. There may be considerable differences between the meteorology recorded at the station and that applicable at the site, particularly if the site is located in a hilly area.

Wind direction data are often recorded with a fairly low resolution. This can result in 'starbursting', whereby distinct 'spokes' radiating away from the odour source are seen. An example of this is shown in Figure 12.10 for a resolution of 10°.

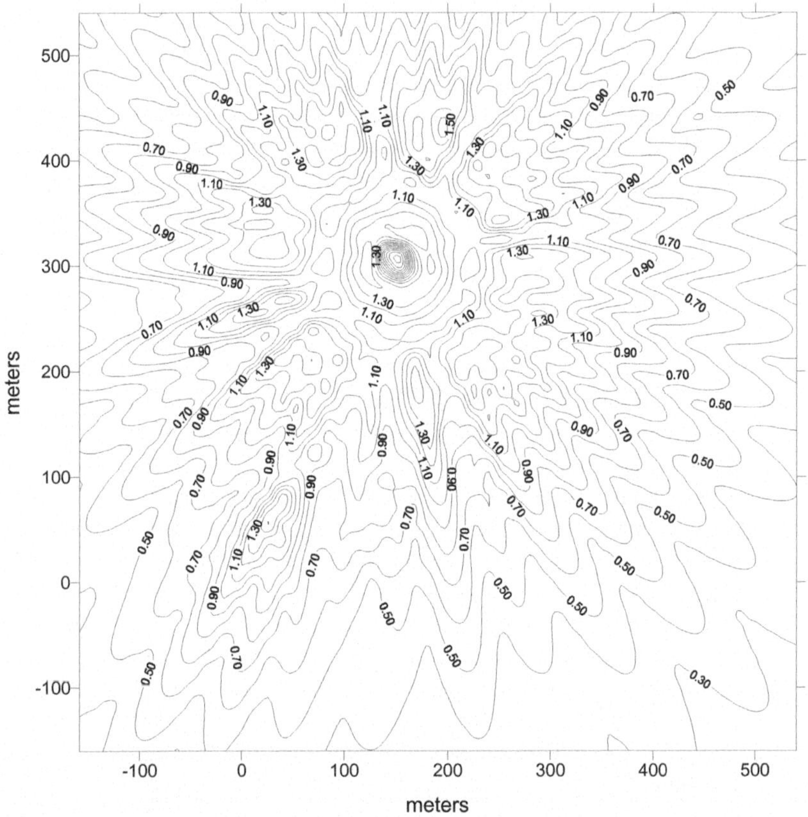

Figure 12.10. Example of 'starbursting' produced by 10° wind direction increment.

Wind velocity data should be handled with care. Meteorological data files prepared for a particular model may have their minimum wind speed set to the minimum wind speed valid for the model. For example, files prepared for the ISCST model will have all recorded wind speeds below 1 m/s set to 1 m/s.

An assessment of the accuracy of dispersion models is summarised in Table 12.6. It is believed that the values quoted here refer to stack emissions, and apply only to the accuracy of the dispersion model itself. Where there is additional uncertainty over emission rates, the possibility of errors is likely to be much greater. It is, therefore, necessary to regard model predictions as an indication of expected concentrations, rather than an absolute prediction of real life.

Table 12.6. Accuracy of model predictions (Petts and Eduljee 1994).

Application	Ratio of predicted to observed concentration	
Peak concentration, not its location, for a release of about 1 hour duration. Flat terrain, steady atmospheric conditions.	0.75 – 1.25	most occasions
Specific hour and receptor point, flat terrain.	0.2 – 2.0 0.1 – 10	50% of occasions most occasions
Maximum hourly concentration from a continuous source, short range, flat terrain.	0.5 – 2.0	
Annual average concentrations at a specific point, short range, flat terrain.	0.5 – 2.0	most occasions
Average concentration over about 1 day, flat terrain, within 100 km.	Simple model 0.12 – 8 one standard deviation Improved model 0.2 – 5 one standard deviation	
Annual average concentrations within 100 km.	0.25 – 4.0	most occasions

12.4 REFERENCES

Hobbs, S.E., Longhurst, P., Sarkar, U. and Sneath, R.W. (1997) Comparison of dispersion models for assessing odour from municipal solid wastes. 4th International Conference on Characterisation and Control of Emission of Odours and VOCs, Hotel Reine Elizabeth, Montreal, Canada, 20-22 Oct 1997.

Petts, J. and Eduljee, G. (1994) *Environmental Impact Assessment for Waste Treatment and Disposal Facilities*. pp. 202-210, John Wiley & Sons, Chichester.

Smith, R.J. (1995). A Gaussian model for estimating odour emissions from area sources. *Mathematical Computer Modelling* **21**(9), 23-29.

USEPA (1992). Screening Procedures for Estimating the Air Quality Impact of Stationary Sources, Revised. EPA-454/R-92-019, U.S. Environmental Protection Agency, Research Triangle Park, North Carolina.

13

Monitoring nuisance and odour modelling

Philip Longhurst

13.1 INTRODUCTION

"*The liberty of the individual must be thus far limited; he must not make himself a nuisance to other people.*" (John Stuart Mill 1806–1873)

Remarkably the human nose not only has 'wide bandwidth' – an extensive detection array, it also has a high sensitivity to many of these chemicals and compound mixtures. Olfactory detection thresholds are often specified in concentrations as low as a few parts per million and in some cases parts per billion. The differing olfactory sensitivities, human responses and cultural interpretations to odour exposure result in *indications of response* being used for this complex chain of human reaction. Regulations to determine and control acceptable levels of odour emission from a process rely on interpreting the exposure and range of human response likely to these odours. This chapter looks at the approaches employed to monitor and predict nuisance and issues associated with the development of standards for municipal processes.

Unlike calibrated sensors, the human response to odours is often inconsistent between reports of the detection, intensity, hedonic properties and character of an odour. Nicell (1994) summarises this sequence as one of odour sensation [threshold detection], discrimination [recognition of the source type], unmistakable perception [the level at which complaints may arise] and finally the degree of annoyance.

Significant proportions of environmental complaints are caused by odour annoyance. In many cases there is a need to determine the extent of the nuisance, particularly where a legal remedy is expected. This requires methods and techniques to characterise, quantify, and define limits for specific odour emissions. Very faint, offensive odours may be unacceptable when compared to pleasant but stronger odours. Methods to assess nuisance need to address not only odour concentrations but also the offensiveness of the odours emitted.

The wide variation in human response to odours results in a need to establish nuisance indicators that represent a probability of annoyance. Each approach must, however, acknowledge the limits to protecting all people within a community from odours when exposure on the one hand is determined to define acceptable limits whilst on the other, not imposing over–stringent controls on site operators.

Maintaining appropriate operating constraints to ensure the public acceptability of operations is vital both for the immediate term as well as being of strategic importance to the ongoing development of businesses. Persistent failures to control odour may prompt complaints which in turn can result in planning restrictions, legal actions, fines, licence modifications or difficulty in achieving development and planning approvals.

13.1.1 Mapping odour emissions to nuisance impacts

Many odour producing operations can be characterised as biologically active, rapidly changing, engineered processes. For such cases attempts to ensure the reliability of operational performance poses a significant management challenge.

Approaches to monitoring annoyance and understanding the issue of amenity effects have been proposed in similar ways to that of other 'failures' or engineering reliability problems that cause environmental impacts (Longhurst and Seaton, 1999). Figure 13.1 shows a conceptual model linking the actions at a site to the cost and revenue implications. Operators need to determine a balance between the investment of resources to enhance operational controls and the amenity impacts that may result in costs in the longer term from restrictions on behalf of the community.

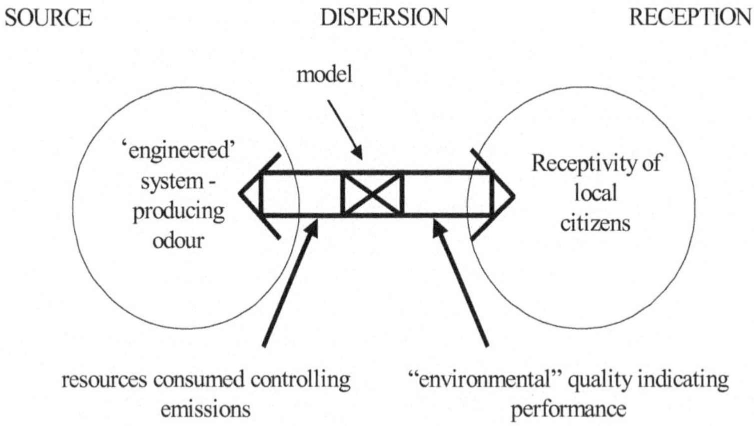

Figure 13.1 Interaction of odour producing 'engineered' systems and the social system.

The two systems describe different disciplines. The disciplines that describe the phenomena of odour within a process include biology, chemistry and civil engineering. Whereas the disciplines associated with the human receptivity to odours are those of physiology, perception, attitude and cultural norms. This concept of receptivity has been applied to a wide range of environmental issues (Lemon and Seaton 1999). What physically separates these two systems in many environmentally sensitive situations is a complex combination of 'transport and transformation'. Dispersing odiferous gases, emitted from operational practices, are transported through the air at a rate dependent on micro-climatic conditions. These gases may be volatile or combine with exposure to each other, other pollution sources, as well as normal and atmospheric gases and sunlight. What, physically, arrives at the nose of the individual in the surrounding area will depend on all these interactions. The arrows and crossed box (in Figure 13.1) denote the mapping function that nuisance monitoring and modelling attempts to represent.

Studies to determine nuisance threshold levels take one of two approaches to develop models of impact to map this function. Either data from reported complaints around a site can be interpreted or alternatively, social enquiry data - often in the form of household surveys - are used (Miedema 2000; Seffelaar *et al.* 1992).

13.2 SPECIFYING ANNOYANCE LIMITS

13.2.1 Dose–effect relationships

Studies of odour annoyance have adopted methods originally applied in air pollution research for toxicity thresholds. Sampling and modelling calculations are used to determine dose–effect relationships where, as in the case of toxic compounds, physiological damage is known to be likely to occur. For toxicity studies, previous evidence of concentrations is extrapolated to concentrations at which chronic exposures are believed likely to be damaging. In such cases, the impact threshold is derived from known physiological deterioration (DOE 1996). By identifying where physiological impact on human health occurs from acute exposure, values can be represented as a frequency and concentrations for a given period of time. In toxicity studies, thresholds for chronic dosing values define all hours as a concentration and percentile, where percentile refers to the number of hours below which a concentration is not exceeded. For example 5 $\mu g/m^3$ of compound 'x' at 98-percentile, i.e. exceeding the specified concentration for only 2% of hours [8,760 × 2% = 175.2 h per annum].

However, unlike epidemiological studies, odour annoyance limits define unacceptable exposure and the likelihood of individuals experiencing annoyance or complaining. Whilst methods continue to be refined, the odour annoyance impact is determined, by predicting concentrations at a given frequency. These concentrations (in European units of ou_E/m^3) are normally specified as a proportion of total hours (per annum). Commonly the 98-percentile is used, though increasingly assessors are asked to consider the 99- or 99.5-percentile. These units represent the anticipated *dose–effect* relationship arising from odour emission, dispersion and the duration of exposure.

13.2.2 Dispersion modelling for olfactometric assessments

The European olfactometry standard (CEN 1997) makes sampling comparisons reliable for odour concentration measurements at source. However, odours dispersed from a source, that strong enough to cause annoyance, are commonly too dilute to measure using dynamic dilution olfactometry. It is also impossible to exclude background odours from a sample taken away from the source. The number of samples that can be taken at any one time and analysed is often limited. In addition, concentrations normally have to be in the range of 500–1000 ou_E/m^3 to achieve the repeatability constraints of 95% confidence limits. Background odours can often be as high as 50 ou_E/m^3 therefore odour samples taken at locations away from the source do not provide reliable measures of off-site odour. To overcome this, site sampling and olfactometric analysis is used to provide emissions data for dispersion assessments. Computational modelling then offers the means to predict the dispersion of odours from a source using local weather data with the source sampled data. Results can be calculated to show the frequency and concentrations for receptors that may cause annoyance.

Where complaints have been received but the source is unknown, *reverse modelling* can be used to identify the location. By plotting emissions for a number of locations the model run can be re-calculated until the receptor location is matched to the point where odour was detected.

Dispersion modelling provides the only effective approach to assess compliance to a defined concentration and frequency standard. The only exception to this is where a single compound or 'marker' is released, such as hydrogen sulphide or isotope markers are used.

13.2.3 Odour parameters and annoyance standards

For each assessment a judgement must be made as to the concentration and frequency at which annoyance may arise.

The use of standards for emission limits and operational controls implies that a common response from receptors can be expected. However, this is very unlikely where the individual threshold sensitivity amongst people differs so greatly. Despite this, the use of concentration measures and a defined frequency when specifying exposure limits for receptors downwind of a facility is well established. Difficulty remains in determining what the concentration and frequency values should be for differing facilities and locations.

In the UK, by default, the Planning Appeals and Public Inquiry system has provided the arena to formalise standards. An early example of this was an Inquiry into the refusal of Planning Permission for a sewage treatment works in Northumberland (DOE 1993). At this Inquiry an hourly average standard of 5 ou/m^3 98-pecentile was accepted. This was based on evidence cited from complaint surveys carried out in The Netherlands.

Despite the outcome of Planning Inquiries debate continues over the setting of appropriate frequency and concentration thresholds, largely for two reasons, the fluctuation of concentrations within hourly averaged met conditions, and the hedonic properties of the odour, i.e. its pleasantness or offensiveness.

Fluctuations in odour concentration are poorly represented within dispersion models that use hourly averaged meteorological data to calculate the rate of change of pollutant concentration and movement. This prompts the reasoning for more stringent standards where it can be argued that the model provides ‘mean’ ratios, whereas the human nose will experience ‘peaks’ in emission that will not be represented by the model output. Estimates can be made to ‘factor up’ the emission values to give a more representative value of the concentration detected by the nose. Commonly a factor of between 5 and 20 is applied. Alternatively more stringent exposure / frequency thresholds can be implemented.

In all cases these standards used to determine annoyance standards principally use only one parameter of odour, namely the threshold concentration.

Odours can be described as having four main parameters:

- Threshold concentration – the lowest concentration at which panellists reliably detect an odour;
- Intensity – the strength of an odour, as perceived be panellists;
- Quality or character – the discriminating qualities of an odour, “what is smells like”;
- Hedonic tone – the pleasantness or offensiveness of an odour.

The threshold concentration of an odour is the concentration at which detection is possible. It is lower than the concentration at which an odour can be identified and forms the basis for specifying annoyance standards. It is the most objective measure of odour, is defined within the CEN standard (1997), but describes only the “detectability” of an odour.

13.2.3.1 Intensity

Intensity is used to refer to the strength of an odour as perceived by a human subject and normally increases with concentration. Human sensory responses (to light, heat, noise, touch and smell) are required to adjust to a wide range of intensities. This wide response rate is recognised in the measurement of sensory impacts by using log-based scales, e.g. decibels. Unlike other intensity scales, odour intensity is subjective relating to the perceived strength of an odour. This is estimated either by using a subjective category scale (such as faint, moderate, strong, etc.) or by comparisons of magnitude (such as sample A is twice as strong as sample B). These subjective or magnitude estimates are then represented on a log scale of concentrations. By adjusting the perceived concentration of a reference gas until it is the same strength as the sample enables a magnitude estimate can also be gained. Two laws are used to reflect this relationship of sensory perception to odour concentration: the Steven's law (Stevens 1957, 1960) and the Weber-Fechner law (Wagenaar 1975):

Steven's law:

$$I = kC^{n} \tag{13.1}$$

Weber-Fechner law:

$$I = a \log C + b \tag{13.2}$$

Where*:*
I = intensity,
C = concentration,
a, b, k, n = constants.

Steven's law is most useful where magnitude or reference scales are used, giving a plot of log intensity against log concentration. Subjective scales, often using 6 or 7 point scales from very faint to extremely strong, employ the Weber-Fechner law to give a linear regression line plot of intensity against log concentration. Each intensity law has limits at low and high concentrations where assessments become unreliable or sensory response is lost, as in the case of hydrogen sulphide, which cannot be detected at the toxic concentrations of 150–250 ppm.

Intensity assessments are particularly valuable where, for example, an estimate is to be made of the change in concentration needed to reduce the perceived detection of an odour from strong to faint.

13.2.3.2 Quality or character

The qualitative terms or 'character' used to describe an odour are entirely subjective. This is the language and analogy used to describe an odour and therefore based upon personal interpretation and shared common experience, e.g. fruity, fishy, eggy, minty, etc. This property is significant in influencing our response to an odour. The ability to discriminate between odours can be tested using pre-prepared "scratch and sniff" sample cards (Doty 1996) though this only offers a test of olfactory function, categorising human performance, and not a means of categorising samples.

13.2.3.3 Hedonic properties of odour

The hedonic tone, or offensiveness or pleasantness of an odour is important when considering the level of annoyance likely from exposure. Yet, much of the standard setting to date has excluded this as being a factor directly considered for industrial processes. Miedema (2000) uses data from field surveys to uncover where emissions from an industry sector are consistently perceived as likely to cause high levels of annoyance. This provides a sound basis for assessing where greater impacts may arise from one source when compared to another of equivalent emission concentration.

13.2.4 Surveys and social enquiry

The analysis of data from field surveys shows that a simple relationship between the 98 percentile and odour concentration gives an approximation as to whether annoyance can be expected. As noted earlier this approach forms the basis of a number of planning decisions within case law where 'unwritten' codes of performance have developed across planning regions.

Whilst the use of percentile measures for exposure has become established as a means of determining annoyance, a limitation of this approach is its inability to take account of the unpleasantness of the odour. An implicit assumption within the existing *dose–effect* method is that all odours are equally as pleasant or unpleasant. This has led to debate over the appropriateness of using the same guidelines established for one development to determine standards for a different process.

Recently, a review of studies by Miedema (2000) developed a method to take account of data indicating the character of the odour from a given industry, thus implying a hedonic quality for the odour. From a comprehensive analysis of data from field surveys, this work shows that a simple relationship exists between the percentage of 'highly annoyed' persons (%HA) and the logarithm of the 98-percentile of the odour exposure concentrations (lgC98).

> "By obtaining 'pleasantness' ratings of odours through laboratory studies it was found that the prediction of %HA improves if the pleasantness of the odour is taken into account. The %HA at a certain level of lgC98 is found to be higher when the odour is less pleasant. This indicates that odour standards may improve if they take the odour pleasantness into account." (Ibid.)

This study helps overcome the difficulties of interpreting concentrations and frequencies predicted for odours off-site as units that indicate acceptable or unacceptable levels of emission. Whilst this work has yet to be tested in planning case law, increasingly the debate over acceptability of predicted emissions is likely to be seen in terms of a combined measure of concentration-exposure and unpleasantness. By taking account of the impact of differing types of operation this development has the potential to improve the relevance of the assessment method for householders likely to experience annoyance.

Despite this odour emissions in some cases are unavoidable. Emissions can arise from multiple causes and it is therefore complex to assess all possible sources. Operators can manage many factors but even incidental releases can heighten awareness of operations and increase the sensitivity of the population to operations and the perceived significance of these impacts. Therefore off-site knowledge, or information on the householders' experience, is valuable to a site operator.

13.3 ANNOYANCE, NUISANCE AND COMPLAINTS

The multiple stages involved when assessing odour annoyance from approximating levels of odour generation; identifying emission sources; calculating dispersion from surface areas; and applying 'averaged' meteorological conditions to assess the potential incidence of annoyance: all introduce the potential for error. For this reason wherever possible, assessments should attempt to employ alternative sources of information to determine the full extent of impact and provide a comparison for what may otherwise be a largely theoretical exercise.

13.3.1 Recording information from the locality

Two sources of information can be routinely recorded from the local community: complaints made to the operators, and reports from site staff and volunteers within the community monitoring the detection of local odours on a daily basis. These records can then be collated and the results used to make comparisons between earlier records. Comparisons may also be possible between differing data sources though often the only data available is that of complaints to the site.

The design of a complaints system can not only influence complaint numbers but can significantly influence the value of the information gained. A reductions in complaint numbers is often seen as an ideal performance measure. This implies that the method of reporting is unchanged and that the reduction is solely due to improved control on behalf of the site operator. This is not always the case. Poorly structured complaint systems can discourage information retrieval, at the risk of concealing problems from a site operator until high annoyance levels are reached within a community.

13.3.1.1 Complaint systems

The effectiveness of a complaints system is determined by how it helps resolve the dissatisfaction of citizens. An effective system will provide:

- Straightforward means for people to make a complaint.
- A procedure for investigation.
- A method of keeping complainants informed about progress.
- Redress where complaints have substance.
- A means of preventing recurrence.
- Feedback to the site to guide resource allocation, establish priorities, plan and ensure quality.

A good complaints system will be:

- Easily accessible.
- Simple to operate.
- Quick, offering prompt action and speedy resolution (with time limits).
- Objective, allowing for independent investigation.
- Confidential in protecting the complainant's privacy.
- Comprehensive in covering all areas of operation that may impact on the locality.

(adapted from: CLA 1992)

13.3.1.2 Complaints to site operators and regulators

Commonly complaints to operators are made through a number of routes, either directly to the site, to an operator's public relations (Liaison Officer), to the local authority, officers responsible for Environmental Health, or an equivalent such as the Environment Agency. Each of these parties need to agree a common reporting structure and agree that both the operator and local authority will be informed of all complaints. Where complaints are recorded by the operator a standard procedure for reports can be designed. The detail of complaint reports should include: the location, complainant's name (where given), the number of people complaining, the nature of the problem, the time of reporting, result of the investigation as to the cause, and weather conditions (possibly recorded from an automatic weather station on-site). Data from these reports can then be used to review the extent to which operational procedures result in odour emissions that cause annoyance from the site.

It is important to note the context from which complaint figures are derived. Unlike routine site monitoring records, complaints are a result of 'direct action' from individuals contacting the company. Whilst the level of annoyance cannot be clearly drawn from this data, the local awareness of odour issues and the evident preparedness of individuals to complain is important. Relevant to this data is whether or not the trend is for more or less people to be affected by an incident or 'complaint events'. Figure 13.2 shows the intensity of events where the number of people complaining (complainants) for any one 'incident' is calculated. If one assumes that for any period people within the locality will complain equally for similar events, then an increase in 'complainants per complaint event' infers a heightened intensity of annoyance whereas a reduction in this scale implies a reduction in intensity. This method becomes a less accurate indicator if the preparedness to complain changes over time. Such a phenomena of 'complaint fatigue' can occur where householders no longer complain, as they believe there is no value in doing so – 'nothing will change'. Reviewing data with significant fluctuations can also be problematic, particularly when it is being used to consider underlying trends.

The detection of intermittent process problems may rely heavily on complaint reports as the only means with which to 'back track' process changes. Here, this data becomes a valuable source of information that benefits the operator if it can be increased. Site surveys or inspection reports can also provide records to compare with complaint data for suspected causes or changes on site. This use of data conflicts with the view that reduced complaint numbers is 'a good thing'. A more appropriate indicator is therefore the ratio of satisfied complainants to complaints.

13.3.2 Issues in regulating nuisance in response to complaints

Where policy makers and planners set out to regulate acceptable levels of exposure to odours the standard setting process relies on a number of assumptions. These commonly form the basis of debate during the investigation of complaints and prosecutions. The issues relate to the determination of standards as well as available evidence and can often only be addressed by approximation and assumption. Common problems are discussed below.

At the planning stage, many sites are given an indication of the upper level of odour exposure that is considered acceptable. Once in place, this standard is difficult to determine from any means other than source sampling. This is problematic and places the burden of sampling costs on the operator to demonstrate compliance. For this reason 'designing out' emissions can be a more effective method of control than proposing an emissions compliance.

There is very limited information on the relative response of people for different odours. Therefore assumptions about new processes cannot rely

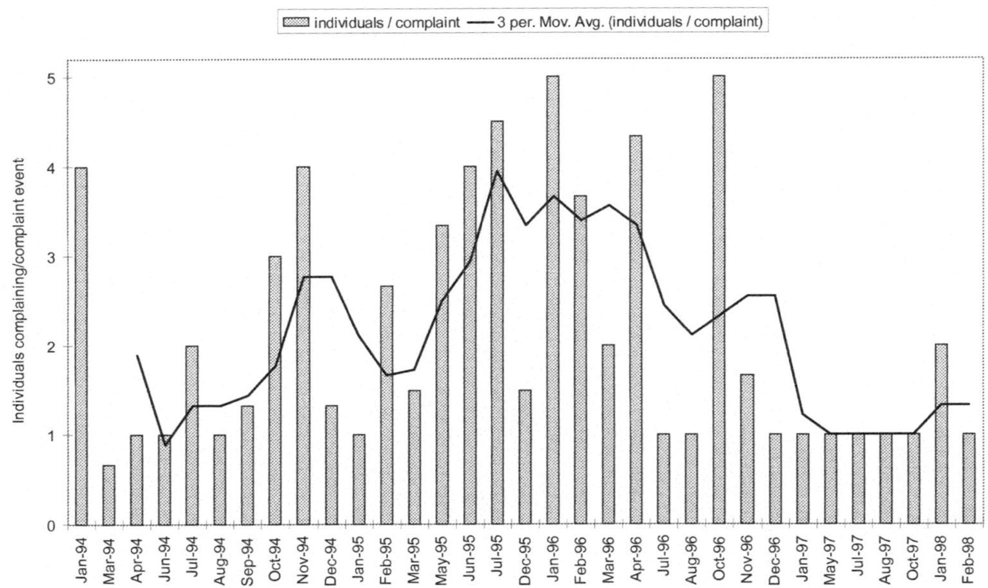

Figure 13.2 Number of individuals complaining per incident for a study site.

directly on data for existing processes unless there is confidence about the odour characteristics.

There can be considerable variation in response across a population, particularly where sections of the population have already been exposed to odours. Process operators need to be alert to the likelihood of 'sensitised receptors'. There is little data available about the range of responses of people to odorants, other than for reference gases such as n-butanol and this does not indicate annoyance. Cultural differences can also result in widely varying responses.

The presence of background odours can be difficult to deal with. It cannot be assumed that an area that has high levels of other odours will be less concerned about potentially odorous processes. The human ability to discriminate between many different odours means that there is not a direct desensitising response for the addition of new odours in an already odour exposed locality.

13.4 ANNOYANCE AND PUBLIC PERCEPTION

The issues concerning odour annoyance are reflected in the wider concerns about environmental quality, uncertainty and risk. Whilst government, industry and the general public acknowledge that protecting and improving the environment matters enormously the decisions about environmental issues are not always open to easy choices. This is because the options and risks are often uncertain. It is clear that uncertainty fundamentally affects how decisions are made. The Economic and Social Science Research Council summarise lessons learnt on environmental change (ESRC 2000). They note:

"Standard decision-making tools rely on quantifiable and objective facts and often fail where there is uncertainty: environmental problems and their solutions arc often complex, value-laden and subjective and will not conform to set assessment criteria.

The gap between 'hard' tools and uncertain issues can lead to 'soft disasters' - environmental and political crises that emerge only slowly but at high costs to society, not least the erosion of public confidence and legitimacy…

Risk assessment and cost-benefit analysis can inform decision-making, but should be supplemented by more qualitative techniques that include a wider range of attitudes and values.

Areas of high uncertainty need to be acknowledged by researchers and policy-makers: 'absence of evidence' of risks is not the same as 'evidence of absence'.

Gaining public trust is not only a matter of new institutions or further research. Rather, openness about risks and legitimate, transparent decision-making is crucial."

These observations are based on global examples of environmental change. However, the observations are consistent with site-specific operations, including the concerns associated with exposure to odours. Scientific evidence on its own can not be considered as a means of achieving public confidence. Neither can the absence of scientific evidence be employed as proof of there being no reason for concern.

Whilst many institutions and governments strive to improve public understanding, the public continue to develop a complex and growing concern over risk from unknown or poorly understood processes – perceived as 'scientific'. This phenomenon referred to, as the 'social amplification of risk' has been a major factor in objections to the development of industrial processes, even where qualitative assessments have been presented on increasingly scientific bases. The definition of these reactions as 'NIMBYism' is commonplace but does not help uncover the reasons for and thus methods to overcome these concerns.

A study by Furuseth (1990) showed that when interviewed, the spatial distribution of householders' concerns did not match the spatial effects predicted for operational sites. This work demonstrated that predictions based solely on the dispersion of odours, noise and dust from sites could not adequately take account of the multiple impacts arising from the site.

In a number of examples where the visibility of sites from the roadside, or vehicles moving to a site, was evident the predicted impacts increased greatly. While the screening of process sites is an obvious solution, more importantly the assumptions about spatial impacts have failed to take into account what Furuseth refers to as the 'non-spatial effects' on the surrounding community. The use of 'visual' space by the community as opposed to the space allocated in a computational model of dwelling locations and dispersion had not been taken into account.

In Furuseth's example the perceived impacts of a visible operation were significant. Similarly, the perception of risk from operations not visible to the public can be as great. In other studies of waste processing sites (Furuseth 1990; Miedema and Ham 1988) the perceived impact and concern over wastes disposal was found to be greater where the nature of operations was unknown and access to information limited. Here, a fear of operations and culture of 'living legends' has had to be overcome in order to gain the confidence of local communities.

For such sites, examples are available where operational sites have known that odour may be caused. For these cases prior warning of this occurrence has not only prevented complaints, but also allayed fears over the control of operations on site. Clearly, the perception of a process as well as exposure to it influences the significance of the impact. Where knowledge of the operations within a community is increased the perceived impact can be reduced.

For these reasons, understanding the cause of complaints and the number of underlying potential complainants, i.e. those who are annoyed but have not yet complained, is of most significance. As noted earlier, a common over-sight is to assume that the total number of complaints is an indication of performance. In all cases, the 'total number of complaints' is a reflection of the accessibility of the complaint system and every individual's motivation to complain. Whereas the performance indicator is the relationship between satisfied and non-satisfied residents, whether complainants or not.

13.5 ODOUR MODELLING AND IMPLICATIONS FOR OPERATIONS AND PLANNING

For the purposes of planning applications, site management, development and control, the application of a *dose–effect* model is the most sensible and scientific basis for assessment. This method provides a sensible starting point and foundation method for understanding the potential for annoyance. The method cannot guarantee a means to overcome objections but can provide a basis for appraising the detailed level of concerns at the planning stage. However, this is distinctly different from a situation where a site operator has to overcome objections from previous impacts, a difficult local history, or significant extensions to an existing operation.

In such circumstances the modelling process will need to demonstrate the efficacy of the approach, the suitability for the location and most importantly the vulnerability or range of sensitivity for the site assessment. Model assessments that take account of the maximum exposure concentrations as well as the percentile values will provide an indication of the vulnerability of locations for short-term exposure and worst conditions.

Allowance for atmospheric conditions, which are difficult to take account of within dispersion models such as short-range dispersion (less than 50 m), and temperature inversions should also be considered. Here, this information is of relevance to the operator in identifying "worst-conditions" and thus the vulnerability of routine and special operations.

Where modelling calculations are used it is important that the sensitivity of results are tested and understood prior to presentation in a planning context or for debate over nuisance. It is widely recognised that the range of variables considered within a modelling run can greatly influence the outcome of each assessment.

Interpretative models that take account of the hedonic quality of emissions from the site (such as that proposed by Miedema, 2000) are likely to provide a more confident basis for determining impacts.

The increased use of information for environmental management standards points towards site operators preparing "odour management plans" that detail the variety of sources, data, sensitive operations and details for information and staff training necessary to maintain appropriate performance.

Within either planning applications or submissions for environmental management systems (EMAS) a number of municipal sites have begun to adopt such an approach. Such a document, based on continuous improvement, helps to ensure continuity of management approach and shared knowledge of the nature of problems across a site. For the majority of instances where "site failures" have resulted in odour complaints, plans that specify the performance of systems for the following have helped reduce these failures; these include:

- Odour control systems;
- Staff guidance and training;
- Complaint reporting and investigation;
- Management of complaint reporting procedures / "hot-lines";
- Process monitoring and control;
- Special incidents and odour control during maintenance;
- Recording site detection of odours.

As patterns of complaint reports are identified, this information can be used not only to inform site improvements but be used to guide future site developments and assessments.

In conclusion, methods to enhance the ongoing application of modelling for planning assessments and site developments will continue to provide clear indications of the likely levels of exposure to odours. These assessments also provide the foundation for planning and legal decisions as to the level of exposure likely to be experienced. However, these factors cannot be directly employed to 'prove' that problems do not occur. The management issues associated with odour control and assessment will continue to be evident but information will be increasingly valuable to operators to improve the targeting of resources and process monitoring.

13.6 REFERENCES

CEN, (1997) TC264/WG2, Odour concentration measurements by dynamic olfactometry. Committé Européen de Normalisation, Brussels.

CLA, (1992) The Local Government Ombudsman. *Devising a complaints system*, The Commission for Local Administration in England, February, 21 Queen Anne's Gate, London SW1H 9BU.

DOE, (1993) Wansbeck District Council - Appeal by Northumbrian Water Ltd.: Additional sewage treatment facilities on land adjacent to Spital Burn, Newbiggin-by-the-Sea, Inspector: Rosser P., File No. APP/F2930/A/92/206240), Inquiry 4-5 & 9-12 March.

DOE, (1996) *Expert Panels on Air Quality Standards, Particles*. Department of the Environment, HMSO, London.

Doty (1996) *The Smell Identification Test*™, Administration Manual, Sensonics Inc., Haddon Heights, New Jersey,

ESRC Global Environmental Change Programme (2000), *Risky Choices, Soft Disasters: Environmental decision-making under uncertainty*, University of Sussex, Brighton, ISBN 0-903622-91-2.

Furuseth, O. (1990) Impacts of Sanitary Landfill: Spatial and Non-spatial Effects on the Surrounding Community. *J. Environmental Management* **31**,.269-277.

Hobbs, S.E., Longhurst, P.J., Sarkar, U. and Sneath, R.W. (2000) Comparison of dispersion models for assessing odour from municipal solid wastes. *Waste Management and Research* **18** (5), 420-428.

Lemon M. and Seaton R.A.F. (1999) Policy relevant research: The nature of the problem. In: *Exploring environmental change using an integrative method* (M. Lemon ed.) Gordon & Breach Science Publishers, The Netherlands.

Longhurst, P.J. and Seaton, R.S. (1999) Employing data on public perception for the strategic management of landfill odour. *Proc. Sardinia 99, Seventh International Waste Management and Landfill Symposium*, S. Margherita di Pula, Cagliari, Italy, 4-8 October.

Miedema, H.M.E. (2000) Exposure-annoyance relationships for odour from industrial sources. *Atmospheric Environment* **34** (18), 2927-2936.

Miedema, H. and Ham (1988) Odour Annoyance in Residential Areas *Atmospheric Environment* **22**, 2501-2507.

Nicell, J.A. (1994) Development of the odour impact model as a regulatory strategy, *International Journal of Environment and Pollution* **4**, 124-138.

Seffelaar, A.M., van der Zalm, C.J.A., Daamen, D.D.L., Dijksterhuis, G.B. and Punter, P.H. (1992) A comparison of odour annoyance survey results Staub – Reinhaltung der Luft **52**, 209-213.

Stevens, S.S. (1957) On The Psychophysical Law, *Psycholoogical Review* **64** (3),153-181.

Stevens, S.S. (1960) The Psychophysics of Sensory Function. *American Scientist* **48**, 226-253.

Wagenaar, W.A. (1975) Stevens VS Fechner: A Plea For Dismissal Of The Case. *Acta Psychologica*, **39**, 225-235.

Part V

ODOUR CONTROL AND TREATMENT

14

Use of chemicals for septicity and odour prevention in sewer networks

Gong Yang and John Hobson

14.1 INTRODUCTION

The use of rising mains to convey sewage has increased steadily in recent years. Rising mains present ideal conditions for development of septic sewage, which could cause odour nuisance and corrosion downstream as well as treatment problems. Consequently, the amount of chemicals used for septicity prevention has risen sharply.

Information about using chemicals for septicity prevention is disseminated mainly by chemical suppliers. Such partly technical and partly commercial materials commonly demonstrate the effectiveness of a particular type of chemical for controlling dissolved sulphide concentration in wastewater. There is a general belief that by eliminating dissolved sulphide, one could also achieve a good degree of odour reduction. However, relatively little data have been seen which illustrates this effect.

 Odours in Wastewater Treatment: Measurement, Modelling and Control edited by Richard Stuetz and Franz-Bernd Frechen. ISBN: 1 900222 46 9

A variety of chemicals have been used for prevention of septicity. Oxygen and less frequently air have been used to prevent septicity. Oxygen promotes aerobic metabolism and prevents the anaerobic metabolism that leads to sulphide and odour formation. Although oxygen can chemically oxidise hydrogen sulphide and perhaps other odorous chemicals, the kinetics is slow unless catalysed (Kotronarou and Hoffmann 1991). Therefore, support of aerobic oxidation of sulphide by oxygen is more important.

Nitrate acts as an alternative electron acceptor for many aerobic microorganisms. In so doing it prevents redox potentials from falling excessively and thus prevents the formation of more chemically reduced products under anaerobic conditions. The most commonly used products are ferric and calcium nitrates supplied in liquid form. Some nitrate products have been given commercial names such as Nutriox® (calcium nitrate) or Anaerite 263® (ferric nitrate).

Both oxygen and nitrate promote alternative microbial activities to anaerobic metabolism. To achieve this, aerobic or anoxic respiration by the microorganisms within the wastewater or sludge must be maintained. As a consequence, a part of the easily biodegradable BOD in the wastewater is destroyed or converted to biomass in the process.

Iron salts, typically ferric chloride and sometimes ferric nitrate, are used to precipitate dissolved sulphide. This reaction eliminates the tendency of hydrogen sulphide to volatilise, which is one of the main causes of odour nuisance and corrosion problems. The effectiveness of iron salts to completely remove sulphide is affected by the pH value in wastewater. The effectiveness of iron salts to remove odour is thought to be affected by the presence of odorous compounds other than hydrogen sulphide. There is no known mechanism by which ferrous salts will prevent the formation of sulphide or other odorous compounds in wastewater, although ferric salt is an oxidising agent.

Stronger oxidising agents, such as hydrogen peroxide, sodium hypochlorite, chlorine dioxide and potassium permanganate have been used for septicity control. These chemicals are able to rapidly oxidise sulphide and other reduced odorous compounds such as mercaptans. Because of the possibility of producing chlorinated by-products, some chlorine-based agents are not viewed favourably in the UK on environmental grounds. They are usually used to treat intensely odorous liquid, such as sludge or return liquors.

Among the chemicals commonly used for the control of septicity, ferric and nitrate salts are applied most widely in the UK. These chemicals are supplied in liquid form, which facilitates simple storage and dosing equipment and allows quick retro-fit to existing process flow sheets. A common disadvantage of this method in general is high costs. Sustained chemical dosing for large wastewater flows can incur a substantial operating cost, thus affecting the overall costs for wastewater treatment.

This chapter presents an introduction to the application of nitrate and iron salts for prevention of septicity and odours in rising mains. The principle of operation for each chemical is described. Wherever possible, olfactometry data are presented along with hydrogen sulphide data to indicate the overall effectiveness of the methods. Practical aspects of the application, such as estimation of chemical demand, control of dosing and costs, are discussed.

14.2 SEPTICITY DEVELOPMENT IN WASTEWATER

The following paragraphs briefly describe the mechanism of septicity formation, which could help the understanding of the principles for its prevention using chemicals.

Microorganisms are present commonly in wastewater. They gain energy through the oxidation of organic matter. Such biological reactions involve the transfer of electrons from electron donor compounds (organic matter) to electron acceptor compounds (e.g. oxygen, nitrate and sulphate). The type of electron acceptor decides which type of reactions take place and hence the end-products of the reactions. When oxygen is utilised as the electron acceptor, organic matter is oxidised to carbon dioxide and water through aerobic metabolism. In the absence of oxygen, nitrate is used as an alternative electron acceptor. Nitrogen gas, carbon dioxide and alkalinity are produced by the so-called anoxic metabolism. In the absence of both oxygen and nitrate, sulphate and carbon dioxide are used as electron acceptors. This results in reactions which produce hydrogen sulphide, mercaptans, volatile fatty acids and other reduced organic compounds. Many of these have strong odours. Accumulation of the reduced metabolic products in wastewater is a sign of septicity development.

Wastewater often contains relatively high concentrations of organic matter measured by BOD but limited dissolved oxygen and little nitrate. Therefore, aerobic or anoxic metabolism could quickly use up all the available oxygen and nitrate, leading to anaerobic conditions. The actual speed of oxygen depletion is a function of aerobic respiration rate and the re-aeration rate. Typically, for domestic sewage the aerobic respiration rate is between 2 and 20 mg/l O_2/h under ambient temperatures (Pomeroy 1990). This implies if no supply of oxygen is available and under suitable temperature, domestic wastewater may remain aerobic only for a short period. Subsequently, anaerobic conditions will prevail.

Establishment of anaerobic conditions in domestic wastewater is not necessary followed by rapid septicity development. In crude sewage outside sewers, for example, the speed of septicity development remain relatively slow long after it has become anaerobic. Figure 14.1 shows the change in hydrogen sulphide concentrations in a typical crude sewage under 20 °C over time. Relatively little sulphide was developed during the first 24 hours.

Since the overall hydraulic retention time for most wastewater transport and treatment systems is less than 24 hours, development of septicity at this slow rate would cause relatively few practical concerns.

However, as we know, sulphide can develop more rapidly in sewerage systems. Figure 14.2 shows the change in dissolved sulphide concentrations over time in sewage in a rising main.

For rising mains, the Pomeroy (1990) formula can be used to predict the sulphide concentrations.

$$\Phi_{se} = M_b * \text{COD} * 1.07^{(T-20)} \tag{14.1}$$

Where:

Φ_{se}	=	sulphide flux from the slime layer into the stream ($g/m^2/hr$),
M_b	=	0.228×10^{-3} under conditions favourable for H_2S build up,
COD	=	chemical oxygen demand of the sewage (mg/l),
T	=	temperature (°C).

or the concentration of sulphide in the sewage:

$$C_t = \Phi_{se} \frac{A}{Q} = \Phi_{se} \frac{2 * L}{u * r} \tag{14.2}$$

Where:

C_t	=	sulphide concentration in sewage (g/m^3),
A	=	Area of internal wall of rising main (m^2),
Q	=	Flow rate of sewage (m^3/h),
L	=	Length of rising main (m),
r	=	radius of rising main (m),
u	=	velocity of flow (m/h).

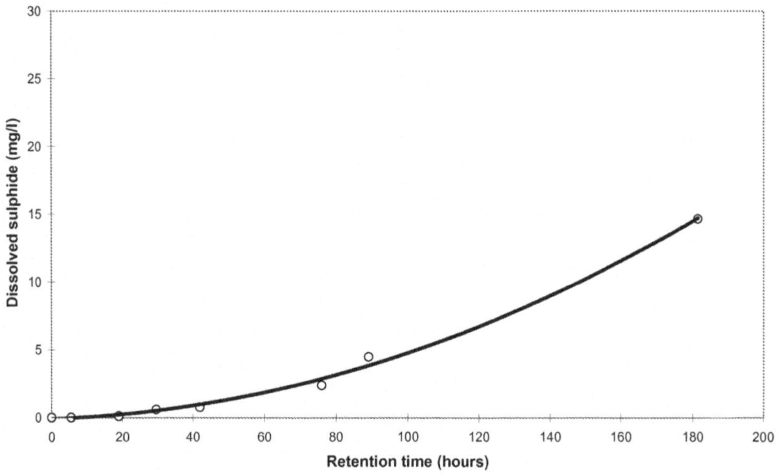

Figure 14.1. Sulphide generation in crude sewage outside sewers.

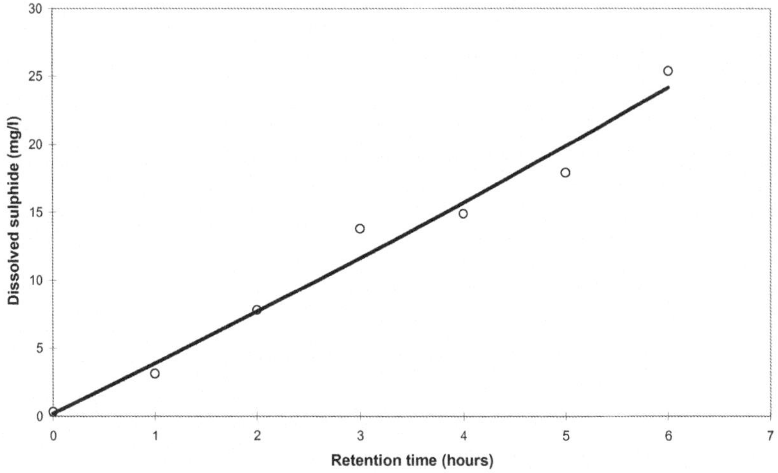

Figure 14.2. Sulphide generation in a rising main.

The composition of the crude sewage used for the two experiments is similar. The difference between the two cases was that a well-developed population of sulphur metabolising microorganisms was present in the rising main as biofilm, but a smaller population of such microorganisms was present in the crude sewage.

This illustrates that two conditions are necessary for rapid development of septicity. First, a strictly anaerobic environment must be established. Secondly, a large population fermentative and sulphur metabolising microorganisms must be present. Significant septicity could not develop quickly if either of conditions is missing. The above is also born out from more practical experiences.

In gravity sewers, fermentative and sulphur-metabolising microorganisms establish in large numbers in the deposit. However, natural aeration provides a continuous supply of oxygen into wastewater. This is often sufficient to satisfy the aerobic respiration demand of the wastewater to maintain aerobic conditions. The absence of anaerobic environment in the bulk of the liquid phase prevents septicity development. When physical blockage or saline intrusion hinders aeration and lead to the exhaustion of dissolved oxygen, septicity develops quickly because both conditions are now fulfilled. This is why odour complaints can often be a good earlier indication to potentially more serious sewer problems, such as partial blockage and saline intrusion.

One the other hand, septicity is often not immediately observed in new rising mains. Even anaerobic conditions may exist due to a lack of oxygen supply, there is not an established population of fermentative and sulphur metabolising microorganisms in new rising mains to catalyse the anaerobic reactions necessary to generate septicity. Since the anaerobic microorganisms grow slowly, it may take several months before significant sulphide production is observed in new rising mains. The capacity of a rising main to produce sulphide can continue to rise for up to a year before reaching its peak.

In summary, development of septicity in domestic wastewater within a short time-scale depends on the presence of a large population of sulphur-metabolising microorganisms and a strictly anaerobic environment. Therefore, septicity can be prevented if at least one of these conditions is withdrawn or avoided.

14.3 CONTROLLING SEPTICITY USING NITRATE

14.3.1 Principle of operation

Addition of nitrate to wastewater prevents the establishment of anaerobic conditions. The procedure involves adding a liquid solution of nitrate into the flow of wastewater upstream of pumping stations. The critical factor to the procedure is controlling the dosage so that a trace, but no more than a trace, of residual nitrate remains in the wastewater. This guarantees enough nitrate is being provided to prevent anaerobic conditions in the wastewater while excessive dosing is avoided, which is both wasteful and could be detrimental to subsequent treatment processes.

Figure 14.3 shows a group of controlled experiments demonstrating this method at work. At the beginning of the experiment, a rising main was filled with fresh crude sewage dosed with different concentrations of calcium nitrate. The sewage was then circulated in the rising main and monitored for dissolved sulphide and nitrate concentrations. As expected, sulphide concentration was not observed as long as a residual nitrate was present in the sewage, but started rising rapidly as soon as the nitrate had been exhausted.

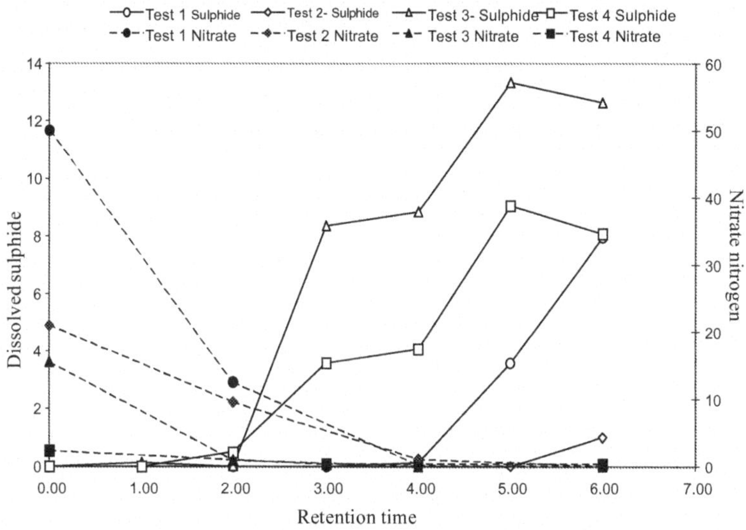

Figure 14.3. Effects of nitrate on sulphide generation in a rising main.

Interestingly, when insufficient nitrate was added, sulphide generation took place at an accelerated rate after nitrate exhaustion. In the worst case, a similar level of septicity had been reached in the wastewater as the control (16 mg/l) to which no nitrate was added. This implies that insufficient dosage may not necessarily achieve partial control of septicity as might be expected. There was no fixed ratio between the expected sulphide concentration and the concentration of nitrate needed. The following paragraphs will provide more explanations.

Excessive nitrate dosage increases the chemical costs unnecessarily. It is argued that high residual nitrate concentrations may impair the performance of some treatment processes, for example, causing floating sludge in primary tanks, when co-settlement of secondary sludge is practised.

14.3.2 Nitrate demand

To maintain a trace of residual nitrate in the wastewater, the amount of nitrate added has to match the nitrate requirement for anoxic metabolism. Typical respiration rates of domestic sewage and biofilm in sewerage systems have been reported to be 2 mg/l/h and 700 mg/m^2/h, respectively (Pomeroy 1990). Assuming the same heterotrophs are also capable of anoxic respiration in the absence of oxygen, these values may be converted to nitrate consumption rates. For equivalent number of electron acceptors, 1 mg/l of nitrate nitrogen can replace approximately 2.86 mg/l of dissolved oxygen. Therefore, the equivalent nitrate consumption rates for sewage and biofilm are 0.7 mg NO_3-N/l/h and 250 mg NO_3-N/m^2/h respectively. Using these values, the nitrate demand by a rising main can be calculated:

$$M_{VSS} = \pi \frac{D^2}{4} L R_{VSS} \tag{14.3}$$

$$M_{VSA} = \pi D L R_{VSA} \tag{14.4}$$

$$M_{Total} = M_{VSS} + M_{VSA} = 0.00025 \pi D L (0.7D + 1) \tag{14.5}$$

Where:

M_{VSS}	=	Nitrate demand by sewage (kg NO_3-N/h),
M_{VSA}	=	Nitrate demand by biofilm (kg NO_3-N/h),
M_{total}	=	Total nitrate demand (kg NO_3-N/h),
R_{VSA}	=	Nitrate respiration rate of sewage (kg NO_3-N/m^3/h),
R_{VSS}	=	Nitrate respiration rate of biofilm (kg NO_3-N/m^2/h),
D	=	Inner diameter of the rising main (m)
L	=	Length of the rising main (m)

The activity of the heterotrophs can be expressed by specific reaction rate, which equals to the overall reaction rate divided by the concentration of volatile solids. For typical domestic sewage, a total suspended solids of 300 mg/l with 80% volatile solids is assumed. Hence the specific reaction rate is around 0.003 mg NO_3-N /mg VSS/h.

A similar calculation may be performed for biofilm. Assuming biofilm has an average thickness of 0.5 mm and a wet density of 1.0 g/cm^3, then each square metre of biofilm surface contains 0.5 kg of wet biofilm. Typical solids contents of biofilm in sewage systems have been shown to be around 5% with 80% of volatile solids (Atkinson *et al.* 1981; Walker and Austin 1981; Characklis and Marshall 1989). Thus each square metre of biofilm surface is associated with 20,000 mg of volatile solids. Hence the specific reaction rate is around 0.013 mg NO_3-N /mg VS/h.

The above values are based on the respiration rates in sewerage systems, but they are within the same range of specific reaction rates found in wastewater treatment systems. For example, specific reaction rates between 0.003 and 0.012 g NO_3-N/g VS/h have been suggested for designing the denitrification stage of activated sludge processes (Wheeldon and Bayley 1981). A specific reaction rate of 0.014 g NO_3-N/g VS/h was found in a denitrifying fluidised bed reactor (Cooper and Wheeldon 1981). In general, well-developed biofilm has higher specific reaction rates than the volatile solids present in crude sewage. But this may change as sewage "ages" in sewerage systems. It has been reported that the respiration rate of "aged" sewage can reach 20 mg O_2/l/h (Pomeroy 1990).

In rising mains less than 300 mm in diameter, there is likely to be more biomass as biofilm than as suspended volatile solids. The nitrate demand is dominated by biofilm. For larger rising mains, biofilm and suspended volatile solids both contribute to the nitrate demand significantly. This is illustrated in Figures 14.4 and 14.5. For example, for a rising main of 400 mm diameter and 3 km long, the estimated total nitrate demand is 29 kg/day at 15 °C. This consists of 6.4 kg by the sewage and 22.6 kg by the biofilm. This corresponds to a chemical consumption of approximately 200 litres per day for a product that contains 150 g NO_3-N/l.

14.3.3 Control requirements

In real sewerage systems, conditions such as temperature and sewage composition change dynamically. As a result, the instantaneous nitrate demand for a particular system does not necessarily equal to the predicted average demand. This means some method of control is needed to respond to these changes. Programmable logical controllers (PLC) are often used for this purpose. Using a PLC, dosing rate may be adjusted to match an empirical dosing pattern triggered by a timer. Such dosing patterns can be designed and tested according to diurnal variations of sewage flow, composition and temperature[1] and other local factors.

---

[1] The following factor is commonly used for temperature adjustment:

$$K_T = \theta^{(20-T)} \qquad (14.6)$$

Where:

K_T	=	Temperature factor,
T	=	Temperature of sewage (°C),
θ	=	Temperature coefficient (1.07).

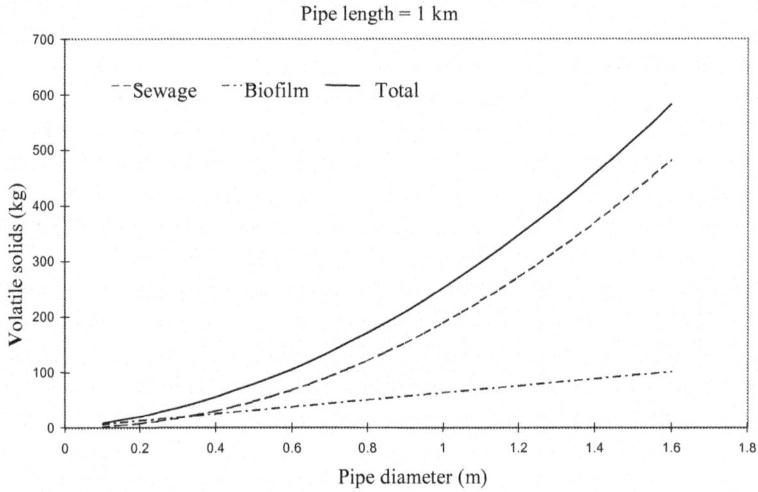

Figure 14.4. Quantities of volatile solids in 1 km of rising main.

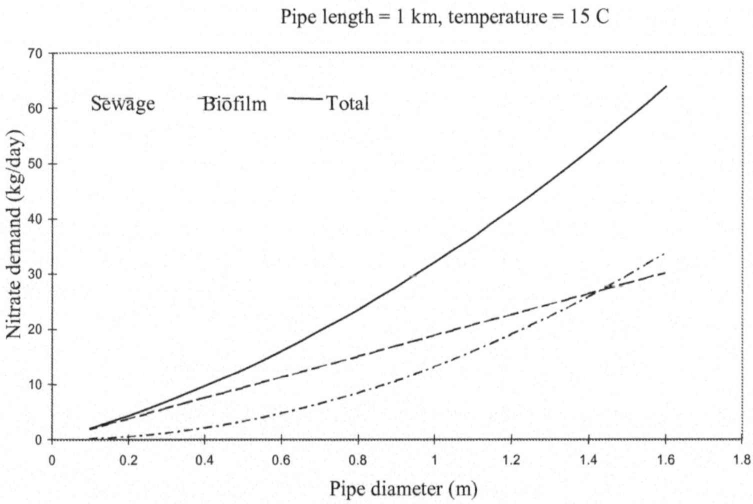

Figure 14.5. Nitrate demands by volatile solids in rising mains.

Because the quantity and activity of biomass in a rising main do not vary quickly with sewage flow rate, the nitrate demand is relatively constant in this respect. This means in theory a more or less fixed nitrate dosing rate may be maintained to prevent septicity when the sewage flow rate fluctuates. In practice, some adjustment of nitrate dosing rate with sewage flow rate may be necessary since variation of sewage flow rate may be associated with variation of other parameters. For example, in rising mains of long retention times, the availability of soluble BOD could be a limiting factor for nitrate consumption. The reaction rate is controlled by endogenous decay or hydrolysis of particulate after soluble BOD is exhausted. The specific reaction rate may decrease by 50 to 80% from the normal rate (US EPA 1991). If the sewage flow rate increases in such rising mains, more soluble BOD is available for normal denitrification, which creates greater nitrate demand. On the other hand, in rising mains of short retention times, the lag phase before the onset of septicity is significant. Decreased sewage flow results in less dissolved oxygen entering the rising main, which is partly responsible for the lag phase. As a result, more nitrate may be necessary to prevent septicity.

It might be possible to create a fully automatic control system to match instantaneous nitrate demands through continuous measurement. For example, such a system may employ a small reactor and a sensor to measure the actual nitrate reaction rates of the sewage upstream the dosing point. These measurements may be used to drive a feed-forward system that controls the chemical dosing pump. A second sensor may used to monitor the residual nitrate concentration downstream of the rising main, which provides feedback calibrations for the feed-forward system.

With either automatic control or fixed pattern control, it is hard to satisfy nitrate demand in sewerage systems with intermittent flow regimes. The nitrate is delivered and transported by sewage flow. When flow stops so does the supply of nitrate. Meanwhile, life goes on in the rising main until nitrate is exhausted.

14.3.4 Effects on odour reduction

Nitrate stops septicity by preventing the establishment of anaerobic environment. Therefore, the formation of all reduced odorous compounds should be prevented. The following example[2] shows this objective has been achieved.

Without nitrate addition, 14–17 mg/l of sulphide was produced in a sewage following 6 hours' retention in a rising main. The odour potential of the sewage also rose from around 5,000 ou/m^3 to over 3,000,000 ou/m^3. Following nitrate dosing, less than 0.1 mg/l of sulphide was produced and the odour potential only rose to around 20,000 ou/m^3. This represents over 99% suppression of both sulphide and odour.

---

[2] Unpublished WRc data.

14.3.5 Costs

The equipment for nitrate dosing includes chemical feed pumps, control cabinet and a chemical storage tank. The capital costs to install such equipment for a medium sized application with 10–20 m^3 chemical storage capacity is around £30,000–£50,000, according to specific quotations from suppliers in 1999.

The operating costs of nitrate dosing are primarily the cost of chemicals. For example, a 3 km rising main with a diameter of 0.4 m may require on average 29 kg of nitrate per day as NO_3-N. This is equivalent to 200 litres per day of a liquid product containing 150 g of NO_3-N per litre. The annual consumption would be 73,000 litres. If the price of this product is £0.30 per litre (an example of quoted prices in 1999); the annual cost would be £22,000.

14.4 CONTROLLING SEPTICITY USING FERRIC

14.4.1 Principle of operation

Septicity in wastewater is closely associated with the formation of sulphide. The later exists as free hydrogen sulphide (H_2S), dissociated sulphide (HS^- and S^{2-}) and metal sulphides (MS). Most detrimental effects of septicity are associated with free hydrogen sulphide. These are eliminated if the free hydrogen sulphide is converted into other forms, particularly to metal sulphides.

Most sulphide compounds with heavy metals have low solubility. Particularly, iron sulphide has much lower solubility than iron hydroxide and iron carbonate. Therefore, iron salts are used to selectively precipitate sulphide in wastewater. In this reaction, the equilibrium concentration of the dissolved sulphide is governed by the solubility product of iron sulphide:

$$K_{\mathrm{FeS so}} = [Fe^{2+}][S^{2-}] \tag{14.7}$$

Iron salts may be used to suppress septicity in two ways. It may be added to septic wastewater to precipitate dissolved sulphide as iron sulphide. Alternatively, it may be added to wastewater before septicity has developed. Any dissolved sulphide which forms subsequently will be precipitated.

When iron salts are added to wastewater before sulphide has formed, insoluble hydroxide and carbonate are produced initially. The equilibrium of both reactions affected by acidity and the concentration of dissolved carbon dioxide:

$$K_{\mathrm{Fe(OH)_2 so}} = \frac{[Fe^{2+}]}{[H]^2[Fe(OH)_2(s)]} \tag{14.8}$$

$$K_{\mathrm{FeCO_3\,so}} = \frac{[Fe^{2+}][CO_3^{-}]}{[FeCO_3(s)]} \tag{14.9}$$

$$= \frac{[Fe^{2+}]}{[FeCO_3(s)]} C_T \left(\frac{[H^+]^2}{K_1 K_2} + \frac{[H^+]}{K_1} + 1 \right)^{-1}$$

Where:

K_{FeSso}	=	solubility products of iron sulphide,
$K_{\mathrm{Fe(OH)2so}}$	=	solubility products of iron hydroxide,
K_{FeCO3so}	=	solubility products of iron carbonate,
K_1, K_2	=	dissociation constants of carbon dioxide,
C_T	=	total concentration of dissolved CO_2.

In typical wastewater[3], iron carbonate is more stable than iron hydroxide at pH values less than 11. For pH values greater than 11 iron hydroxide is more stable, as shown by Figure 14.6.

In practice, the total concentration of dissolved iron salt and the total concentration of dissolved sulphide are move relevant than the concentrations of specific ionic species. To estimate these, the conditional solubility product (P_s) of iron sulphide may be used:

$$P_s = Fe_T S_T = \frac{K_{\mathrm{FeSso}}}{\alpha_{Fe} \alpha_s} \tag{14.10}$$

Where:

P_s	=	conditional solubility products of FeS,
Fe_T	=	total concentration of dissolved iron,
S_T	=	total concentration of dissolved sulphide.

[3] The value of C_T is around 0.001 to 0.01 M

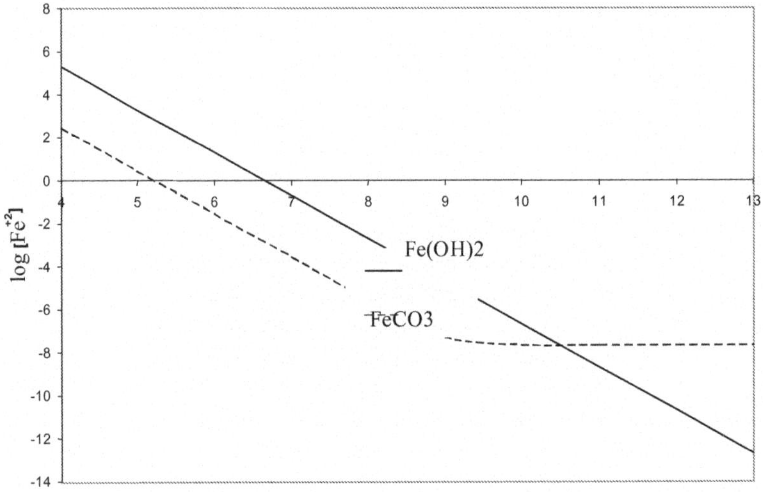

Figure 14.6. Solubility of iron hydroxide and iron carbonate at various pH values.

The values of α_{Fe} and α_s may be calculated by:

$$\alpha_{Fe} = \left(1 + \frac{K_{Fe1}}{[H^+]^2}\right)^{-1} \tag{14.11}$$

$$\alpha_S = \left(1 + \frac{[H^+]}{K_{S2}} + \frac{[H^+]^2}{K_{S1}K_{S2}}\right)^{-1} \tag{14.12}$$

Where:
K_{Fe1} = solubility product of ferrous hydroxide,
K_{S1}, K_{S2} = the dissociation constants for sulphide.

For a range of total dissolved iron concentrations, the total dissolved sulphide concentration is plotted against the pH value in Figure 14.7. This indicates that the dissolved sulphide concentration can be reduced to virtually zero by maintaining a small dissolved iron concentration. This can be achieved most effectively at the pH range between 7 and 8.

The above analysis is simplistic compared with the reality because complex formation of iron hydroxides and iron sulphides are not considered. Nevertheless, the result of the analysis is consistent with the fact that little extra iron is needed to completely precipitate dissolved sulphide. The analysis also confirms that the effectiveness of precipitation is highly sensitive to the pH value.

14.4.2 Ferric demand

The following experiment proves that ferric iron salts can precipitate dissolved sulphide stoichometrically. A septic sewage that contains 9.2 mg/l of dissolved sulphide was used in the experiment. Between half to two times of the stoichiometric amount of ferric chlorite was added to 10 litre sub-samples of this sewage. The dissolved sulphide concentrations in the sewage were measured after 30 minutes. The results are shown in Figure 14.8. This experiment confirms that the iron salt reacts selectively with sulphide.

When an iron salt is added to wastewater contains no dissolved sulphide, iron carbonate, hydroxide or other complexes form initially. It is possible that some fraction of this iron may be recovered irreversibly and is then unavailable for reacting with sulphide that subsequently forms. When ferric chlorite was continuously dosed to fresh sewage prior to a rising main, 120% of the stoichiometric amount was required in order to completely react with the anticipated sulphide formation in the rising main.

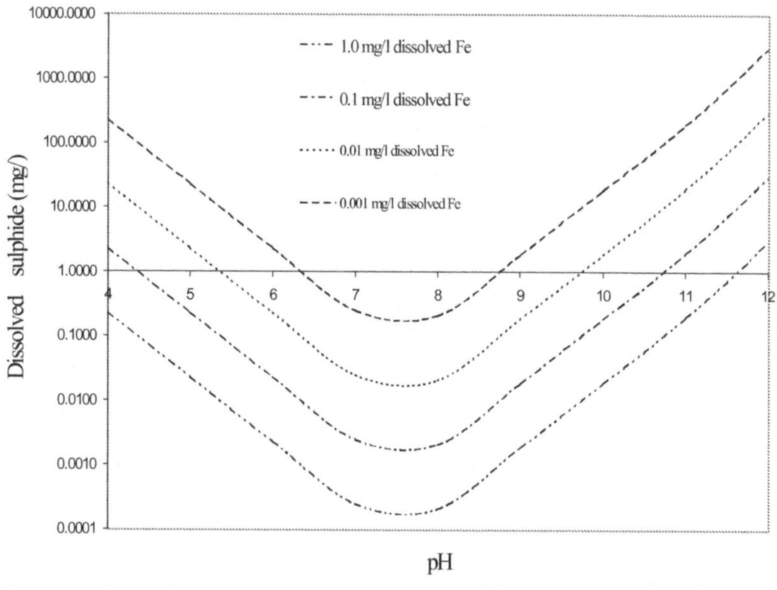

Figure 14.7. Effects of pH and iron concentration on residual sulphide.

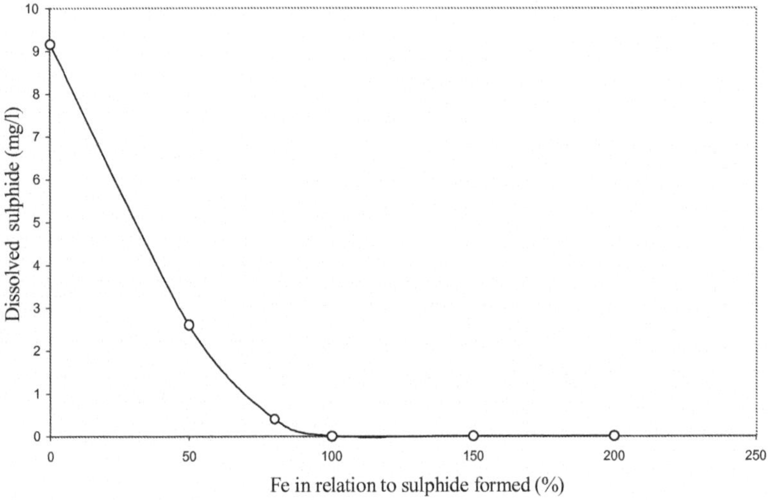

Figure 14.8. Precipitation of sulphide by ferric chlorite.

To make an estimate of ferric demand, the first step is to establish how much sulphide has formed or is expected to form. Once the actual or anticipated sulphide concentration is known, the ferric demand may be predicted by:

$$C_{Fe} = \alpha \times \beta \times C_{H2S} \text{ and } C_{Fe} < \alpha * \beta \times C_s \tag{14.13}$$

Where:

C_{Fe}	=	iron salt concentration required as Fe ions (g/m^3),
α	=	constant, equals the molecular weight of iron (56) divided by / molecular weight of sulphide (34),
β	=	equals 1 if iron salt is added after sulphide has formed and up to 1.5 if iron salt is added before sulphide formation,
C_s	=	total available sulphur in the wastewater, a value of 15 to 20 mg/l is typical for domestic sewage,
C_{H2S}	=	sulphide concentration (mg/l).

Ferric is utilised most efficiently when added to wastewater already having a moderate sulphide concentration. A strongly saturated solution is instantly created, prompting rapid formation of iron sulphide precipitates. This is characterised by instant changes in the colour of wastewater to dark grey and black. If ferric is added to wastewater which contains no sulphide, a saturated solution will develop gradually when sulphide is produced. Precipitation may not take place until a certain degree of supersaturating is achieved. The level of supersaturating may be different for each wastewater since this depends on the concentration and the surface nature of suspended solids, which are involved in heterogeneous nucleation during the initial stage of precipitation. Extra ferric salts may be required to overcome the delay of precipitation in systems where a low concentration of sulphide is formed. The actual ferric demand is best determined by laboratory tests or by trial and error.

14.4.3 Control requirements

Like nitrate dosing, ferric dosing is typically controlled by adjusting the feed pump triggered by a timer according to a fixed pattern. The particular pattern for each application is determined by experience or trial and error.

The dosing system usually includes a chemical storage tank and chemical feed pumps which deliver the chemical to a application point. Liquid ferric chloride stock solutions normally require no dilution, so feed pumps can draw directly from the storage tank. The storage tank is sized according to the feed rate requirement and delivery constraints. The system needs to be protected against corrosion.

The corrosive nature of the chemical requires some extra caution on the control system. Ferric dosing causes a small decrease in the pH value of the wastewater, but this usually remains neutral. However, localised low pH value is possible with poor mixing or under unexpected conditions. For example, if the wastewater pump stops unexpectedly and the ferric feed pump continues operating, concentrated ferric solution could accumulate near the delivery point, usually just upstream of the main wastewater pump. This could cause damage of pipeline section and the wastewater pump unless these are corrosion resistant. A flow fail switch can be used to avoid this situation by preventing the chemical feed pump operating while there is no wastewater flow.

14.4.4 Effects on odour

There are no known mechanisms by which iron salts can prevent the formation of sulphide or other odorous compounds. Iron salts is not known to react with organic sulphur compounds. Therefore, it is expected that ferric dosing may not always completely suppress odour. However, experiments using iron salts in a rising main showed some unexpected results for odour suppression.

Figure 14.9 shows the relationship between hydrogen sulphide and odour suppression following different amount of ferric was added to a septic sewage. The level of odour suppression was almost proportional to H_2S suppression. Similar results were obtained when ferric was dosed to fresh sewage to prevent the effects of septicity.

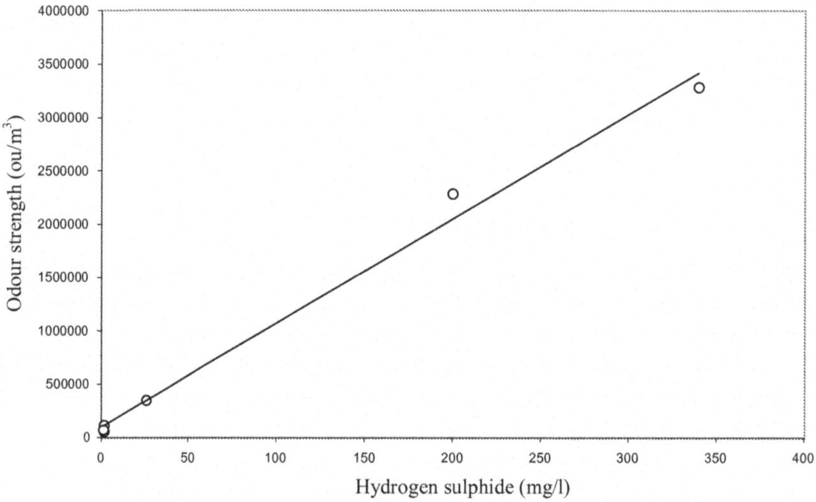

Figure 14.9. Correlation between H_2S and odour following ferric dosing.

A closer examination of Figure 14.9 yields a linear relationship between odour potential and hydrogen sulphide concentration with a slope of 9800 ou/m^3/ppm. In other words, 9.8 ou/m^3 was removed for each ppb of H_2S removed. Similar, but more scattered relationships between odour and H_2S were also found for digester gas[4] and for septic sewage in rising mains without chemical treatment[5], as shown in Figures 14.10 and 14.11. Such relationships usually break down when the odours are of a non-septic nature (e.g. from secondary treatment) or when the hydrogen sulphide concentration is very low.

One simple explanation for the correlation shown in Figures 14.9–14.11 is that hydrogen sulphide has a threshold odour concentration between 0.1–0.25 ppb. Accordingly, once a given concentration of hydrogen sulphide is removed, an equivalent concentration of odour is also removed. However, this explanation contradicts with the most widely believed value of 0.5 ppb for the threshold odour concentration of hydrogen sulphide (Vincent and Hobson 1999).

---

[4] Measured from digesters at 8 sewage treatment works in the UK.

[5] Measured during a pilot-scale trial to evaluate odour and sulphide development in rising mains.

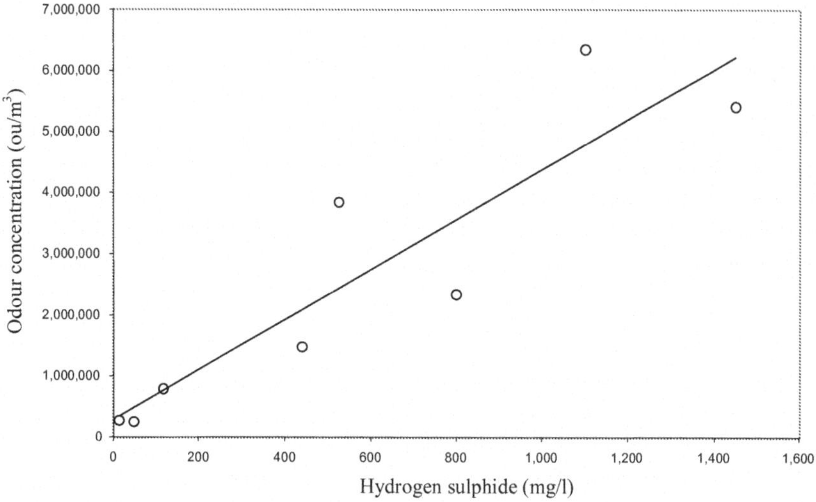

Figure 14.10. Correlation between H_2S and odour in digester gas.

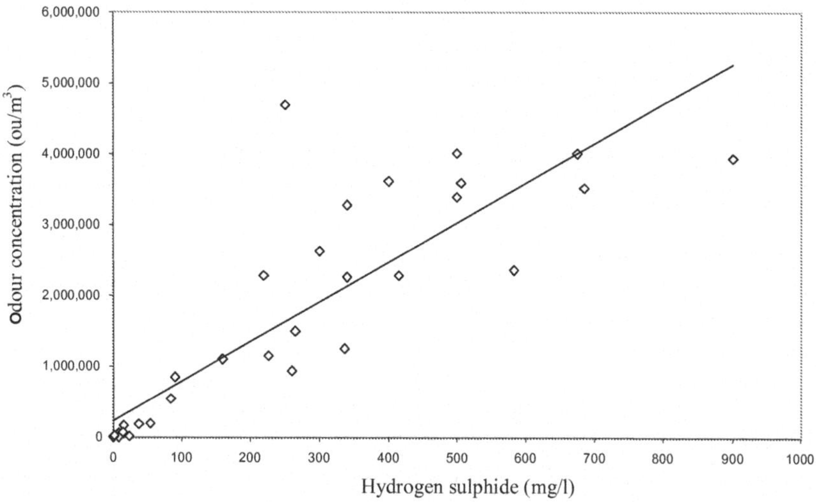

Figure 14.11. Correlation between H_2S and odour in septic sewage.

It is possible that other odorous compounds to have a synergistic effect on hydrogen sulphide. In other words, their presence together with hydrogen sulphide may enable the human nose to detect hydrogen sulphide at a lower concentration than that with the presence of hydrogen sulphide alone. When hydrogen sulphide is dominant, the synergistic effects of other odorants may be more important than their own contribution to odour. Thus proportional reduction of odour would occur with the reduction of hydrogen sulphide until nearly all the hydrogen sulphide is removed. Subsequently, the other odorants become dominant for the perceived odour concentration. This hypothesis can explain the observations well but more direct measurements are needed for its substantiation.

14.4.5 Costs

The equipment for ferric dosing includes chemical feed pumps, control panel and a ferric solution storage tank. The capital costs to install such equipment for a medium sized application with 10–20 m^3 storage capacity is similar or slightly higher than a nitrate dosing facility of the same capacity (due to corrosion resistance requirement).

The operating costs of septicity control by ferric dosing can be estimated through the cost of chemicals. For example, a 3 km rising main with a diameter of 0.4 m may require an average ferric dosing rate of 200 kg per day (as ferric chloride). The annual consumption would be 73,000 kg. For an estimated price of £200 per tonne, the annual chemical cost would be £15,000.

14.5 CONTROLLING SEPTICITY USING FERRIC NITRATE

Nitrate and ferric suppress septicity through different mechanisms and acting in sequence. Therefore, apparent extra value may be achieved when they are supplied as a single chemical. This offers an incentive for selecting ferric nitrate instead of calcium nitrate or ferric chloride for septicity control.

In theory, if 55% of the nitrate demand by a rising main is met by ferric nitrate, 45% of the iron requirement is provided too. Based on this calculation, one may expect 55% dosage of ferric nitrate to provide 100% protection. Unfortunately, it did not turn out like this in practice. As indicated earlier in this chapter, when insufficient amount of nitrate is added to a rising main, sulphide production may start at accelerated rates after the nitrate is exhausted, so part of the gains through nitrate is lost. There is also a possible delay of action by ferric following nitrate exhaustion, as explained earlier and shown in Figure 14.12. This could result in possible break through of odour. Therefore, it is not possible to achieve the theoretical saving by switching from calcium nitrate to ferric nitrate. A moderate saving of 5 to 15% may be possible. When ferric nitrate was applied to meet 100% of the nitrate demand, it offered more reliable odour suppression than either calcium nitrate or ferric chloride[6].

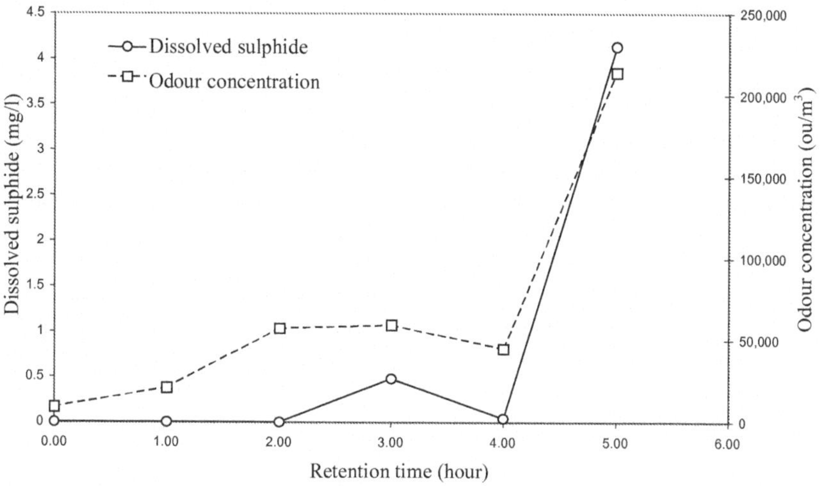

Figure 14.12. Use of ferric nitrate to suppression odour.

14.6 CONTROLLING ODOUR BY pH ADJUSTMENT

Because the free hydrogen sulphide concentration in solution decreases as the pH value increases, chemicals for pH adjustment (e.g. lime and sodium hydroxide) may be used for odour control. The proportion of free dissolved sulphides in solution is governed by the following reactions, which can be expressed as a function of pH.

$$H_2S \Leftrightarrow HS^- + H^+] \ \{K_1 = 7.94 \times 10^{-8}\} \tag{14.14}$$

$$HS^- \Leftrightarrow S^{-2} + H^+] \ \{K_2 = 1 \times 10^{-12}\} \tag{14.15}$$

$$H_2S(\%) = \frac{100}{1 + \frac{K_1}{10^{-pH}} + \frac{K_1 K_2}{10^{-2pH}}} \tag{14.17}$$

[6] Unpublished WRc data.

The above relationship suggests that the free hydrogen sulphide concentration in solution is highly sensitive to the pH value. For example, if the pH value is increased from 6.5 to 8.5, the free hydrogen sulphide concentration in solution would decrease by a factor greater than 20.

Figure 14.13 shows H_2S concentrations measured through odour potential sampling (Hobson 1995). Before sampling, the pH value of the sewage was adjusted over the range of pH value between 6.4 and 8.8. The predicted hydrogen sulphide concentrations were calculated using Equation 14.14, assuming the measured and predicted values have a common baseline for hydrogen sulphide concentration at the highest pH value.

The observed changes of hydrogen sulphide concentration with the pH value were less dramatic than predicted. Further measurements revealed that this discrepancy was more predominant with liquid in systems with lower hydraulic turbulence[7].

The following hypothesis was used to explain the observation. If the loss of CO_2 from boundary layer of the liquid due to the mass transfer to atmosphere was not replenished at the same rate by the diffusion of CO_2 in the bulk liquid, a dynamic equilibrium could be established. This dynamic equilibrium would allow a higher pH value in the boundary layer than that in the bulk of the liquid. This in turn, would cause a lower free sulphide concentration in the boundary layer than that in the bulk of the liquid. Consequently, less hydrogen sulphide than predicted according to the bulk free sulphide concentration would be emitted to the atmosphere at low pH values. In practice, this implies that in order to achieve a given percentage reduction of hydrogen sulphide emission, more dramatic pH adjustment than the theory prediction is required.

The effect of pH value on odour is a matter of the nature of the odour. If hydrogen sulphide is a major constituent of the odour, then a similar relationship could be expected. Figure 14.14 shows the relationship between hydrogen sulphide and odour concentration when the pH value changes from 6.5 to 8.5. In this case, a rough linear relationship existed between the measured odour and hydrogen sulphide concentrations.

---

[7] Unpublished WRc data

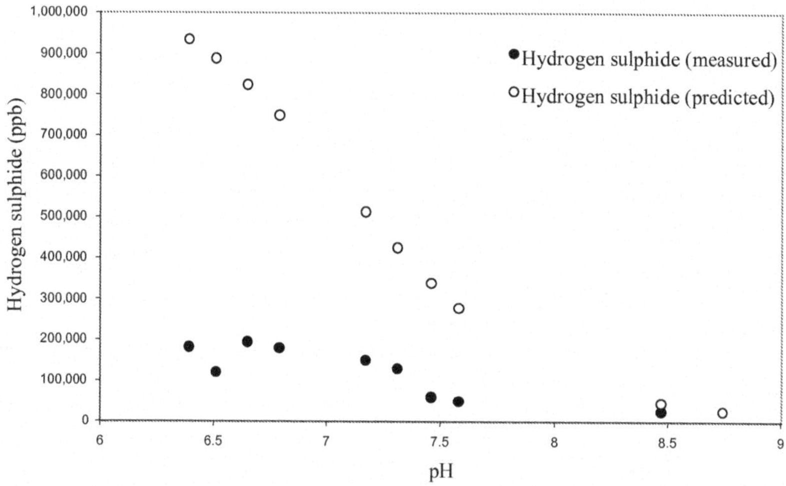

Figure 14.13. Effects of pH value on free hydrogen sulphide in solution.

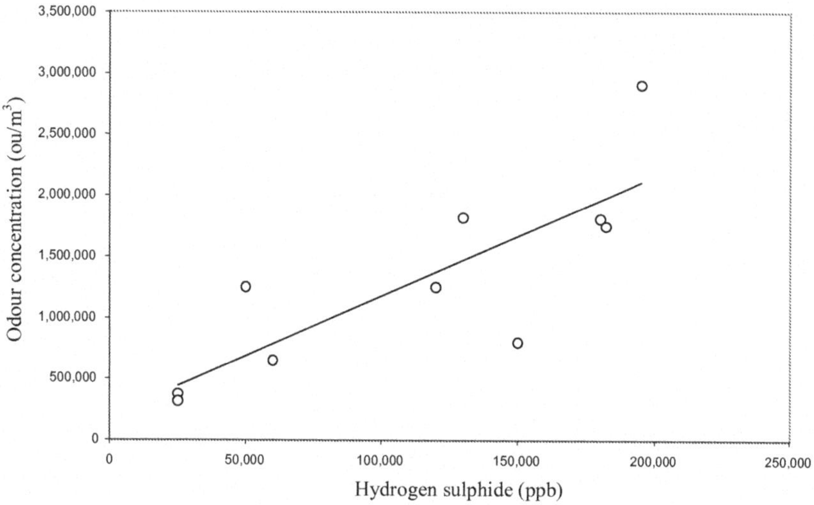

Figure 14.14. Effects of pH on hydrogen sulphide emitted.

14.7 REFERENCES

Atkinson, B., Black, G..M. and Pinches, A. (1981) The characteristics of solids supports and biomass support particles when used in fluidised beds. In: *Biological Fluidised Bed Treatment of Water and Wastewater*, (P.F. Cooper and B Atkinson, eds.) Ellis Horwood Ltd, London.

Characklis, W..G. and Marshall, K. G. (1989) *Biofilms*. John Wiley & Sons, London

Cooper, P.F. and Wheeldon, D.H.V. (1981) Complete treatment of sewage in a two-fluidised bed system. In: *Biological Fluidised Bed Treatment of Water and Wastewater*, (P.F. Cooper and B Atkinson eds.), Ellis Horwood Ltd, London

Hobson, J. (1995) The odour potential - a new tool for odour management. *J. Chart. Inst..Water Enviro. Manag.* **9**, 458-463.

Kotronarou, A. and Hoffmann, M.R. (1991) Catalytic autoxidation of hydrogen sulphide in wastewater. *Enviro. Sci. Techno.* **25**, 1153-1160.

Pomeroy, R.D. (1990) The problem of hydrogen sulphide in sewers. The Clay Pipe Development Association (A. Boon ed.) 2nd edition.

Tchobanoglous, G. and Burton, F. L. (1991) *Wastewater Engineering: Treatment, Disposal and Reuse,* Metcalf and Eddy Inc., McGraw-Hill Inc., New York.

US EPA. (1991) *Nitrogen Control*. Technomic Publishing Company Inc., Lancaster.

Vincent, A. and Hobson, J. (1999) Odour Control. *CIWEM Monographs on Best Practice* No 2, Terence Dalton Publishing, London.

Walker, I. and Austin, E.P. (1981) Use of Porous biomass supports in a pseudo-fluidised bed for effluent treatment. In: *Biological Fluidised Bed Treatment of Water and Wastewater*, (P.F. Cooper and B Atkinson, eds.), Ellis Horwood Ltd, London.

Wheeldon, D.H.V. and Bayley, R.W. (1981) Economic studies of biological fluidised beds for wastewater treatment. In: *Biological Fluidised Bed Treatment of Water and Wastewater*, (P.F. Cooper and B Atkinson, eds.), Ellis Horwood Ltd, London.

15

Process covers for odour containment

Lawrence Koe

15.1 INTRODUCTION

Sewage treatment facilities all over the world typically have multiple process chambers, tanks and building structures linked together by channels and various pipeworks. To contain the odorous gases that are emitted at these facilities, it is common for the major processes to be covered up or enclosed within a building structure such that the odorous air that is released can be efficiently trapped and pumped to appropriate odour control systems for treatment, prior to discharge to the ambient environment.

A wide variety of cover types and configurations are in use today. Many of these covers have been aesthetically designed to blend nicely with the surrounding architecture. A careful consideration of various factors such as cover material, durability, corrosion-resistance, shape and size as well as costs is

 Odours in Wastewater Treatment: Measurement, Modelling and Control edited by Richard Stuetz and Franz-Bernd Frechen. ISBN: 1 900222 46 9

required before final selection of process covers for odour containment can be made.

15.2 COVER MATERIALS

15.2.1 Properties

Considering that corrosive gases are emitted from the covered processes at a wastewater treatment plant, cover materials used should therefore be corrosion resistant and be durable for the aggressive environment. A variety of cover materials are available for selection by the design engineer. The most common materials include concrete, wood, fabric, aluminium and fibre reinforced plastic (FRP). A list of manufacturers of process covers is in Table 15.1.

Table 15.1 List of manufacturers of process covers.

Company and location	Comments
Temcor, USA (www.temcor.com)	Manufacture clear span covers made of aluminium for tanks, digestors, basins and reservoirs
Conservatek Industries, Inc., USA (www.conservatek.com)	An ISO 9001 certified company that engineers, manufactures and installs aluminium domes and covers
Geomembrane Technologies, Inc., Canada (www.gti.ca)	Design, install and maintain structures and floating cover systems for municipal wastewater plants
ILC Dover, Inc., USA (www.ilcdover.com)	Manufacture tank covers to contain odour and VOC emissions from wastewater processes
Thermacon Enviro Systems, Inc., USA (www.thermacon.com)	Design and manufacture specialty products for water and wastewater industries
American Grating, USA (www.amgrating.com)	Manufacture molded and pultruded fibreglass gratings and covers

15.2.1.1 Concrete

Concrete is typically used as low level cover in the form of precast prestressed concrete planks. The concrete planks are usually keyed into and grouted together to form a permanent cover over areas that do not emit significant corrosive gases such as effluent channels and basins. As concrete planks are heavy, they are usually not designed to be removable. In addition, corrosion can quickly impair the strength of the concrete member and it may be necessary to line the underside of concrete covers with a corrosion-resistant lining such as high density polyethylene or other suitable plastics.

15.2.1.2 Wood

Wood has been a favoured material for low-level covers particularly at old sewage treatment works. It is relatively light and easily shaped to cover a variety of channel shapes and tanks. Unfortunately, wood has a short lifespan often subject to weathering. Whilst it may be possible to protect wooden covers with appropriate plastic lining and painting system, such covers still deteriorate rapidly in the aggressive sewage environment. The short lifespan makes wood rather unattractive as a permanent cover material.

15.2.1.3 Fabric

Fabric is typically used as material for high level covers. Its use as low level cover is hampered by the difficulty in providing access over the covers and by safety considerations.

The PVC coated polyester often lined with polyvinyl fluoride for corrosion protection, is a very popular fabric cover. The flexibility of the fabric enables various free-form shapes to be designed, which can blend in nicely with the surrounding aesthetics. Although fabric covers are longer lasting than wood, there is still a need for regular (perhaps every 10–15 years) replacement and vendors have been known to offer several years of guarantees on fabric durability.

In use, the fabric can be stretched over a supporting framework which is typically made of corrosion-resistant material such as aluminium or steel. Figure 15.1 illustrates a geodesic dome spaceframe that could act as support for the fabric covers.

A clear advantage of fabric covers is that there is considerable choice in the range of colours available. White is typically selected because it is light translucent, thus providing excellent visibility during the day. Internal lighting would be needed for night use.

15.2.1.4 Aluminium

Aluminium is a very popular choice as cover material for sewage treatment processes. It has a high strength:weight ratio and being light can easily be configured as covers over various-shaped tanks and channels. With the correct choice of alloys, aluminium is also known to be very corrosion resistant. Low copper alloys should be used to avoid stress corrosion and cracking. The corrosion resistance of the aluminium is due to the formation of oxidic layers on the aluminium surface. There have been instances where splash surfaces of aluminium (areas subject to sewage splashing) exhibit corrosion. Hence, aluminium should be recommended in situations where it is unlikely to be splashed by sewage.

Figure 15.1. Geodesic framework.

Dome covers in the form of modular units of geodesic design can be moulded to suit a variety of spans. Components are standardised and being factory produced, quality control is very good. Inspection hatches, access openings etc. can be easily provided either at the time of installation or at any other convenient time. The track record for the use of aluminium covers is rather impressive. There have been many cases of its use at sewage treatment facilities worldwide, and even after 25 to 50 years of usage, aluminium covers have been known to show no significant signs of corrosion.

15.2.1.5 Fibre reinforced plastic

Fibre reinforced plastic is another very popular material used as cover material at sewage treatment plants. It has a high strength:weight ratio and its low weight makes it suitable as both low and high level covers. FRP typically manufactured with selected resins and with the appropriate resin choice, the material is very economical to use and can be shaped to suit a variety of cover shapes and sizes.

There are many FRP fabricators in the world and reasonably high quality control can be expected as FRP covers can be factory manufactured. There are of course instances of FRP deterioration, particularly for the older works where many FRP covers had been incorrectly designed and inappropriate resin for UV protection has been used. Sufficient technical knowledge is now available to ensure that FRP covers are properly detailed during design and that premium-grade chemical resistant resins are used. Free-spanning FRP domes, exceeding 50 m diameter, have been constructed at sewage works.

FRP is produced by reinforcing a relatively low strength resin matrix with a very high strength, high modulus glass fibre. The tensile strength of the glass fibre is in the region of about 2400 MPa which is nearly 10 times that of steel. Manufacturers provide different amounts of glass fibres positioned in various directions such that different strength and elasticity can be obtained. Hence, different laminates will result in different strength and elasticity and the directional positioning of the laminates causes directional strength to be obtained.

Un-idirectional e.g. roving, uni-directional cloth laminate, provides very high strength and modulus values in one direction with relatively low values in the other direction. Glass content in the laminate is in the range of 60% to 80% by weight. Bi-directional e.g. woven rovings and cloth, provides equal strength in two directions and glass content is about 40% to 60% by weight. A random e.g. chopped strand mat will provide equal strength in all directions and glass content is in the range of 25% to 40% by weight.

Resins such as isophthalic polyester and vinyl ester are suitable for FRP that requires corrosion resistant properties. For an aggressive environment, vinyl ester resin, though more expensive than isophthalic polyester, should be used as it has excellent corrosion resistance as well as inherent toughness. The cheaper isophthalic polyester is ideal for mildly corrosive conditions. To protect against UV deterioration, resins will require the addition of ultra-violet absorbers and light stabilizers. Without the presence of appropriate UV inhibitive additives, FRP covers will break down over time and structural restoration work would be required. Proper UV protected FRP covers can be expected to last for some 20–25 years.

The most common configuration of FRP covers is the dome-shaped cover and typically, pie-shaped roof segments are designed to span on to a central compression ring. Figure 15.2 gives an example of a segmented FRP roof cover and Figure 15.3 shows FRP segments over a tank.

FRP has been used in various configurations as low and high level covers. For aeration basins, it is common to find flat covers made out of modular segments spanning on to FRP supporting beams. Ideally, the flat covers are slightly curved to improve stiffness and undersides of the covers are likely to have stiffening sections. A ribbed profile is also often designed for FRP covers so that each segment can span from the centre of the tank to the edge wall.

Figure 15.2. An example of segmented FRP roof cover.

Figure 15.3. An example of FRP segments over individual tanks.

15.2.2 Costs

The covers should be cost effective based on the required life of the covers. The total cost of the covers is affected by the following factors:

(1) the cost of the raw material;
(2) local availability of material;
(3) ease of transportation;
(4) ease of installation;
(5) wastage during installation.

For some materials there may be costs associated with maintenance of the covers if corrosion protection coatings need to be replaced at a regular intervals. The relative costs of various covers are summarised in Table 15.2.

Table 15.2. A comparison of the cost of various materials.

Materials	Wood	Concrete	Fabric	Aluminium	FRP
Relative Cost	Low	Low	Moderate	High	High

15.2.3 Aesthetics

The covers installed at wastewater treatment plants will be increasingly visible in the future due to the reduction in the buffer zone and high rise development closer to the boundary of plant sites. It is therefore important that the covers blend in well with the surroundings, with the existing works facilities and with any future extension works. Aesthetics is influenced by:

(1) The size and shape of the cover systems influencing aesthetic style and form.
(2) The material from which the covers are made influencing visual texture and reflectivity.
(3) The colour of the covers in relation to surrounding natural colours and the colour of other man-made structures.

15.3 COVER CONFIGURATION

15.3.1 Types

A variety of cover configurations are available. These can range from low-level covers which are designed to be placed immediately above the liquid surface (with minimal headspace) to high-level covers, which typically form a secondary enclosure over the low-level primary covers. It is also possible to have dual high/low level cover configuration.

15.3.1.1 Low-level covers

Low-level covers typically consist of a deck which spans between tank walls and is placed just above the liquid surface to minimise the volume of odorous headspace between the liquid and the cover. Access for maintenance is provided through hatches or removable cover sections. Such covers are favoured whenever there is a need to minimise the volume of odorous air that is required to be contained and treated, although the concentration of emitted volatile within the headspace will be higher for the same mass rate of gaseous emissions. Some disadvantages of low-level covers do exist. For example, visual inspection of the process unit is difficult and collection of sample for routine testing is often hampered. For some wastewater treatment processes e.g. aeration units, the visual observation of liquid appearance, presence of foams etc. provide the plant operator with an indication of process performance. Observation hatches do not necessarily provide the same degree of visual appreciation of the liquid surface. Maintenance and cleaning of covered tanks is also difficult. Safety precautions for entering enclosed tanks are necessary and often, tank maintenance is ignored when procedures are cumbersome and troublesome. Figure 15.4 shows an example of a low-level cover.

15.3.1.2 High-level covers

High-level covers are usually designed to span over liquid surfaces such that there is sufficient headspace to allow entry of personnel or maintenance equipment. In several cases, the cover comprises a fixed dome that spans from fixed low walls built independently of the existing tank structure. For example, existing travelling bridges and machinery at a settling tank need not be altered but can be accommodated within the tank structure. Figure 15.5 shows an example of a high-level cover built to enclose an existing primary settling tank.

Figure 15.4. An example of low-level covered tanks.

Figure 15.5. An example of high-level covered tank.

Access to high-level covered tanks for normal inspection or maintenance will be via doors built at the side wall of the covered tank. In the event that large items of machinery need to be moved in and out of the tank, such access can be done by removing sections of the cover (the design can be specified to allow for this).

To prevent fugitive emissions during the opening of doors, covered tanks are designed with a slight negative pressure so that air flow direction will be inwards.

15.3.1.2 Dual-low and high-level covers

In many instances, some major odorous processes such as those at a treatment plant's headworks or the primary settling tanks, are designed to be covered by both low- and high-level covers. The low-level cover, which is placed just above the process liquid surface, acts as the primary containment enclosure while the high-level cover acts as an additional enclosure to contain any obnoxious air that happens to leak out from the primary cover e.g. during maintenance when access hatches are opened. Such a dual system is costlier but will provide the highest security against fugitive emissions. Vent air for the low-level cover is drawn from the space between the two covers while fresh air from the outside is vented into the space via operable dampers around the perimeter. Fresh air flow rate is adjusted to ensure adequate ventilation and a net inflow of fresh air. A typical configuration is shown in Figure 15.6.

Figure 15.6. Dual low- and high-level covers.

It is possible for the low-level cover to be designed to rotate with travelling bridges or to be fixed. If the cover is rotating, then the framework needs to be supported on wheels on top of the process tank wall. Fixed covers, however, allow for better flexibility in accessing the covered process units and ductwork will be less obtrusive.

15.3.2 Application

The sewage works has multiple process tanks, channels, chambers and building structures, which require enclosures to contain and allow collection of odorous gases. The enclosures are discussed based on the following four broad categories.

15.3.2.1 Building enclosures

To contain and control odorous gases, openings on the exterior of these buildings must be eliminated, such that the interior space is placed under a slight negative pressure, while maintaining adequate ventilation. The materials selected for closing up the buildings for odour control are those most appropriate given the aesthetics, operational requirement and the nature of the surrounding cladding.

15.3.2.2 Channels, chambers and tanks

Low-level covers generally are used although small enclosures over things such as the stairwell leading up from the pump room in the inlet works will be more appropriate. Aluminium is preferred for such covers, which are subjected to pedestrian traffic due to its robustness.

15.3.2.3 Circular settlement tanks

High-level cover is in the form of a fixed dome spanning from new shallow founded walls built independently of the existing tank structure. This is the simplest alternative for covering these tanks as the existing travelling bridges and drives will not need to be altered. Low-level covers can either span between the existing tank wall and the tank centre, or they can be suspended from the nodes of a supporting framework. The system has the lowest capital cost and is most suitable when the need for maintenance access is infrequent or when odour production rates are low so that the risk of significant fugitive odour emissions is minimised.

The dual high/low-level covers, which are based on the provision of a high level dome together with a low-level cover just above the liquid surface are usually preferred even though the system has the highest capital cost. The system provides the greatest certainty that there will be no fugitive emissions. Fabric, aluminium and FRP are suitable for high-level covers and only aluminium and FRP are suitable for use as low-level covers.

15.3.2.4 Rectangular aeration tanks

Both high- and low-level covers are suitable for covering the aeration tanks. Low level covers were cheaper than high-level covers based on both capital and operating costs of the odour control system. Aluminium is the most robust but aluminium covers are susceptible to corrosion when splashed by sewage or sewage spray. FRP covers are safe and are easier to produce in a profiled appearance than aluminium.

15.4 CRITERIA FOR SELECTION

15.4.1 Evaluation criteria

Selection of material for covers typically requires the consideration of the following factors:

(1) strength and stiffness;
(2) durability;
(3) weight;
(4) cost;
(5) aesthetics;
(6) quality control;
(7) operational requirements.

15.4.1.1 Strength and stiffness

Covers must obviously have sufficient strength to span freely over process tanks. For example, over primary and secondary settling tanks, covers are designed to span over the length or diameter of the tank without any intermediate support. Support to tank covers are potential corrosion spots and there could be obstruction to air and liquid flow within the covered process unit. Covers must also have sufficient stiffness so that deflection of the member is kept within tolerable and acceptable limits.

15.4.1.2 *Durability*

Considering the presence of corrosive gases within covered tanks, it is necessary for covers to be durable, both physically and chemically. During the construction and life time of the cover, it must be able to withstand physical impacts, pedestrian traffic, wind forces as well as self weight and maintenance loads. Covers must be resistant to corrosion both from external forces such as weathering and UV radiation, as well as internal attack on the underside due to sewage gases.

Hydrogen sulphide, mercaptans and organic volatiles are present in sewage air. These gases easily react within enclosed space to form corrosive substances such as sulphuric or organic acids. Cover materials must hence be durable chemically against attack by these chemicals.

Typically, processes at the headworks of a sewage plant e.g. inlet chambers, grit channels and primary sedimentation tanks, emit corrosive sulphuric acid due largely to the presence of gaseous hydrogen sulphide, while, downstream processes such as sludge processing units, tend to emit volatile fatty acids. Hence, appropriate cover material selection must consider the types of corrosive gases present in the different unit operations.

15.4.1.3 *Weight*

It is desirable to select materials that have high strength:weight ratios. Cover materials that are light in weight will not place unnecessary overburden on supporting structures. For covering existing structures, it is vital to ensure that cover materials are not too heavy to result in structural deformity or cause eventual collapse of the existing support system.

15.4.1.4 *Cost*

Covers must be cost effective for the design life of the covers. Designers must be aware of the local availability of certain types of cover and of the assembly cost of covers.

15.4.1.5 *Aesthetics*

Covers are highly visible and hence should be designed with an appropriate configuration and material such that the covers will blend nicely with the surrounding buildings. High-level roof covers of settling tanks have a significant visual impact and it is desirable to ensure that their shapes and colour are aesthetically appealing. Figures 15.7–15.9 show various forms of covered structures that have been designed to be aesthetically pleasant.

Figure 15.7. An example of a curved shaped covers over aeration basins.

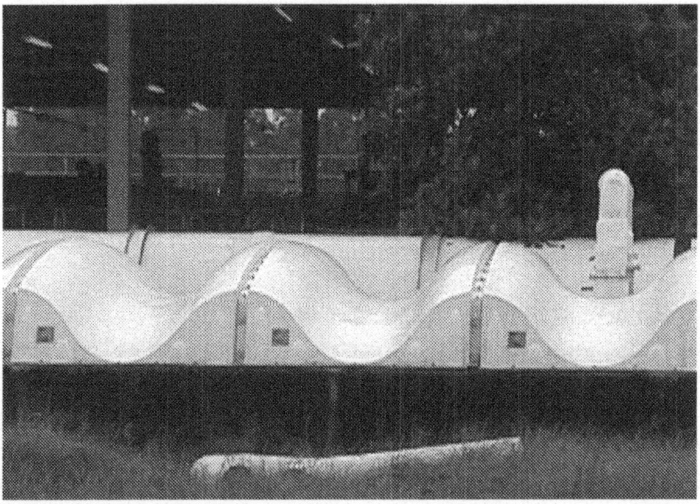

Figure 15.8. An example of curved-shaped covers over a channel.

Figure 15.9. An example of a nicely designed aluminium covers over primary settling tank.

15.4.1.3 Quality control

Good quality control particularly in the manufacturing process is important to ensure that cover shapes and sizes are fabricated to fit properly over the intended process. A track record by the manufacturer is hence essential. In addition, covers that are easily fitted up on site will allow for better quality control during the construction process thus resulting in a better fitted cover.

15.4.1.4 Operational requirements

Covers must be able to contain the sewage gases underneath them. In general, a slight negative pressure should be present within the enclosed headspace so that air flow will be from the direction outside the cover into the covered space. Designers typically, allow slight gaps between covers and walls or between adjacent cover segments to facilitate such air flow.

The covers must also be suitable for inspection and access hatches to be constructed for maintenance purposes. Alternatively, cover sections are designed to be removable so that repair and maintenance can be conveniently carried out.

15.4.2 Applicability

Some sewage treatment processes such as sludge dewatering units (belt filters, centrifuges etc.) and dissolved air flotation units, are often housed within building structures. Cover materials selected to enclose up such buildings for odour control must be appropriately selected taking into consideration aesthetics and operational requirements. Enclosures must blend in to the surrounding building and if operational requirements call for natural lighting, then glazing can be used. Sometimes, concrete is used as lower walls while upper walls and roof segments may use lightweight material with appropriate glazing.

15.5 BIBLIOGRAPHY

Kissell, J.R. and Ferry, R.L. (1995) *Aluminium Structures: A Guide to Their Specifications and Designs*. John Wiley & Sons, Inc., New York.

Sharp, M.L. (1993) *Behavior and Design of Aluminium Structures*. McGraw-Hill, Inc., New York .

American Society of Testing Materials. (1987) Degradation of Metals in the Atmosphere. PCN 04-965000-27, American Society for Testing and Materials, Pennsylvania.

United States Environmental Protection Agency. *Odor and Corrosion Control in Sanitary Sewerage Systems and Treatment Plants*, EPA/625/1-85/018.

Barret, A.E. (1989) Geodesic-dome tank roof cuts water contamination, vapor losses. *Oil and Gas J.* (July).

Bray, W.H. (1999) Putting the lid on odors and VOCs. *Environmental Protection* (July).

16

Chemical odour scrubbing systems

Tom Card

16.1 INTRODUCTION

This chapter addresses the technologies that are available for the chemical control of odours in ventilation air from wastewater treatment processes or other odour sources.

16.1.1 Overview of available systems

Chemical scrubbing technologies rely on the intimate contact of a scrubbing chemical with the odorous gas stream. The two leading technologies are packed tower systems and atomised mist systems.

16.1.1.1 Packed towers

Packed towers remove odorous compounds from wastewater treatment plant ventilation air by providing an opportunity for the compounds to absorb into a liquid solution. Classically, sorption is the primary removal mechanism for

 Odours in Wastewater Treatment: Measurement, Modelling and Control edited by Richard Stuetz and Franz-Bernd Frechen. ISBN: 1 900222 46 9

packed tower scrubber. However, oxidation in liquid solution can greatly enhance the adsorption and therefore enhance removal. In addition, there may be some instances where gas phase oxidation may be a significant removal mechanism.

The most common configuration for packed towers is a vertical shell with gas flow going up through packing and the liquid scrubbing solution going down through the packing. The gas and liquid pass over packing material to promote a large interfacial area. Liquid solution is usually circulated over the packing by pumping from a collection sump in the bottom of the tower. Chemicals are added to the scrubbing solution either in the sump or in the recirculation piping. For best performance, a portion of scrubbing solution is continually wasted to remove the accumulated contaminants from the liquid solution. Figure 16.1 presents a schematic of a typical vertical counter-current packed tower system.

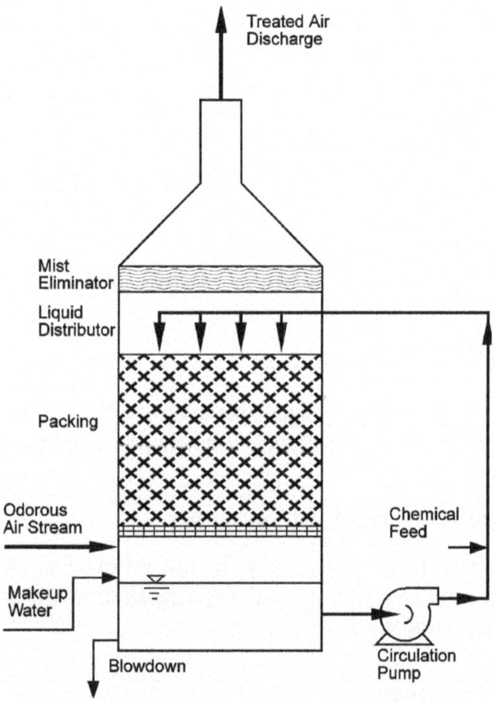

Figure 16.1. Vertical counter-current packed tower schematic.

The other most prevalent configuration is a cross-flow system. In this system, gas flows horizontally through a packed bed and the liquid scrubbing solution is sprayed over the packing media. Figure 16.2 presents a schematic of a typical cross-flow scrubber system.

Packed towers do an excellent job of removing hydrogen sulphide, but can have problems with some organic compounds that are not water soluble. They are generally the lowest cost chemical scrubbing technology for applications with over 14 m^3/s (30,000 $ft^3/min.$) air flow rate. They are also the most easy to operate of the liquid scrubbing systems. The scrubbing solution is usually sodium hydroxide and sodium hypochlorite, although many solution chemistries can be used. There is some concern about chlorine emissions from large systems that use chlorine based scrubbing solution. Packed towers can either be purchased as packaged systems from system vendors or purchased as individual components. The packaged systems normally are the lowest cost systems. Operating costs average between \$US 2.00 and \$US 8.00 per kg of sulphide removed. Figure 16.3 shows a typical large installation at a wastewater treatment facility.

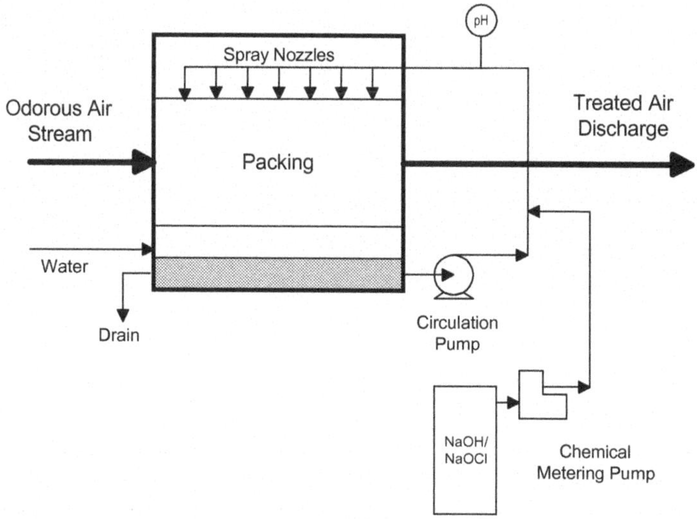

Figure 16.2. Schematic of cross-flow packed scrubbing system.

A specialised packed tower system is a catalytic air oxidation system. These are packed towers that contain a proprietary liquid catalyst that will oxidise the sulphide to sulphur. The sulphur is a by-product that must be removed and disposed appropriately. They cost about \$US 1.32 per kg (\$US 0.60 per pound) of sulphide removed, but have high capital costs. They start to become cost effective at sulphide concentrations above about 100 ppmv (parts per million by volume or mole). These systems only remove hydrogen sulphide. If there are other odorous compounds present, they must be removed by another scrubbing technology.

Figure 16.3. Example large packed tower installation at a wastewater treatment facility.

16.1.1.2 Mist chamber systems

The other leading technologies for chemical scrubbing systems are mist systems. Historically these have been low cost systems using relatively course sprays in a fibreglass or plastic box. In about 1975 these systems were upgraded using high performance atomising nozzles that use compressed air to produce droplets on the order of 10 microns in diameter. Figure 16.4 shows an air flow schematic of a typical atomised mist system.

Atomised mist systems have a larger interfacial area than packed towers, but have less liquid solution to remove absorbed pollutants. They cost the about same as packed towers for applications for air flows up to 14 m^3/s (30,000 $ft^3/min.$) and 25 ppmv of hydrogen sulphide. Above these levels packed towers are normally cheaper. They have been shown to remove some air toxic and hydrocarbon compounds in addition to odorous compounds. However, with a few exceptions, they can only operate using chlorine based solutions and there can be some concern about the chlorine emissions from large systems. They are more complicated than packed towers and require more maintenance. There are very few vendors for atomised mist systems. There are no practical theoretical methods to design these systems, so they are usually sized and configured based on past experience. Atomised mist systems will normally occupy more space than packed towers. The operating cost is between $US 5.00 and $US 8.00 per kg of sulphide removed. Figure 16.5 shows a typical atomised mist system contact tower.

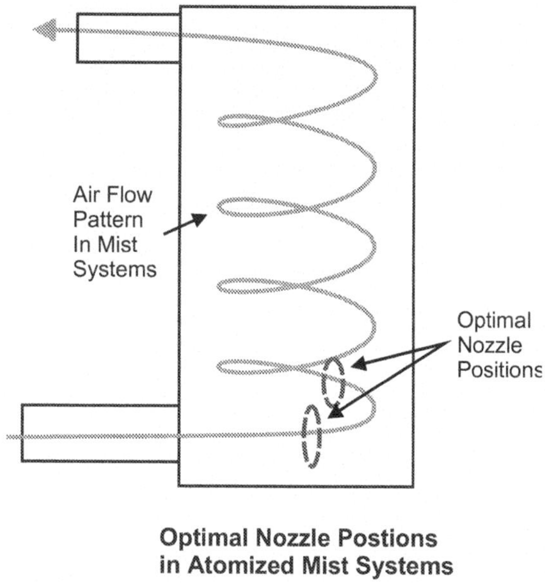

Figure 16.4. Atomised mist system air flow schematic showing optimal chemical solution spray nozzle locations.

16.2 CHEMISTRY OF WASTEWATER TREATMENT ODOURS

This section addresses the types of odorous compounds that can occur in wastewater treatment ventilation streams and how the chemistry of those compounds can affect the design of chemical scrubbing systems.

Figure 16.5. Typical atomised mist scrubbing system.

16.2.1 Hydrogen sulphide (H_2S)

Hydrogen sulphide is the most prevalent, but by no means the only, compound of concern for wastewater treatment odours. It is most often associated with the ventilation of air from contact with raw sewage, plant headworks, primary clarifiers, and digester gas.

16.2.1.1 Volatilisation

Hydrogen sulphide follows Henry's Law at low gas phase concentrations.

$$X_G = X_L H_C \tag{16.1}$$

Where:
X_G = the gas phase concentration of hydrogen sulphide,
X_L = the liquid phase concentration of hydrogen sulphide,
H_C = the Henry's Law Coefficient.

Table 16.1 presents the Henry's Law Coefficients at various temperatures for hydrogen sulphide and ammonia. The use of Henry's Law Coefficients has to be completed with some care for both hydrogen sulphide and ammonia because the X_L value only represents the un-ionised fraction and will be a function of the solution pH.

Table 16.1. Henry's Law Coefficients (atm per liquid phase mole fraction) for hydrogen sulphide (Tchobanoglous and Burton 1991)) and ammonia (Kohl and Riesenfeld 1979).

Temperature (°C)	Hydrogen sulphide	Ammonia
10	367	
20	483	0.738
30	609	
40	745	1.851
60	1,030	4.30

16.2.1.2 Ionisation

Hydrogen sulphide will ionise when dissolved in water. The reactions are:

$$H_2S \rightarrow H^+ + HS^- \quad (16.2)$$
$$HS^- \rightarrow H^+ + S^= \quad (16.3)$$

Figure 16.6 shows the distribution of these ionic species as a function of solution pH. This graph shows that at pH values below 7 most of the total sulphide in solution is in the un-ionized form. This is the form that can volatilise from solution. At pH values above 7 most of the sulphide is in the ionic form that is not volatile.

16.2.1.3 Chemical oxidation

As mentioned in the preceding paragraphs, many chemicals can chemically oxidise sulphide. Oxygen can also oxidise sulphide, but it is a very slow reaction unless a catalyst is present (chelated iron is very powerful catalyst for the oxidation of sulphide). Chlorine, hydrogen peroxide, and potassium permanganate can rapidly chemically oxidise sulphide. The chlorine can be in gaseous form, as hypochlorite, chlorite, or even chlorine dioxide.

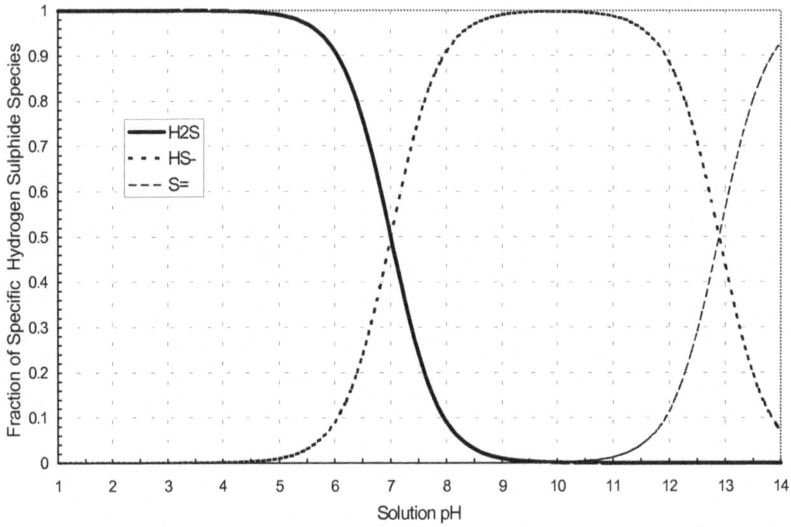

Figure 16.6. Distribution of hydrogen sulphide ionic species as a function of solution pH.

The chemical oxidation reaction of sulphide using oxygen is

$$2S^{=} + 2O_2 \rightarrow SO_4^{=} + S^o \quad (16.4)$$

The chemical reaction with chlorine (White 1992) is either

$$H_2S + Cl_2 \rightarrow 2HCl + S^o \quad (16.5)$$

or

$$H_2S + 4Cl_2 + 4H_2O \rightarrow 8HCl + H_2SO_4 \quad (16.6)$$

The chemical reaction with sodium hypochlorite is either

$$H_2S + NaOCl \rightarrow NaCl + S^o + H_2O \quad (16.7)$$

or

$$H_2S + 4NaOCl \rightarrow 4NaCl + H_2SO_4 \quad (16.8)$$

The chemical reaction with hydrogen peroxide is

$$H_2S + H_2O_2 \rightarrow S^o + 2H_2O \quad (16.9)$$

or

$$S^{=} + 4H_2O_2 \rightarrow SO_4^{=} + 4H_2O \quad (16.10)$$

It is generally believed that at lower oxidant concentrations and at lower pHs, the reaction will tend to only go to elemental sulphur, thus saving considerable chemical expense. This is, in general, only true for the peroxide reaction and not for the chlorine reaction.

For the chlorine reaction that oxidises sulphide to sulphate the stoichiometric requirement is 8.9 kg of chlorine per kg of H_2S. In most cases, only a stoichiometric amount is required. Chlorine costs between $US 200 and $US 400 per 1,000 kg depending on quantity and location. Sodium hypochlorite costs between $US 400 and $US 800 per 1,000 kg.

Hydrogen peroxide will react with sulphides at a stoichiometric rate of 1 to 4 pounds peroxide per pound sulphide, depending on whether the oxidation reaction is to elemental sulphur or to sulphate. Hydrogen peroxide will cost between $US 500 and $US 2,000 per 1,000 kg.

16.2.2 Ammonia (NH_3)

Ammonia odours are most commonly associated the biosolids handling processes, particularly anaerobically digested sludge or almost any biosolids that undergo a dramatic pH increase as part of stabilisation process. Ammonia chemistry is very similar to H_2S except for the exceptions as noted below.

Ammonia ionises when it dissolves in water according the following relationship:

$$NH_3 + H_2O \rightarrow NH_4^+ + OH^- \qquad (16.11)$$

Note that the behaviour of ammonia as a function of solution pH is the exact opposite of hydrogen sulphide. Figure 16.7 shows that in low pH (acidic) solutions most of the ammonia is in the non-volatile ionic form. Ammonia also oxidises with chlorine (and other oxidants, but to a lesser extent, as well).

The reaction with chlorine is (White, 1992):

$$NH_3 + 3HOCl \rightarrow NCl_3 + 3H_2O \qquad (16.12)$$

Therefore one mole of ammonia will consume three moles of chlorine if the reaction goes all the way to nitrogen trichloride (trichloroamine).

16.2.3 Organic odours

The usual organic odours of concern are organic reduced sulphur compounds, such as methyl mercaptan, dimethyl disulphide, carbonyl sulphide; and organic nitrogen compounds like amines, indole, and skatole. Many of these compounds are both resistant to oxidation and not very water soluble, both of which provide challenges for packed towers. Most of the time chlorine-based systems are required to remove these types of compounds. In addition, many of the reactions with organic sulphur compounds are both reversible and produce odorous intermediate products.

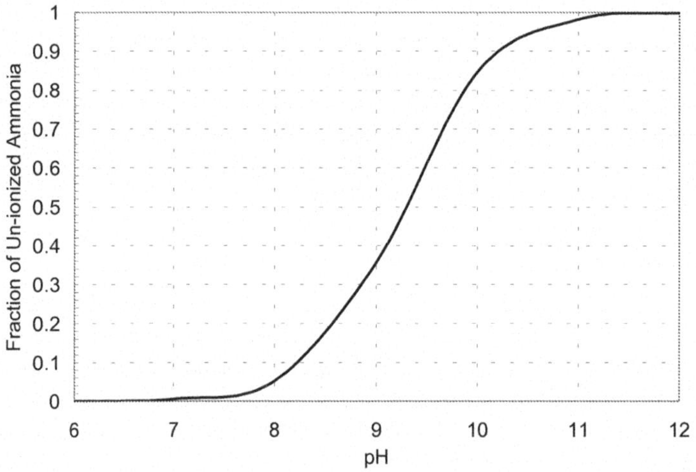

Figure 16.7. Fraction of unionised liquid phase ammonia as a function of solution pH.

16.3 DESIGN OF PACKED TOWER SCRUBBERS

Packed towers are one of the primary types of equipment used to control odours at wastewater treatment facilities. They are very flexible and can be configured to reliably and efficiently remove most common odour causing compounds at wastewater treatment facilities.

16.3.1 Configuration and chemical selection

Typical situations that packed towers are used in are provided with some discussion below. Chemical selection is critical to operating both a functional and economic system. Figure 16.8 shows a graph of the cost of a system operating on caustic only and one operating on caustic with an oxidant for various inlet H_2S concentrations. The particular economics of this comparison are site specific, but the general trend is universal.

16.3.1.1 Using sodium hydroxide (caustic) only

This is probably the first and one of the most common applications for packed towers used for odour control. Previous work has shown that it is most economical to remove high hydrogen sulphide concentrations using caustic alone with a high blowdown (scrubber solution wasting) rate. Under normal circumstances, using caustic only becomes the most economical at inlet hydrogen sulphide concentrations over 25 to 100 ppmv. The exact economic threshold is a function of chemical costs, makeup water costs, blowdown disposal cost, and inlet carbon dioxide concentration. In this configuration about 10% of the inlet carbon dioxide is removed, which can be a major economic burden. It is almost never economical to remove H_2S with caustic alone if the concentration of H_2S is below 25 ppmv. In addition, this configuration can only remove between 90% and 95% of the H_2S, and it only removes H_2S, no other odorous compounds. This results in the requirement for a second stage for most situations. Although this is still a popular technology for high concentration H_2S control, iron catalytic systems and autotrophic biofilters are significantly lower cost at H_2S concentrations above 100 ppmv.

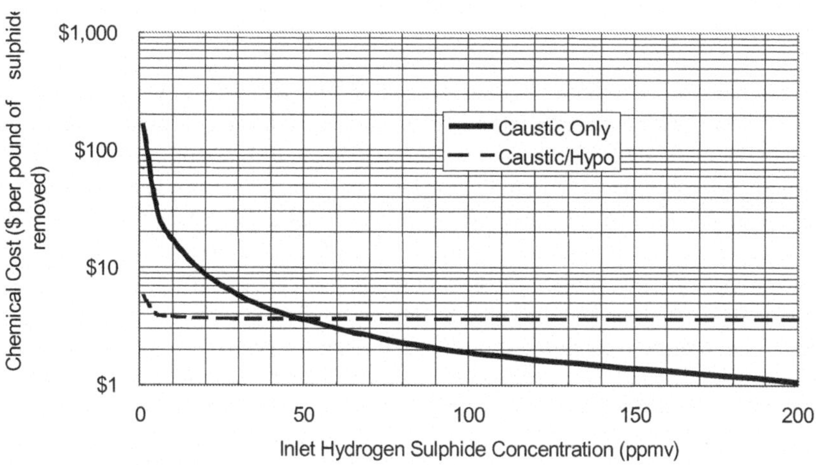

Figure 16.8. Example operating cost (US$) as a function of inlet sulphide level for both caustic only and caustic/hypo systems.

16.3.1.2 Using sulphuric acid only

Where the ammonia concentration is over 5 to 50 ppmv it is often more economical to remove ammonia in a packed tower that has a low pH (usually 3) sulphuric acid solution circulating in it. Oxidant scrubbers, particularly scrubbers using sodium hypochlorite can effectively remove ammonia, but at a very high cost. In addition, high ammonia concentrations can reduce the performance of hypochlorite scrubbers on other compounds of concern. It is thought that this may occur because the ammonia reaction is both very fast and consumes substantial quantities of chlorine resulting in localised low oxidation-reduction potential areas in the scrubber.

A variant of this technology is the Ammonia Removal and Recovery Process (ARRP) developed by the North American engineering consulting firm CH2M HILL in the mid 1970s. This process utilised a packed tower ammonia stripper that was coupled to a packed tower ammonia adsorber. The adsorber was operated so that it produced a ammonia salt that was concentrated enough to sell as fertiliser. Stand alone ammonia odour scrubbing systems can be operated to produce high concentration ammonia salt blowdown that would be suitable for recovery as fertiliser.

16.3.1.3 Using an oxidant with pH control

Probably the most common configuration of packed tower used to control odours from wastewater treatment plants is using sodium hypochlorite (bleach) with or without the addition of caustic for pH control. Using bleach alone is normally adequate when the inlet odours are 10 ppmv of H_2S or less. Above that level, the pH drop in the tower will produce substantial chlorine odour. The addition of caustic to maintain a pH of 8 to 9 will dramatically reduce the chlorine odour. The amount caustic required to accomplish this is usually around 10% of the bleach flow. At sustained H_2S concentrations above 25 ppmv, alternative technologies or additional stages to reduce the inlet concentration should be considered. For very hard to oxidise odours, sodium hypochlorite with acid has been used as a first stage process. This produces gas phase chlorine that will oxidise non-water soluble compounds. Caution is urged when implementing this type of system due to severe corrosion and safety concerns. This type of system is almost always followed by a high pH system to remove the chlorine and any residual odours.

Hydrogen peroxide is also widely used in packed towers. With peroxide, caustic is almost always used to increase the pH to around 9. Peroxide has a very low volatility and will only oxidise compounds in the liquid phase. This configuration has had performance problems with organic odours.

For severe organic odours that are not water soluble, chlorine dioxide has been used as an oxidant. Chlorine dioxide is very volatile and provides ample opportunity for gas phase reaction. However, it is necessary to assure that unreacted chlorine dioxide is not released from the system.

16.3.2 Novel packed tower systems

16.3.2.1 Iron catalyst

At least two vendors supply packed tower systems that utilise an iron catalyst solution that oxidises the sulphide to elemental sulphur using either oxygen in the gas stream or an oxygen sidestream (for digester gas applications). These systems use fluidised plastic ball media because of plugging concerns. The elemental sulphur is recovered from the solution as a liquid slurry. The sulphur can be recovered or disposed of. This type of system will typically have operating costs less than $US 2.00 per kg of sulphur removed for systems that have over 100 ppmv of gas phase sulphur. This system has been used at large scale for wastewater treatment odour control in Hawaii.

16.3.2.2 Liquid phase hypochlorite catalyst

A catalyst was developed in Great Britain that dramatically increases the oxidative power of hypochlorite in the liquid phase. In this system, the scrubbing solution is passed through a catalyst bed as it is recirculated through the tower. This catalyst will increase the removal of organic sulphur compounds and some VOCs.

16.3.2.3 Dissolved ozone

Some work has been done using ozone in combination with packed towers. This has been accomplished by scrubbing a combined odorous and ozone laden stream in a packed tower that circulates water. The thought is that the packed tower will given the ozone time to react in solution. Since ozone is not very soluble, this may have limited applicability.

16.3.2.4 Ultraviolet light enhanced

A large ultraviolet light enhanced packed tower system is being installed in Stockton, CA (Kundidzora and Reichenberger 1999). This system uses UV light in two separate systems. The first pre-conditions the odorous gas and the second is used to promote oxidation in the recirculating scrubber solution. This system is not operational at the time of this writing. However, it has been successfully pilot tested.

16.3.3 Example applications

For complex odour streams, packed towers can be sequenced with different chemical compositions for more comprehensive odour removal. As an example, it is common to use multiple stages to control complex odours. During design of packed tower systems it is important to characterize the odour thoroughly in order to design the system appropriately.

16.3.3.1 Hydrogen sulphide with ammonia

When ammonia is present in significant quantities (10–100 ppmv) it should be removed first by an acid scrubber stage. This configuration will be much cheaper to operate (it costs about $US 0.65 per kg of ammonia removed in an acid stage and about $US 6.50 per kg in a hypochlorite system) and the second stage (caustic) will work much more effectively if the ammonia is removed.

16.3.3.2 Hydrogen sulphide with organic reduced sulphur compounds

If significant organic reduced sulphur exists (over 100 ppbv) then it is advisable to have a first stage with lower pH (7 to 8) with hypochlorite only. This will dramatically increase the removal of organic reduced sulphur compounds.

16.3.4 Types of packed powers

The most efficient type of packed tower is the vertical counter-current configuration. Packed towers can also be configured in cross-flow or co-current. These types of configurations are less efficient, but the efficiency loss may be offset by the profile difference, which may allow these units to fit into some process configurations more efficiently.

16.3.4.1 Compact multi-stage packed tower systems

Currently, compact multi-stage systems are very popular. They provide flexibility and streamlined installation. Normally these systems are three stages and have all components pre-installed on the compact unit. These systems can be installed at a prepared site in a very short time (sometimes less than two weeks). Figure 16.9 shows a schematic of an example system. These systems can have identical chemistry in each stage or the chemistry can be adjusted in each stage for optimal removal or lowest cost.

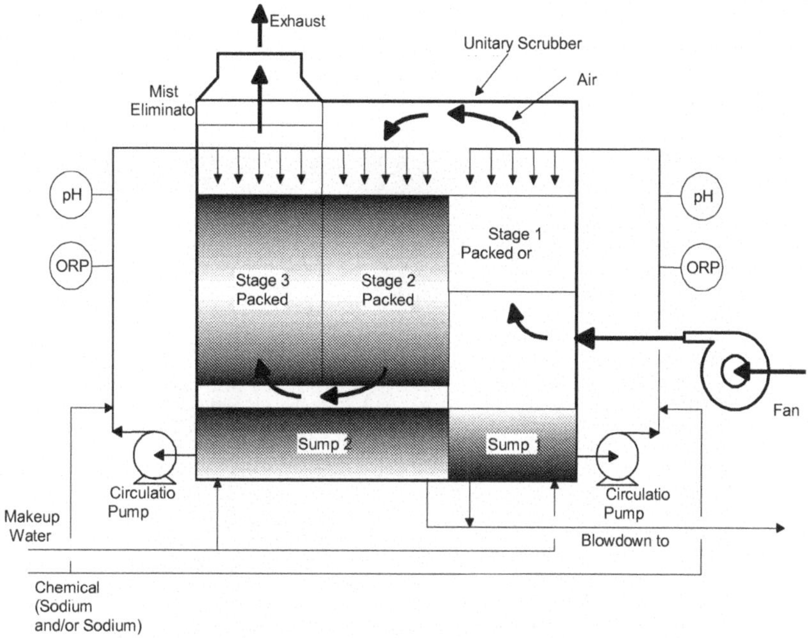

Figure 16.9. Compact packed tower system schematic.

16.3.5 Packed tower components

16.3.5.1 Shell

The shell is constructed of either steel, polyvinyl chloride (PVC), or fibreglass (FRP). In wastewater odour control, the shell is almost always FRP and occasionally PVC. Most systems today use a fibreglass laminate based on the use of a vinyl ester resin. This type of system is chemically resistant to all chemicals used in wastewater odour control. Chlorine or sodium hypochlorite are the most chemically severe service and require the use of a vinyl ester resin system.

16.3.5.2 Packing

There are several packing manufacturers that make packing that is appropriate for wastewater odour control. Some of the major packing vendors include Jaeger, LanTec, Norton, Glitsch, and Ceilcote. Most odour control applications do not need extremely high performance packing. Where space constraints are severe, structured packing can be used that is about 30% less volume than random packing. Figure 16.10 shows some of typical packing shapes used for wastewater odour control.

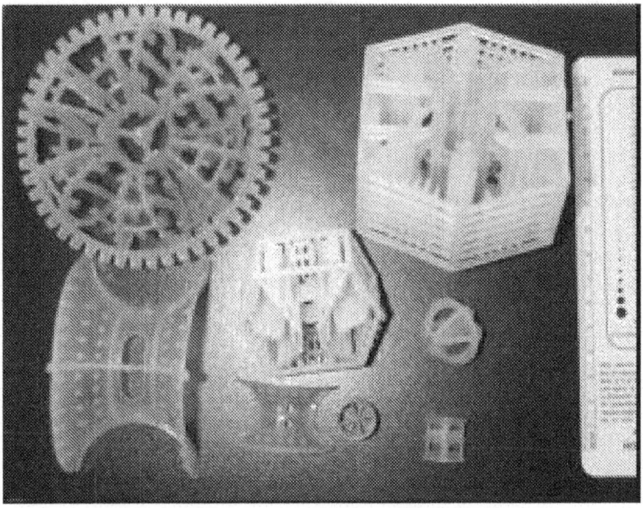

Figure 16.10. Typical packing used in packed tower systems.

16.3.5.3 Tower internals

The packing is usually supported on fibreglass grating. It is usually a good idea to select a grating that has an opening area of 90% of the total area. For high performance applications gas injection plates can be used. These are quite expensive but can provide more than 100% open area by using large waffle-type corrugations.

Bed limiters are only required for high gas velocities (greater than 2.5 m/s) or when there is a requirement for maintenance personal to walk on top of the packing. These are almost always FRP grating.

Most odour control system use nozzles for liquid distribution. For large systems, multiple nozzles can be used. Gravity distribution systems provide better distribution and are cheaper to operate (lower head loss) but are much more expensive than nozzles. Figures 16.11–16.13 show the various types of gravity distribution systems.

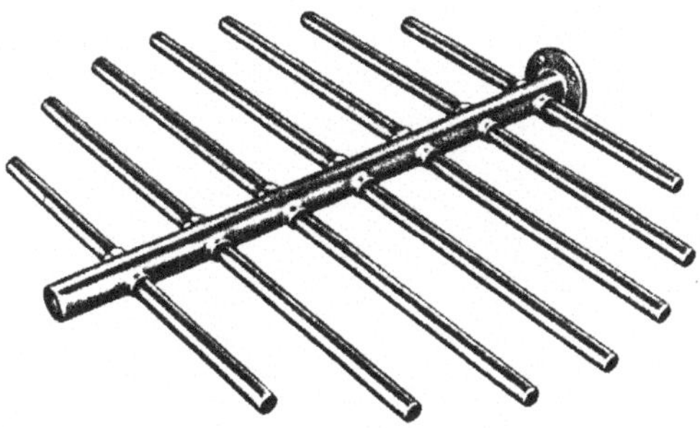

Figure 16.11. Ladder distributor (Courtesy of Norton Products, USA).

Figure 16.12. Weir-trough distributor (Courtesy of Norton Products, USA).

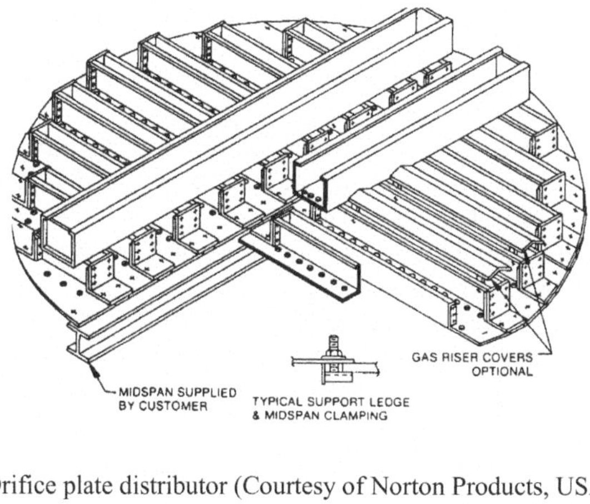

Figure 16.13. Orifice plate distributor (Courtesy of Norton Products, USA).

16.3.5.4 Mist elimination

Mist elimination is almost always necessary. Mist can be reduced by using any of the three common types of mist eliminators:

- Mesh pad;
- Small random packing;
- Chevron.

Mist eliminators are often a maintenance problem. Mesh pads have the highest performance, but are also the most maintenance intensive.

16.3.5.5 Pumps

With the exception of once through systems, the scrubbing liquid must be circulated over the packing using a pumping system. Most systems use American National Standards Institute (ANSI) B73.1 chemical process pumps made out of either high-performance stainless steel (Alloy 20) or fibre-reinforced plastic (FRP). A less expensive approach is to use all plastic pumps, but they do not have the same service life as the chemical process pumps. One of the main problems with pumps is the seal. The caustic/hypochlorite solution creates a severe service environment for a seal. In order to reduce maintenance effort, and to increase seal life, it is better to operate all pumps on line and not leave pumps on standby.

An alternative approach that resolves the seal issue is to configure the systems such that seal-less vertical pumps can be used to circulate the scrubbing solution.

16.3.5.6 Water make-up

All systems need to have the scrubbing solution removed (blowdown) so that the scrubbed contaminants are removed from the system. This is always done by adding water to the scrubbing solution and removing an equal amount of the circulating liquid. This is better if done continuously rather than in a batch mode. The amount of water required is dependent on solution chemistry. Water-only and caustic-only scrubbing systems require the most blowdown. Oxidising systems (those with hypochlorite or peroxide) require the lowest water blowdown rates. If there is significant hardness in the make-up water, softening is recommended to reduce scaling in systems that will operate with a high pH.

16.3.5.7 Chemical addition

Many chemicals can be used to enhance the performance of packed towers. For acid gases (hydrogen sulphide), a high pH or basic solution is normally used.

For basic gases (ammonia), a low pH or acid solution is normally used. Chemical oxidants (chlorine, hydrogen peroxide, potassium permanganate, etc.) can dramatically increase performance and decrease water use.

16.3.5.8 Controls

Most control system utilises a pH sensor to pace chemical make-up. In addition to this it is possible to enhance the control system using an oxidation–reduction potential (ORP) and/or and exhaust gas (either the odour causing substance or vapour phase residual chlorine) analyser. ORP control is more complicated than pH control and requires experience and patience to achieve success.

16.3.6 Practical design issues

16.3.6.1 Sizing

For most packed tower systems the most economical sizing point occurs at approximately 1.5 m/s (300 feet per minute) superficial gas velocity and a gas to liquid ratio (m^3/m^3) of 400. However, units can be sized to be much more compact than this, but at a significant operational cost penalty (i.e. fan head loss). The maximum gas velocity that is practical in odour control packed towers is about 3 m/s (600 fpm). Most odour control systems that operate well are sized between 1 and 2 m/s.

16.3.6.2 Volume of packing required

For normal tower systems, 1 m^3 of packing per 0.5 m^3/s (1 ft^3 of packing per 30 cfm) gas flow rate will provide 99% removal of hydrogen sulphide.

16.3.6.3 Liquid circulation rate

The optimum liquid circulation rate is normally 170 l/min liquid per 1 m^3/s gas rate (1 gpm per 50 cfm). The circulation rate should never be less than 85 l/min liquid per 1 m^3/s gas (1 gpm per 100 cfm).

16.3.6.4 Packing selection

The most common random packing materials used in wastewater odour control systems are Lan-Pack®, Jaeger Tri-Packs® and Ceilcote Tellerettes®. Either of these packing materials is usually more than adequate for wastewater odour control. A new ultra-low-headloss packing called Q-Pac® is now available that reduces tower headloss as much as 75%. When using low headloss packing the concerns about good gas and liquid distribution become more critical.

The highest performance packing available is structured packing. It has not seen wide use in odour control because of the cost (2 to 3 times more expensive than random packing). It is typically only used in areas of severe space constraints.

16.3.6.5 Internals

The most important internals are the liquid distribution system and the packing support. Towers that have high liquid rates and low gas rates do not need sophisticated supports or distributors. However, as gas rates increase and/or liquid rates decrease, internals become a very important issue. The highest performance liquid distributor is a gravity distributor of either a weir trough or orifice injection plate type. Orifice plates are not normally used in odour control because of their increased propensity to plug. Spray nozzles are appropriate for small towers that have low gas rates and high liquid rates. The energy to circulate the liquid scrubbing solution is substantially higher with spray nozzles, although they are several times cheaper than gravity systems.

Whenever the gas flow rate exceeds 2 m/s (400 fpm), then a gas injection support plate should be used. This has a wavy cross section and allows gas into the packing with minimal headloss due to the large open area.

16.3.6.6 Tower construction

Again, with high gas rates, the gas inlet conditions are critical. Gas should never enter the tower at velocities over 7.6 m/s (1,500 fpm), and 5 m/s (1,000 fpm) should be a design goal.

16.3.6.7 Ductwork

The most important aspects of duct design is to direct the gas flow into the tower symmetrically and to slope into the tower so that the scrubber solution that sprays into the ductwork flows back to the tower.

16.3.6.8 Fans

Fans can be configured in either an induced draft or forced draft mode. Forced draft will keep the fan out of the scrubbing solution, but will pressurise the tower and may cause leaks. Induced draft keeps the tower at a negative pressure, but the fan will always be in the spray of the scrubbing liquid. This can be critical if chlorine is used, but is not a concern with the other chemicals. Induced draft will also reduce plume and drift problems, since the fan blades tend to coalesce water droplets.

16.3.6.9 Chemical injection

There are several possible locations in the scrubber liquid circulation loop to inject the chemicals. Common locations include:

- Directly into the tower sump,
- Into the pump suction,
- Into the pump discharge.

Injecting into the pump suction allows the chemicals to achieve very good mixing. However, the injection point must be coordinated with the pH measurement point in order to have a stable control system. The highest performance control systems can inject immediately before the pH probe and will work very effectively if the loop is tuned correctly and the solution is well mixed. However, it is generally not recommended to go this route unless the situation demands it. Allowing the chemicals to accumulate in the sump and providing for ample lag time in the control system will allow for a functioning system with lower performance controls.

16.3.6.10 Pumps

The biggest problem with pumps is the seal system. Pumps that are not in operation tend to experience corrosion of the mechanical seals. It is preferable to keep all pumps on line in service to reduce seal damage. The use of vertical pumps eliminates the seal problem, but vertical pumps have more overall maintenance problems than horizontal pumps. Suction piping design is critical to avoid cavitation and vortexing. Entrance velocities should be kept below 0.6 m/s (2 fps) in the piping in the tower sump.

16.3.7 Operations issues (optimisation)

16.3.7.1 pH and blowdown

For towers that use caustic only, the optimisation of pH and blowdown is critical. Figure 16.14 shows the relationship between pH and blowdown for packed towers operated at the Orange County Sanitation Districts in Fountain Valley, CA. In general, as the blowdown increases, the pH can decrease. If the costs for water and caustic are known, then the system can be optimised for low cost operation.

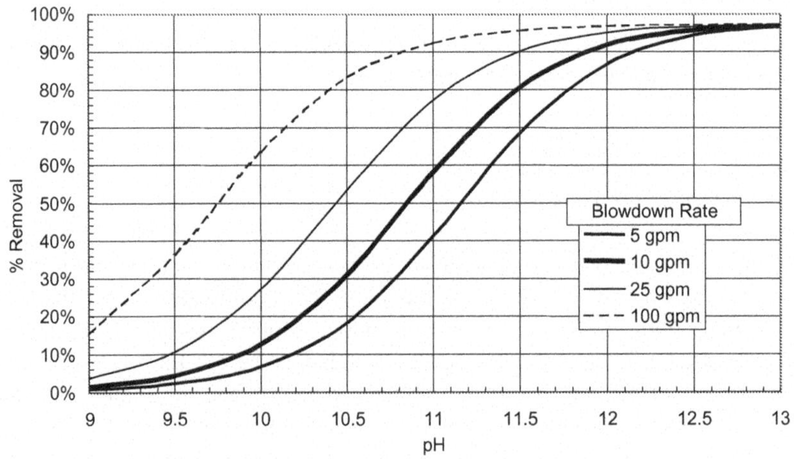

Figure 16.14. Percentage H_2S removal as a function of pH and blowdown rate.

16.3.7.2 Liquid rate

Performance increases slightly as liquid rate increases. The primary advantage of higher liquid rates is that if the liquid rate falls below the packing wetting rate, due to poor liquid distribution or plugging, performance decreases quickly to zero. The wetting rate for conventional packings occurs at approximately a gas to liquid ratio of 80 –1,000 (m^3/m^3). Operating at a gas:liquid ratio of 400 provides an ample safety margin so that packing wetting should not be a concern.

16.4 PACKED TOWER THEORY

This section addresses the design practice and theory for packed tower systems. Some of the equations are semi-empirical and therefore must be presented in the non-metric units that they were derived in.

16.4.1 Theoretical background

There are two distinct (although theoretically identical) methods for analyzing and predicting the performance of packed tower systems (Sherwood *et al.* 1975). Figure 16.15 shows the nomenclature used for this analysis.

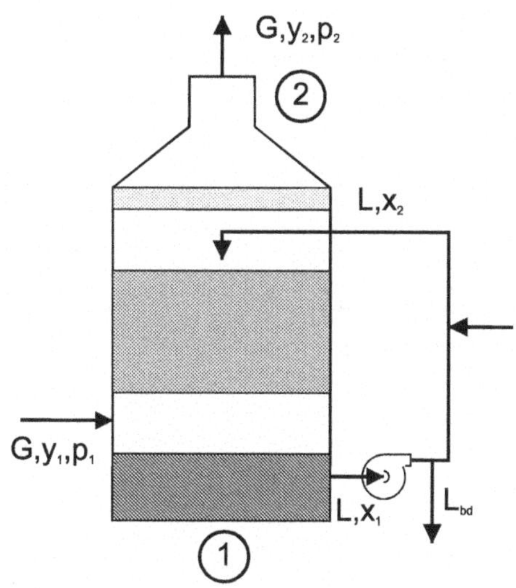

Figure 16.15 Packed tower schematic and nomenclature.

The first method is

$$k_g A = \frac{N}{VP\Delta Y_{LM}} \tag{16.13}$$

Where:
$k_g A$ = mass transfer coefficient lb-moles/(ft^3-hr-atm),
N = lb-moles per hour transferred,
V = tower packing volume (ft^3),
P = system pressure (atm).

$$\Delta Y_{\mathrm{LM}} = \frac{\left(Y_i - Y_i^*\right)_1 - \left(Y_i - Y_i^*\right)_2}{\ln\left[\frac{\left(Y_i - Y_i^*\right)_1}{\left(Y_i - Y_i^*\right)_2}\right]} \tag{16.14}$$

Where:
ΔY_{LM} = log mean concentration difference,
Y_i = gas phase mole fraction of constituent i,
Y^*_i = equilibrium gas phase mole fraction of constituent i,
1 = the bottom of the tower,
2 = the top of the tower.

The other method involves the concepts of number of transfer units and the height of transfer units.

$$Z = \text{HTU} * \text{NTU} \tag{16.15}$$

Where:
Z = packing depth (ft),
HTU = height of a transfer unit (ft),
NTU = number of transfer units.

$$\text{HTU} = \frac{G}{k_g A} \tag{16.16}$$

Where:
G = molar air flow rate (#mole/(ft^2-hr)).

$$\text{NTU} = \frac{\ln\left[\frac{y_1 - Hx_2}{y_2 - Hx_2}\left(1 - \frac{1}{A}\right) + \frac{1}{A}\right]}{1 - \frac{1}{A}} \tag{16.17}$$

Where:
y = gas phase concentration (mole fraction),
H = Henry's Law Coefficient (atm/mole fraction),
x = liquid phase concentration (mole fraction),
1 = the bottom of the tower,
2 = the top of the tower.

The adsorption factor (A):

$$A = \frac{L}{HG} \tag{16.18}$$

Where:
L = molar liquid flow rate (#mole/(ft^2-hr)).

This can be solved explicitly for removal efficiency by the following algebraic manipulations:

$$\left(1-\frac{1}{A}\right)\mathrm{NTU}=\ln\left[\frac{y_1-Hx_2}{y_2-Hx_2}\left(1-\frac{1}{A}\right)+\frac{1}{A}\right] \tag{16.19}$$

$$\exp\left[\left(1-\frac{1}{A}\right)\mathrm{NTU}\right]=\frac{y_1-Hx_2}{y_2-Hx_2}\left(1-\frac{1}{A}\right)+\frac{1}{A} \tag{16.20}$$

Now assigning:

$$E=\exp\left[\left(1-\frac{1}{A}\right)\mathrm{NTU}\right] \tag{16.21}$$

$$F=\frac{1}{A}=\frac{HG}{L} \tag{16.22}$$

$$F_b=\frac{1}{A_b}=\frac{HG}{L_{\mathrm{bd}}} \quad F_b=\frac{1}{A_b}=\frac{HG}{L_{\mathrm{bd}}} \tag{16.23}$$

Incorporating both recirculating and non-recirculating systems where:

$$R=\frac{Q_{\mathrm{c}}-Q_{\mathrm{bd}}}{Q_{\mathrm{c}}}=1-\frac{L_{\mathrm{bd}}}{L} \tag{16.24}$$

Where:
Q_{c} = the liquid circulation rate,
$Q_{\mathrm{b}d}$ equals the liquid blowdown rate.

$$x_1=\frac{G}{L_{\mathrm{bd}}}(y_1-y_2) \tag{16.25}$$

$$x_2=x_1R=\frac{G}{L_{\mathrm{bd}}}(y_1-y_2)\left(1-\frac{L_{\mathrm{bd}}}{L}\right) \tag{16.26}$$

$$hx_2 = \frac{hG}{L_{\text{bd}}}(y_1 - y_2) - \frac{hG}{L}(y_1 - y_2) = (F_b - F)(y_1 - y_2) \tag{16.27}$$

$$T = \frac{E-F}{1-F} = \frac{y_1 - (F_b - F)(y_1 - y_2)}{y_2 - (F_b - F)(y_1 - y_2)} \tag{16.28}$$

This reduces to:

$$\frac{y_2}{y_1} = \frac{1 + F_{\text{b}}(E-1) - FE}{E + F_{\text{b}}(E-1) - FE} \tag{16.29}$$

Where:
F_{b} is based on the blowdown rate,
F is based on the circulation rate.

For once through scrubbers ($F_{\text{b}} = F$), this reduces to:

$$\frac{y_2}{y_1} = \frac{1-F}{E-F} \tag{16.30}$$

and if the adsorption factor, A, is very large ($F >> 0$) as in the case of oxidants:

$$\frac{y_2}{y_1} = \frac{1}{E_m} = \frac{1}{\exp(\text{NTU})} \tag{16.31}$$

When the adsorption factor, A, is very large ($F >> 0$), but blowdown is very small

$$\frac{y_2}{y_1} = \frac{1 + F_{\text{b}}(E_m - 1)}{E_m + F_{\text{b}}(E_m - 1)} \tag{16.32}$$

or

$$\frac{p_2}{p_1} = \frac{\frac{1 - F_b}{E_m} + F_b}{1 - F_b\left(\frac{1}{E_m} - 1\right)} \tag{16.33}$$

Where:
y_1 = mole fraction of gas in,
y_2 = mole fraction of gas out,
p_1 = partial pressure of gas in,
p_2 = partial pressure of gas out.

$$F_b = \frac{HG}{L_{bd}} \tag{16.34}$$

Where:
L_{bd} = molar liquid blowdown rate.

$$E_m = e^{NTU} \tag{16.35}$$

K_gA values range from 12 to 36 lb-moles/(ft^3-hr-atm) for H_2S removal depending on oxidant concentration and packing type. For lower pH hypochlorite solutions the apparent K_gA can be as high as 60 lb-moles/(ft^3-hr-atm) due to the presence of vapour chlorine. K_gA values for ammonia range from 10 to 20 lb-moles/(ft^3-hr-atm) depending on packing type. K_gA values for carbon dioxide range from 2 to 4 lb-moles/(ft^3-hr-atm) depending on packing type. K_gA also has some dependance on liquid rate. Most K_gA values are reported for liquid rates of 5,000 pounds per ft^2 per hour. They can be adjusted using a 0.175 power law. For example:

$$\frac{K_g A_1}{K_g A_2} = \left(\frac{L_1}{L_2}\right)^{0.175} \tag{16.36}$$

Note that this relationship collapses when the minimum packing wetting rate is reached. For most packings the minimum wetting rate occurs at a volumetric gas to liquid ratio of 800 to 1,000. The above relationship should only be used for gas to liquid ratios of 600 or less.

16.4.1.1 Ionisation

Acids and bases (and their salts) will ionise in aqueous solutions. As an example, one of the more common odorous compounds ionises as

$$H_2S \Leftrightarrow H^+ + HS^- \Leftrightarrow 2H^+ + S^= \tag{16.37}$$

This ionisation reaction is essentially instantaneous. The equilibrium concentration of each of the species as a function of pH can be calculated from the following equations. The total amount of hydrogen sulphide species in solution is represented by

$$C_t = [H_2S] + [HS^-] + [S^=] \tag{16.38}$$

Where:
$[H_2S]$ = the concentration of unionised dissolved gas (mol/l),
$[HS^-]$ = the concentration of the bisulphide ion (mol/l),
$[S^=]$ = the concentration of the sulphide ion (mol/l).

The equilibrium condition of the following reaction:

$$H_2S \Leftrightarrow H^+ + HS^- \tag{16.39}$$

is represented by the following equation

$$K_1 = \frac{[H^+][HS^-]}{[H_2S]} \tag{16.40}$$

Where:
K_1 is the first ionisation constant and the hydrogen ion concentration is calculated from the pH by

$$[H^+] = 10^{-pH} \tag{16.41}$$

The equilibrium of

$$HS^- \Leftrightarrow H^+ + S^= \tag{16.42}$$

is represented by

$$K_2 = \frac{[H^+][S^=]}{[HS^-]} \tag{16.43}$$

Where:
K_2 is the second ionisation constant.

Values of ionisation constants for compounds of interest are shown in Table 16.2.

Table 16.2. Ionisation constants (at 20 °C) for odour control chemicals (Sorum 1967).

Compound	Formula	Ionisation Constant(s)
Hydrogen Sulphide	H_2S	$K_1 = 1.0 \times 10^{-7}$ $K_2 = 1.3 \times 10^{-13}$
Hypochlorous Acid	HClO	$K_1 = 3.2 \times 10^{-8}$
Ammonium Hydroxide	NH_4OH	$K_1 = 1.8 \times 10^{-5}$

The portion of the total liquid phase sulphide concentration that is unionised is:

$$\alpha_O = \frac{[H_2S]}{C_t} = \frac{1}{\left(1 + \frac{K_1}{[H^+]} + \frac{K_1 K_2}{[H^+]^2}\right)} \tag{16.44}$$

The portion that is partially ionised is:

$$\alpha_1 = \frac{[HS^-]}{C_t} = \frac{1}{\left(\frac{[H^+]}{K_1} + 1 + \frac{K_2}{[H^+]}\right)} \tag{16.45}$$

The portion that is completely ionised is

$$\alpha_2 = \frac{[S^=]}{C_t} = \frac{1}{\left(\frac{[H^+]^2}{K_1 K_2} + \frac{[H^+]}{K_2} + 1\right)} \tag{16.46}$$

Figure 16.6 shows how the concentration of the ionised species of hydrogen sulphide is a function of pH.

16.4.1.2 Oxidation

Hydrogen sulphide is readily oxidised by any strong oxidant (chlorine, hydrogen peroxide, and potassium permanganate). The reaction is almost instantaneous with chlorine and can have a half-life as long as 5 to 10 minutes with hydrogen peroxide. The reaction with oxygen is quite slow unless a catalyst (iron) is present. The performance of chlorine is generally superior in packed tower systems because of the combined gas phase liquid phase reactions in chlorine systems. This is due to the vapour pressure of chlorine in liquid solutions.

16.4.1.3 Pressure drop

Jaeger provides pressure drop information for No. 2 Tri-packs® in a graphical form. This can be approximated by

$$\log(\Delta p) = \left(a_1 + \frac{b_1 L}{1000}\right) + \left(a_2 + \frac{b_2 L}{1000}\right)\log(G) \tag{16.47}$$

Where:
Δp = pressure drop (inches of water column per foot of packing depth),
L = liquid rate (#/hr-ft^2),
G = gas rate (#/hr-ft^2),
a_1 = -8.2828,
b_1 = 0.0897,
a_2 = 2.2342,
b_2 = -0.0171.

The flooding point is approximated by:

$$G_f = a_3 \exp\left(\frac{L}{b_3}\right) \tag{16.48}$$

Where:
G_f = gas rate at flooding (#/hr-ft^2),
a_3 = 5,536.2,
b_3 = - 42,000.

16.4.2 Example of mass balance

Packed Tower Analysis
Schematic Mass Balance

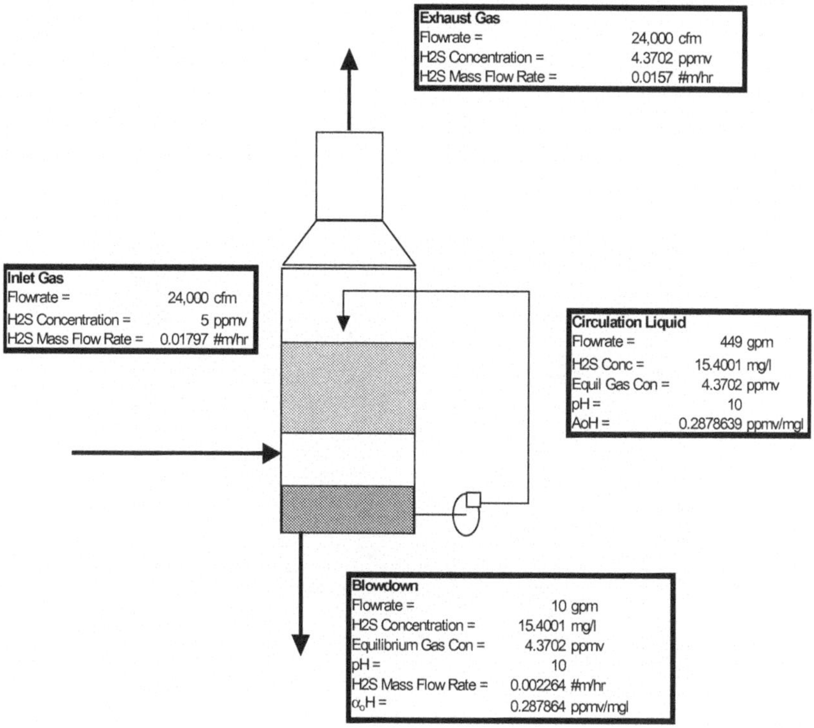

Key Tower Parameters

pH =	10
Blowdown =	10 gpm
% Removal =	13%
Kga =	18 #moles/ft3/hr
Tower Dia =	10 ft
Packing Dpth =	10 ft
Gas Loading Rate =	306 cfm/ft2
G/L =	400 ft3/ft3

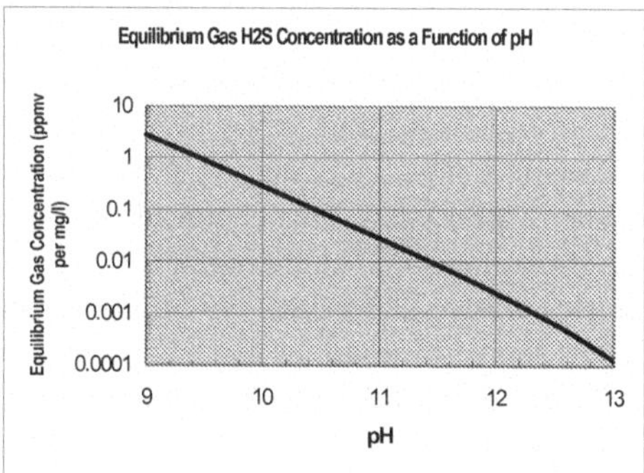

16.5 DESIGN OF MIST SYSTEMS

There are two types of atomised mist systems that are currently available. One type uses compressed air to accelerate a liquid stream to supersonic velocities fragmenting the stream into droplets around 10 microns in diameter (Figure 16.16). This nozzle is a cylinder about 6 cm in diameter by 10 cm long and has liquid nozzle clearances of about 10 microns. The other nozzle type uses compressed air to accelerate the liquid stream into two jets that impact each other (Figure 16.17). This technology has nozzle clearances of about 5 mm, but has a much larger droplet size. The whole nozzle assembly for the impact technology is about 1 m long.

Figure 16.16. Supersonic nozzle technology.

Atomised mist technology has a much larger gas/liquid interface area than packed towers but has a much lower liquid rate. The typical liquid flow rate through a nozzle is 4 l/min. Normally one nozzle is used for each 4 m^3/s (10,000 cfm) of air flow rate. The liquid system is almost always once through with the excess chemical wasted. This chemical dosing is normally controlled by sensing pH and sometimes ORP in the liquid waste line.

This technology almost always utilises a chlorine compound as the oxidant, to take advantage of possible gas phase reactions. It is common to use multiple stages operated at different pH values to increase scrubbing performance on recalcitrant compounds.

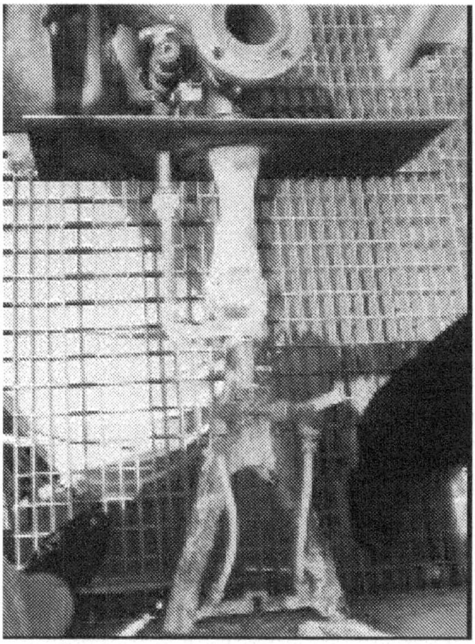

Figure 16.17. Impact nozzle technology.

The systems are sized for about 10 seconds contact time for odours similar to about 10 ppmv of H_2S. For levels up to 25 ppmv H_2S equivalent, about 15 seconds of contact time is necessary. For more concentrated odours contact time will have to be field verified.

Each nozzle will require about 0.036 m^3/s (80 cfm) of compressed air at between 3 and 7 atm (40 and 100 psi_{gas}) depending on the nozzle type. Generating the compressed air is a major cost.

The design of the contact chamber to avoid short circuiting is an art. Normal criteria are to have tangential entrance and exit with velocities less than 7.6 m/s (1,500 fpm). Contact chambers have had either up-flow or down-flow designs and it doesn't seem to matter. The process is always co-current because of the small droplet size. Mist carryover out of the system has always been a concern with these systems.

There is no practical analytical design methodology for mist systems. They are sized based on past experience with similar applications.

The Washington Suburban Sanitary Commission in Silver Spring, Maryland spent over ten years optimising atomised mist technology to control odours from biosolids composting. Their system relied on a multi-stage multi nozzle approach. A schematic of their process is shown in Figure 16.18.

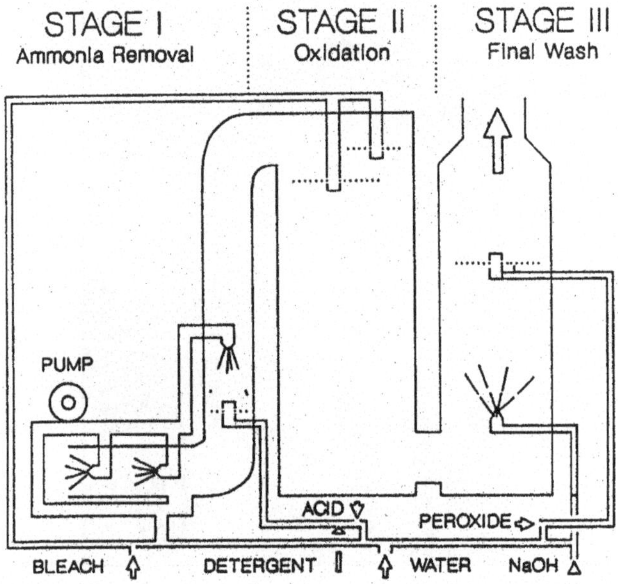

Figure 16.18. WSSC atomised mist system schematic (Hentz *et al.* 1992).

16.6 ESTIMATING COSTS FOR CHEMICAL ODOUR CONTROL

The purchase cost for the two leading liquid scrubbing technologies, packed towers and atomised mist systems, are very similar with some exceptions. The maximum size of a single train atomised mist system is usually about 14 m^3/s (30,000 cfm). For larger air flows, multiple systems must be installed. Packed towers can handle up to 27 m^3/s (60,000 cfm) per tower, and much of the support equipment can be common to a set of towers. Therefore packed towers have a much better economy of scale than mist systems.

Packed tower systems have many vendors, with over 40 vendors actively producing systems. Atomised mist systems have only two vendors. The actual cost to construct a mist system is lower than a packed tower system, however, so far this economy has translated into increased profit margins on mist systems, not lower costs. Therefore, for systems under 14 m^3/s (30,000 cfm), packed towers and atomised mists systems are essentially the same price for the same performance level. Figure 16.19 presents the approximate costs for these systems as a function of air flow rate.

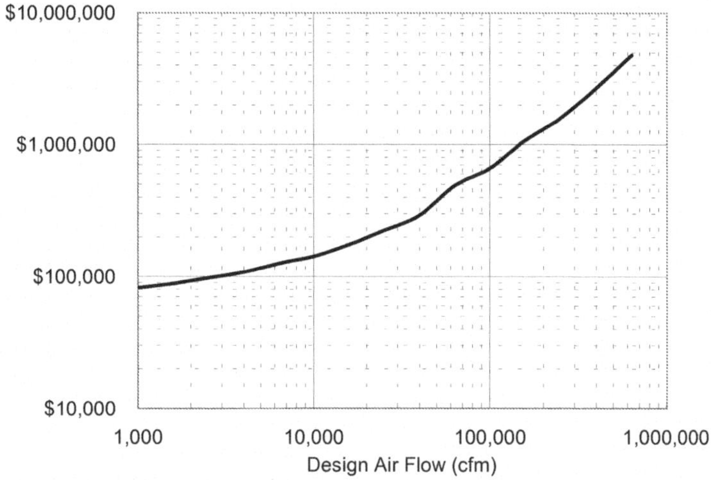

Figure 16.19. Capital costs (US$) of single stage packed tower scrubbing systems (equipment only) for various air flow rates.

Operating costs consist of chemicals, water, electrical power, replacement parts, and labour. The chemical usage is usually close to stoichiometric for hydrogen sulphide and ammonia removal, but can be much higher for other odorous compounds. Water costs are minimal when an oxidant is used, but can be quite large for caustic only towers. Note also that water must often be softened to reduce scale build-up on packing. Electrical power is consumed by the fan and circulation pumps.

16.7 REFERENCES

Hentz, L.H., Murray, C.M., Thompson, J.L., Gasner, L.L., and Dunson, J.B. (1992) Odor Control Research at the Montgomery County Regional Composting Facility. *Water Environ. Res.* **64**, 13–18.

Kohl, A and Riesenfeld, F. (1979) *Gas Purification*, 3rd Ed., Gulf Publishing Corporation, Houston.

Kundidzora, E., and Reichenberger, J. (1999) Cost-Effective WWTP Odor Control with UV Light. Proc. WEFTEC, New Orleans, October 9-13.

Sherwood, T. G., Pigford, R.L., and Wilke, C.R. (1975) *Mass Transfer*. McGraw-Hill, New York.

Sorum, C.H. (1967) *Introduction to Semimicro Qualitative Analysis*, 4th Edition. Prentice-Hall, Inc. Englewood Cliffs.

Tchobanoglous, G. and Burton, F. L. (1991) *Wastewater Engineering: Treatment, Disposal and Reuse,* Metcalf and Eddy Inc., McGraw-Hill Inc., New York.

White, G. C. (1992) *Handbook of Chlorination and Alternative Disinfectants*, 3rd. Ed. Van Nostrand Reinhold, New York.

17

Adsorption systems for odour treatment

Amos Turk and Teresa J. Bandosz

17.1 INTRODUCTION

The surface of a solid always accumulates a concentration of molecules from its gaseous or liquid environment; this phenomenon is called *adsorption*. The "surface" includes all accessible areas, and can therefore be extensive for solids that incorporate an inner network of pores, including those with diameters down to molecular dimensions. Such solids are known as *adsorbents*. The removal of adsorbed matter (*adsorbates*) from a solid is called *desorption*.

Adsorbents are useful in odour control because they serve as media for removing odorous gases and vapours from air streams by concentrating and retaining them, thus facilitating their subsequent disposal or their conversion to odourless products. Adsorbent systems also serve to recover valuable chemicals, but this function does not ordinarily apply to wastewater operations.

edited by Richard Stuetz and Franz-Bernd Frechen. ISBN: 1 900222 46 9

Adsorbed odorants, if they are stable and relatively unreactive in air (butyric acid, for example) may simply remain on the carbon indefinitely. Others, such as reduced sulphur compounds, including hydrogen sulphide, are more or less rapidly oxidised to products that are frequently less odorous, and sometimes not odorous at all. In many instances, the oxidation products are higher in molecular weight and more strongly adsorbed and retained. Different adsorbed odorants, being concentrated and in close proximity to each other on the surface or in the pores of the adsorbent, may interact and others, like styrene or acrylates, may polymerise. Most of these actions favour effective deodorization.

An adsorbent may be previously impregnated with a reagent that is selective for removal or destruction of specific odorants, or with a catalyst that speeds up a desired reaction, usually oxidation by air. Alternatively, the adsorbent itself may provide the catalytic activity. For practical deodorization objectives in wastewater applications, adsorbent impregnations are a mixed blessing: the total capacity for physical adsorption of the usual wide range of odorous vapours is reduced by the sacrifice of surface area and pore volume occupied by the impregnant. The net result is usually disadvantageous.

Adsorbent systems for odour control in wastewater applications generally consist of static beds of granular materials in vertical cylindrical columns (see section on equipment and systems below). Accordingly, the adsorption process starts where the air stream enters the column, progresses in the direction of air flow, and continues until odour "breaks through" at the exit end. The section of the bed in which the adsorption process is taking place (after the exhausted section but before the fresh section) is called the *adsorption zone*. The adsorbent is considered to be exhausted when the breakthrough odour level or gas concentration has reached some arbitrarily selected value. Frequent questions about the odour control performance of an adsorbent bed are: How efficient is it? What is its capacity? How long will it last? At what temperature range does it operate? What is the effect of the humidity of the air on its efficiency and capacity?

Efficiency: Since the depth of an adsorbent bed for wastewater odour control is typically several adsorption zones, the effluent air for most of its life is generally odourless. For such systems, the initial efficiency of 100% is not an issue in the selection of an adsorbent.

Capacity: The capacity of an adsorbent bed for a particular odorous gas or vapour depends on the properties of the adsorbent, the concentration of odorant in contact with it, the ambient temperature and humidity, the presence of other adsorbable gases and vapours, the physical design of the bed, and the rate of air flow through it. A typical value for a 1-metre bed of unmodified activated carbon with a linear air flow of 25 cm/s may be about 10% of the mass of the dry bed. In equation 17.1 , that would correspond to 0.1 for the value of S.

Service life:

$$t = 6.7 \times 10^6 SW/EQMC \qquad (17.1)$$

Where:
t = service life (time to breakthrough) (hr),
S = proportionate saturation of adsorbent at breakthrough (fractional)
W = weight of adsorbent (kg),
E = fractional average adsorption efficiency over time of service (usually close to 1),
Q = volume rate of air flow through adsorbent bed (l/s),
M = average molecular weight of adsorbates (g/mol),
C = average concentration of influent absorbable vapours (ppm by volume).

An average molecular weight for such adsorbates is about 100. Substituting this value in the equation, and assuming S = 0.1, yields the rough approximation:

$$t \cong 6700\ W/QC \qquad (17.2)$$

Carbon towers used for odour control in wastewater treatment plants have employed up to about 10,000 kg of carbon for air flows of about 5000 l/s. The least predictable and most variable of these factors is C, the influent VOC concentration. In practice, such activated carbon odour control systems have served for up to a year or two, or sometimes even more.

Temperature: In general, the capacity of adsorbent decreases with increasing temperature. As a rule of thumb, 50 °C is often cited as an approximate practical upper limit for odour control by physical adsorption with activated carbon. Many odorants, however, especially including H_2S, are oxidised on the carbon after they are adsorbed. Impregnated carbons utilise chemical reactions, such as acid–base neutralisation, to enhance their capacity for specific odorants. Neutralisation is very rapid in any aqueous environment, and oxidation, like chemical reactions in general, is favoured by higher temperatures.

Humidity: Activated carbon is said to be "hydrophobic", which means that its non-polar nature causes it to bind preferably to odorous VOCs rather than to water vapour, which is much more strongly polar than organic compounds. It is this property of activated carbon that makes it suitable for removing organic impurities even from liquid water. Most of the moisture that carbon does adsorb from air is gradually displaced by the less polar or non-polar vapours that are subsequently adsorbed (Turk and Van Doren 1953). Liquid water, which may condense on the carbon when the temperature of a humid air stream drops, does slow down the diffusion of organic vapours to the carbon surface, and is therefore disadvantageous. However, such moisture is rapidly re-vaporised when dryer air flow is resumed.

17.2 ADSORBENTS

17.2.1 Characteristics of activated carbon and other granular media

Activated carbons are widely used as adsorbents, separation media, and catalyst supports (Bansal *et al.* 1988). They are obtained by carbonisation followed by activation at high temperature of organic precursors such as wood, bituminous or anthracite coal, petroleum pitch, peat, or coconut shells. Activated carbons used for odour control in wastewater treatment plants are predominantly based on coal from China and North America. Carbon from coconut shells originates mostly from southern Asia. Vendors include companies that produce the activated carbons, or that purchase it from a manufacturer and modify it chemically or physically, or that simply resell it. Some vendors do all three, and others change their functions from time to time. Table 17.1 lists some of the largest carbon manufacturers.

Table 17.1. List of activated carbon suppliers.

Company	Internet address
Calgon Carbon	www.calgoncarbon.com
Chemiviron carbon	www.chemivironcarbon.com
Norit	www.norit.com
Pica	www.picausa.com
Waterlink Barnebey Sutcliffe	www.waterlink.com
Westvaco	www.westvaco.com

Activation is a process where agents such as steam or carbon dioxide are used to create pores in carbonised char. Those pores range from a few to several hundred ångstroms in diameter and are active in the adsorption process (Figure 17.1) (Donnet *et al.* 1994). It is believed that they are slit shaped. Activated carbons are characterised by high surface area, up to about 2000 m^2/g, high pore volume (over 1 cm^3/g), and a high degree of surface structural and chemical heterogeneity. Chemical heterogeneity is the result of the presence of atoms other then carbon in the activated carbon matrix (Boehm, 1966; Puri, 1970; Leon y Leon and Radovic 1992). The origin of those hetero-atoms is in the nature of organic precursors (nitrogen, sulphur), in the chemistry of activation process (oxygen, phosphorus) and in the storage conditions (oxygen). The most common hetero-atom is oxygen, which is present on the carbon surface in the form of acidic, basic, or neutral organic groups such as carboxylic acids, lactones, phenols, pyrenes, carbonyls, esters, etc (Figure 17.2) (Fanning and Vannice 1993).

Figure 17.1. Schematic structure of an active carbon (Donnet *et al.* 1994).

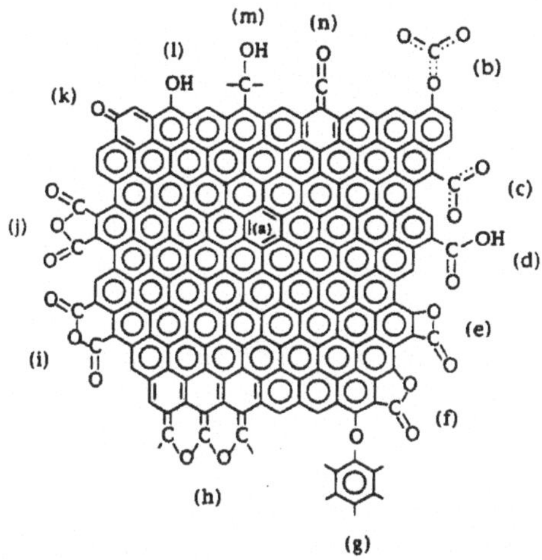

Figure 17.2. IR-active funtionalities on carbon surafces: (a) aromatic C=C stretching; (b) and (c) carboxyl-carbonates; (d) carboxylic acid; (e) lactone (4-membered ring); (f) lactone (5-membered ring); (g) ether bridge; (h) cyclic ethers; (i) cyclic anhydride (6-membered ring); (j) cyclic anhydride (5-membered ring); (k) quinone; (m) alcohol and (n) ketene (Fanning and Vannice 1993).

Physical sorption on activated carbons results from dispersive or specific interactions of sorbate molecules with the carbon surface (Gregg and Sing 1982). The dispersive forces are increased by the presence of small pores which enhance adsorption energy as a result of interactions of sorbate molecule with more than one surface (Everett and Powl 1976). In addition, surface chemistry has a significant influence on specific interactions, among which the strongest are hydrogen bonding and interactions between Lewis acids and bases (Leon y Leon and Radovic 1992).

The activated carbon surface is known to catalyse various chemical reactions, particularly oxidation in the presence of air, as well as interactions between sorbates and previously impregnated reagents (Bansal *et al.* 1988). This effect is due to the presence of small micropores (Everett and Powl 1976), the presence of functional groups and inorganic impurities (Puri 1970) and to the electrical conductivity of carbonaceous materials important for electron transfer (Leon y Leon and Radovic 1992). As a result, some properties useful in wastewater odour control such as the adsorption/oxidation of hydrogen sulphide, are greatly improved when activated carbon is used (Bandosz 1999; Adib *et al.* 1999a,b, 2000a, Bandosz *et al.* 2000).

Capacity of other materials such as alumina, silica, zeolites or various inorganic oxides for removal of hydrogen sulphide from wastewater effluents is some one tenth that of activated carbons (Steijns and Mars 1977). Some oxides as alumina or zirconia do show a high conversion of H_2S (but only half of that observed for carbon). The enhanced activity of those oxides is attributed to the presence of Lewis acidic centres related to the coordination of aluminium and other metal ions in the oxide framework. These sorbents have surface areas and pore volumes about half of those of carbons and the surface is not so heterogeneous as in the case of carbonaceous sorbents (Brinker and Scherer 1990).

17.2.2 Activated carbon: impregnated, catalytic, gas-injected, or unmodified

17.2.2.1. Impregnated activated carbons

Activated carbons impregnated with caustics (NaOH or KOH) are the most often used adsorbents of hydrogen sulphide in wastewater treatment plants. Both NaOH and KOH react with atmospheric CO_2, forming the corresponding carbonates. These basic compounds facilitate the removal of H_2S (Turk *et al.* 1992). The action proceeds further as the activated carbon catalyses the oxidation of hydrogen sulphide by atmospheric oxygen to yield a variety of products, including elemental sulphur and more highly oxidised forms.

Other impregnated sorbents for removal of hydrogen sulphide from gas streams are carbons impregnated with heavy metal salts such as copper sulphate or lead acetate. These media, when spent, are usually classified as hazardous materials because of their content of heavy metals. Among non-carbonaceous sorbents, activated alumina impregnated with potassium permanganate has also been used for oxidative removal of H_2S, but its severely limited capacity for physical adsorption of VOCs from humid air makes it impractical for application to wastewater effluents. Recently, a new catalytic carbon, Centaur®, has been introduced by the Calgon Carbon Company. It is made by impregnating low temperature char with urea followed by its activation at around 800 °C (Matviya and Hayden 1994). This treatment introduces basic nitrogen species into the activated carbon matrix. Owing to Centaur's® high relative microporosity (the ratio of the micropore volume to the total pore volume is over 80%) the resulting pyridine-like species are highly dispersed in the carbon's micropores. The hydrogen sulphide, upon its dissociation in the adsorbed water films, is oxidised to isolated sulphur radicals, which are then further oxidised to sulphur oxides and on to sulphuric acid (Adib *et al.* 2000b) (Figure 17.3). The consequent selectivity of conversion of H_2S to H_2SO_4 is almost 100%, making it feasible to regenerate Centaur® carbon by washing it with water (Hayden 1995). From wastewater plants that contain relatively low concentrations of hydrogen sulphide, the use of Centaur® carbon may provide only a marginal advantage.

17.2.2.2 Catalytic carbons

It has been demonstrated recently that unmodified activated carbons can provide enough capacity to efficiently remove hydrogen sulphide from effluent gas in sewage treatment plants (Bandosz *et al.* 2000). This is an important finding since application of caustic-impregnated materials is associated with many disadvantages. They are: (1) their ignition temperatures are lowered by the exothermic reaction taking place on the carbon surface. When such carbon is allowed to stand as a thick bed with access to air for some time before the full cooling air flow is started, a gradual temperature rise may lead to self-ignition. The result is not a conflagration or explosion, but a glowing center that can be extinguished with water; (2) hydrogen sulphide is mainly converted to elemental sulfur (Bandosz and Le 1998). Carbons with deposited sulfur are exhausted for the removal process and they cannot be regenerated *in situ* using inexpensive methods such as washing with water. According to the mechanism of the oxidation process (Turk *et al.* 1989) their activity lasts only until the caustic is exhausted and then sulfur and salts deposited on the surface block the pore structure where sorption of hydrogen sulphide can occur (Bandosz and Le 1998).

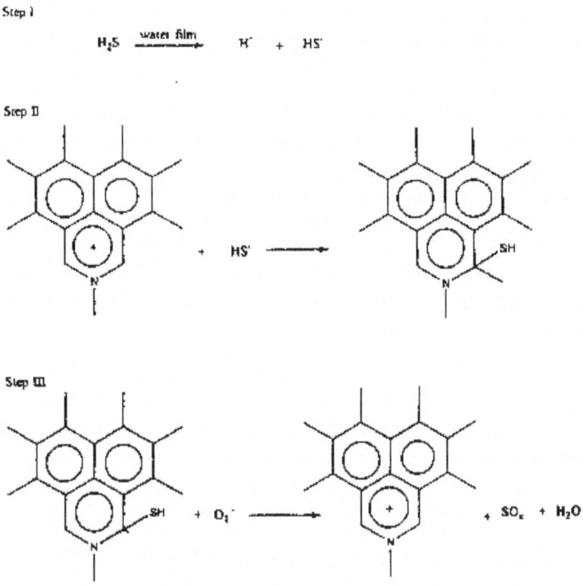

Figure 17.3. Proposed pathway of H_2S oxidation on nitrogen modified activated carbon (Adib *et al.* 2000b).

17.2.2.3 Gas-injected carbons

Ammonia (NH_3), which catalyses the oxidation of hydrogen sulphide to sulfur by atmospheric oxygen, can be injected continuously into the air stream preceding the carbon (Turk *et al.* 1989). At the same time, methyl mercaptan is oxidized to dimethyl disulphide which, owing to its almost doubled molecular weight, is more effectively retained by the carbon. A rate of injection that provides an ammonia concentration of about 10 ppm, which is well below ammonia's odour threshold of about 50 ppm, has been found adequate. Since ammonia is lighter than air and is not retained by the carbon, it is continuously displaced by air flow and does not behave like an impregnant, and therefore does not diminish the carbon's capacity for adsorption of odorous VOCs. Furthermore, ammonia, being non-toxic and highly soluble in water, is not an air pollutant. It has also been found that the useful life for hydrogen sulphide of an exhausted caustic-impregnated carbon can be extended by about one third by the retrofit of an ammonia injection system. For plants where the presence of ammonia cylinders is undesirable, injection of ammonia water may be substituted.

17.2.2.4 Unmodified carbons

The above mentioned disadvantages of caustic-impregnated carbons directed the attention of researchers to unmodified carbons as alternative sorbents for odour control in wastewater treatment plants (Bandosz *et al.* 2000). The advantages of unmodified activated carbons are as follows: (1) capacity for physical sorption is undiminished; (2) the unmodified activated carbon surface with its incorporated hetero-atoms can act as an oxidation catalyst (Adib *et al.* 2000b), (3) the deposition of inorganic salts is limited and (4) their costs per pound are significantly lower than these of impregnated carbons or patented catalytic carbons. Furthermore their densities are lower than those of caustic impregnated carbons, which contain significant moisture. Another important factor, which can make the application of unmodified carbon feasible in, is the frequent low concentration of hydrogen sulphide in wastewater effluents. It follows that even the slow kinetics of the reaction does not present an obstacle for this particular application (Bandosz *et al.* 2000).

17.2.3 Role of surface chemistry

Research on unmodified carbons as hydrogen sulphide adsorbents has shown that surface pH (surface chemistry) largely influences this performance and can even be incorporated into the specifications for selection of an activated carbon for wastewater odour control (Adib *et al.* 1999a,b.). The pore volume of carbons along with the local pH in the pore system has a significant effect on the efficiency of hydrogen sulphide dissociation and its oxidation to various sulphur species (Figure 17.4). A moderately low average pH of the carbon surface suppresses the dissociation of H_2S and the creation of hydrogen sulphide (HS^-) ions. The latter, when present in only low concentration in small pores, are oxidised to sulphur oxides and then to sulphuric acid. On the other hand, a pH in the basic range promotes the dissociation of H_2S, yielding a high concentration of HS^- which is then oxidised to sulphur polymers such as, for example, S_6 or S_8. When the pH is very low only physical adsorption can occur. This raises a question about the limits of acidity of the carbon surface. The problem is important not only when the sorption capacity is considered but also for the feasibility of regeneration of the spent material. The ideal situation requires both high capacity, which is promoted by a basic environment and the presence of sulphuric acid which, in the case of unmodified carbons, is usually created at a moderate pH level (Adib *et al.* 2000a). The ideal "compromise" would be a pH that yields a high capacity for hydrogen sulphide removal and promotes ultimate oxidation to sulphuric acid. Figure 17.5 shows the dependence of the normalised capacity (in mg of H_2S per unit pore volume of carbon) upon the pH of the carbon surface. The method of pH evaluation is simple, fast, and inexpensive and for some purposes it is good enough as a specification of the manufactured product. The threshold value derived from the analysis of H_2S adsorption/oxidation on activated carbon of various origin was indicated to be about 5.0 (Adib *et al.* 2000a).

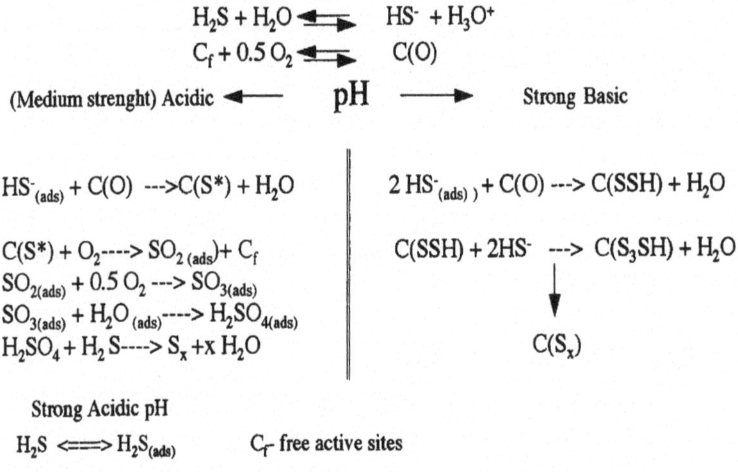

Figure 17.4. Mechanism of H_2S adsorption/oxidation on activated carbons at various pH levels (Adib *et al.* 1999a).

The justification for the existence of the threshold is based on the proposed mechanism of hydrogen sulphide adsorption/oxidation on unmodified carbons (Adib *et al.* 1999a) (Figure 17.4). It involves H_2S adsorption on the carbons surface (17.3), its dissolution in water film (17.4), dissociation of H_2S in adsorbed state in water film (17.5), and surface reaction with adsorbed oxygen (17.6).

$$H_2S_{gas} \xrightarrow{K_H} H_2S_{ads} \qquad (17.3)$$

$$H_2S_{ads} \xrightarrow{K_S} H_2S_{ads\text{-}liq} \qquad (17.4)$$

$$H_2S_{ads\text{-}liq} \xrightarrow{K_a} HS^-_{ads} + H^+ \qquad (17.5)$$

$$HS^-_{ads} + O^*_{ads} \xrightarrow{K_{R1}} S_{ads} + OH^- \qquad (17.6)$$

$$HS^-_{ads} + 3\,O^*_{ads} \xrightarrow{K_{R2}} SO_{2\,ads} + OH^- \qquad (17.7)$$

$$H^+ + OH^- \longrightarrow H_2O \qquad (17.8)$$

Where H_2S_{gas}, $H_2S_{ads\text{-}liq}$, and H_2S_{ads} correspond to H_2S in gas, liquid and adsorbed phases, respectively; K_H, K_S, K_a, K_{R1}, and K_{R2} are equilibrium constants for related processes (adsorption, gas solubility, dissociation, and surface reaction constants); O^*_{ads} is dissociatively adsorbed oxygen and S_{ads} represents sulphur as the end product of the surface oxidation reaction.

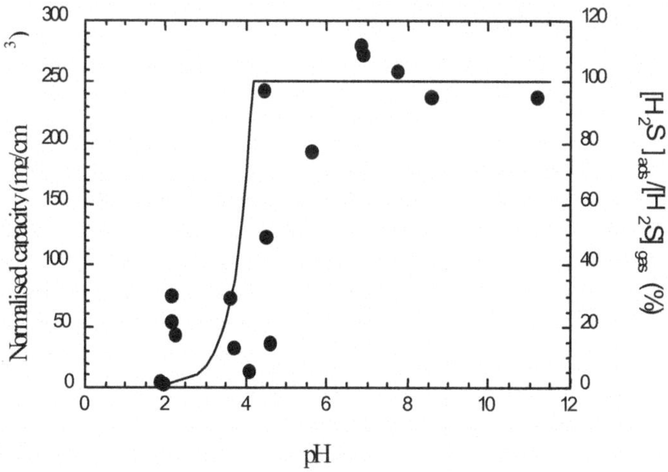

Figure 17.5. Dependence of normalised capacity (in mg of H_2S per cm^3 of pore volume) on pH of activated carbon surface.

The surface reactions (17.6 and 17.7) are considered as the rate-limiting steps of H_2S oxidation (Ghosh and Tollefson 1986). The following expression was proposed to calculate HS^-_{ads} concentration:

$$\text{Log}\,(HS^-_{ads}) = \log(K_S) + \log(K_H) + \log(K_a) + pH + \log(H_2S_{gas}) \qquad (17.9)$$

Substituting equilibrium constants derived from specific experimental conditions into equation (17.9) leads to the following expression (Adib *et al.* 2000a):

$$\text{Log}\ (HS^{-}_{ads}) = -4.2 + pH + \log\ (H_2S_{gas}) \tag{17.10}$$

This equation suggests that for carbon having a pH equal to or larger than 4.2 the concentration of HS^- in the adsorbed state will be equal to or larger than its concentration in the gas phase which is required for the effective removal of hydrogen sulphide. Figure 17.5 shows that the graphical estimation of the pH threshold is around pH equal to 5. The discrepancy is due to the fact that the K constants for equation (17.8) were calculated using simplified estimations (Adib *et al.* 2000a).

17.3 OPTIONS FOR REGENERATION OR DISPOSAL OF SPENT ADSORBENTS

The options for spent unmodified carbon are diagrammed in Figure 17.6 (Turk *et al.* 1991). Carbon that has been used for odour control in wastewater treatment plants is generally free of wastes that are classified as "hazardous," and discarding it in a municipal landfill may therefore be permissible. If it is to be reused, it must be removed to a facility for reactivation, usually with steam at around 700 to 900 °C, and where the furnace exhaust is scrubbed and/or incinerated before release to the environment. No one is interested in recovering odorous sewage gases. On-site reactivation is not used because the furnaces needed are not consistent with the ambient temperature odour control adsorbers.

Caustic-impregnated carbon may also be discarded, the considerations being the same as for unmodified carbon. Alternatively, it may be regenerated in place by a series of washings with strong caustic solutions and water, between air dryings (Farmerie 1985). The soakings take about a week, and we are not aware of any wastewater facility that has elected to carry this out more than once. Alternatively, the spent carbon can be washed with water to remove soluble salts, after which it is easier to reactivate. Another option is to extend its life by using the ammonia injection system, preferably after water washing (Turk *et al.* 1989).

Any activated carbon that catalyses the oxidation of H_2S completely to sulphuric acid suggests the possibility of its *in situ* regeneration with water. Centaur's® manufacturer claims that it can be regenerated to at least 80% of its original capacity by using 1046 litres of water per kilogram of spent adsorbent

(Hayden 1995) or 30 L/kg (Calgon Carbon Manual). As indicated, the two sources differ significantly in the amount of water needed.

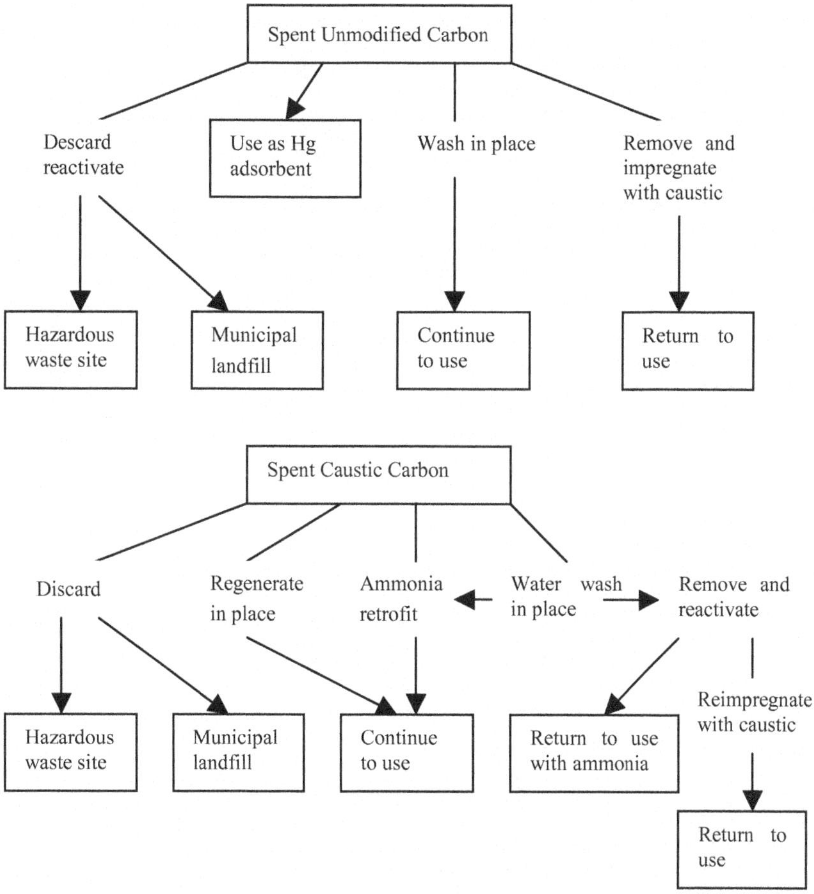

Figure 17.6. Options for regeneration/disposal of spent carbon.

The oxidation of H_2S in unmodified carbons yields a mixture of sulphur and sulphuric acid in various ratios depending on the type of carbon (Adib *et al.* 1999a,b, 2000a). Whether interim washing with water to extend the carbon life is worthwhile must be determined on an *ad hoc* basis. The performance of coconut-based carbon after four adsorption/ cold water regeneration cycles is shown in Figure 17.7. With an increasing number of regeneration cycles the breakthrough time decreases and the amount of deposited sulphur increases, causing the fouling of the catalyst due to the blockage of active sites (Bagreev *et al.* 2000a,b).

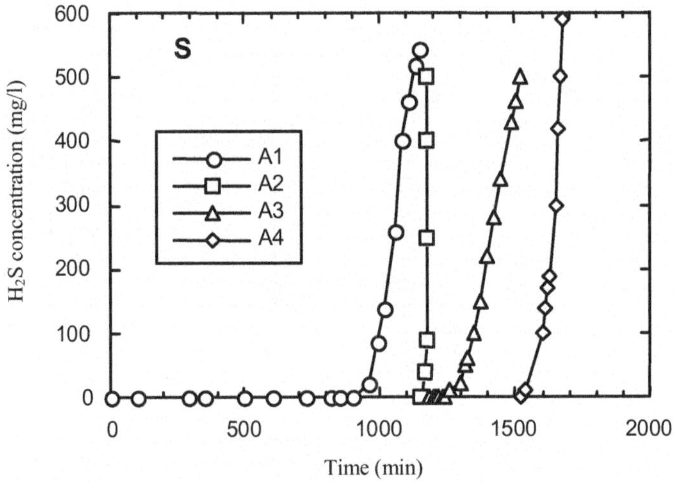

Figure 17.7. H_2S breakthrough curves obtained on cold water washed coconut-based carbon after 4 adsorption/regeneration cycles.

Ammonia-injected carbon can also be regenerated in place by the caustic-soaking procedure mentioned above. Otherwise, it can be treated essentially as spent unmodified carbon.

Spent carbon after H_2S adsorption with adsorbed/incorporated sulphur can find an application as adsorbents of mercury vapours owing to the high ability of mercury to form insoluble sulphides. So far the experimental results published have presented an enhanced capacity for mercury removal after sulphurisation of activated carbon surfaces and incorporation of organic sulphur compounds (Liu *et al.* 2000; Chang 1981).

17.4 CHARACTERISTICS OF CARBON BEDS

If an adsorbent bed were no deeper than the adsorption zone, then the deodorization would be effective only initially, and would begin to deteriorate as soon as it started to operate. Typical activated carbon bed depths in wastewater treatment plants are about one metre, with a linear air flow of about 0.26 m/s (≅ 50 ft/min) for a total contact time of some 4 seconds. The number of adsorption zones encompassed in this interval depends in part on the packing density and particle size distribution of the adsorbent granules, but should well reach 10 or more. Figure 17.8 shows a dual bed arrangement designed to provide a large capacity in a single vessel without imposing the excessive resistance to air flow that a single bed of double thickness would impose. Each bed is provided with three equally-spaced sampling ports. Such design imposes a strict requirement of balance of air flows through the two beds to insure uniform exposures to contaminants (Turk *et al.* 1993).

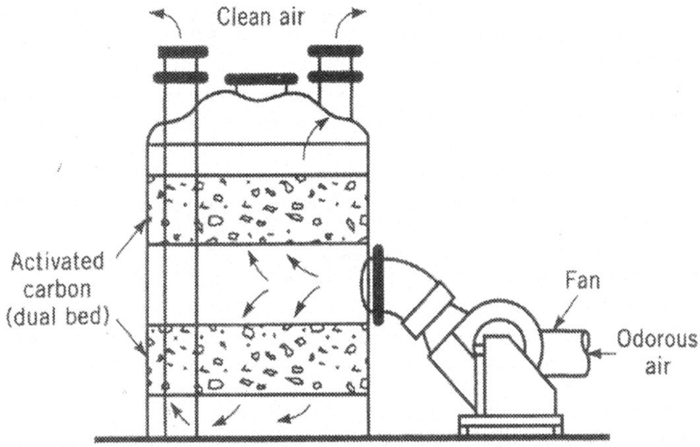

Figure 17.8. Illustration of North River dual bed vessel.

Other types of systems include thin bed (2-3 cm) adsorbers such as are used in air conditioning applications, adsorbents disposed on inert carriers, and fluidized or other moving adsorbent beds. The more stringent requirements normally imposed by wastewater odour control applications generally preclude the use of such alternatives.

As discussed in the preceding section and shown in Figure 17.6, some of the options available for spent carbons involve reuse of the carbon after suitable treatment or modification. It is therefore helpful to consider the recoveries that can be attained in such transformations.

Recoveries expressed in terms of mass of base carbon require analyses of the various components shown in Figure 17.9. In fact such mass balance exercises are rarely, if ever, carried out with activated carbons used for odour control in sewage operations. Recovery may also be expressed in terms of the ratios of volume of treated carbon to that of the virgin product. Losses of carbon volume require make-ups if the original vessel is to be refilled. However, we are not aware of any published data that bear on this question.

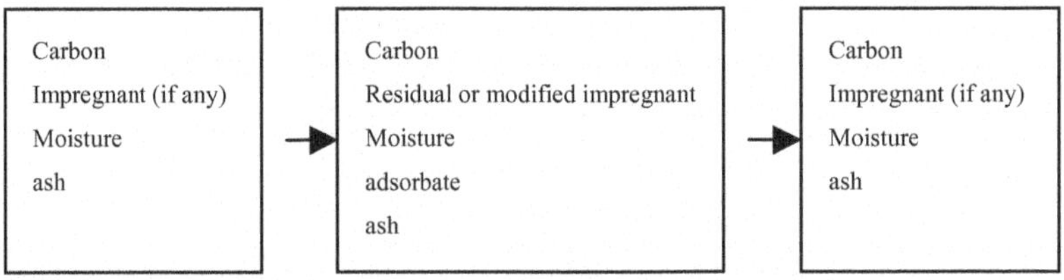

Figure 17.9. Components of carbon recovery systems.

17.5 CONTROL OF HYDROGEN SULPHIDE

The selection of activated carbon as a means of purifying air discharged from wastewater treatment plants is often predicated on its effectiveness in controlling hydrogen sulphide. Because thick bed carbon system may operate for many months, laboratory tests are greatly accelerated, and preferably should be completed within a working day. The test procedure now in general use world-wide challenges the carbon with humidified air at ambient temperature containing 1% (10,000 ppm v/v) of H_2S until a breakthrough concentration of 50 ppm of H_2S is detected. The breakthrough capacity is then calculated from the flow rate, the input H_2S concentration, time to breakthrough, and volume of the carbon test column, and is expressed as mg of H_2S per ml of carbon. A frequently specified capacity for wastewater treatment plants is about 0.14 g/ml. The test was originally designed and is applicable to comparison between different caustic-impregnated carbons, where the neutralisation of H_2S is practically instantaneous and its subsequent oxidation is catalysed by the high pH of KOH or NaOH. When unmodified activated carbon is used, oxidation of H_2S is slower and the reaction zone is broader, and the accelerated test underestimates the carbons' capacity.

A modified test has been developed recently (Adib *et al.* 1999a). It uses a lower challenge gas concentration of H_2S, smaller volume of carbon, and lower flow rate of air. Under such suitably modified conditions, the tested advantage of caustic impregnated over unmodified carbon disappears (Table 17.2). Of course, a fully valid test must be carried out under "real-life" conditions in a wastewater treatment plant at full scale. Such a test was conducted at the North River Water Pollution Control Plant in New York City. Comparisons between caustic and unmodified carbons over a span of 18 months (Bandosz *et al.* 2000) have shown that the latter continue to maintain capacity for removal of H_2S after the caustic carbons have been spent (Figure 17.10). The full-scale tests at North River continue.

Table 17.2. Comparison of the H_2S breakthrough capacity measured using two different tests.

Carbon	ASTM test	CCNY test
Caustic:		
7383C-B1	0.002	0.002
7383C-B2	0.093	0.080
Virgin:		
7383F-B2	0.020	0.021
WVA-1100	0.014	0.079
Maxsorb	0.003	0.026
S208C	0.029	0.055
Centaur®	0.066	0.068

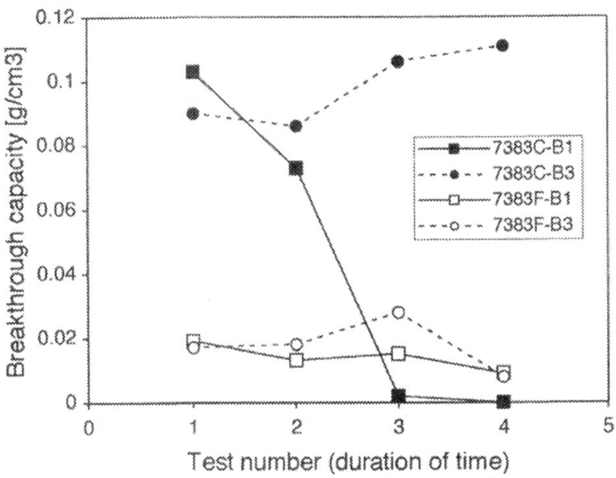

Figure 17.10. Changes in H_2S breakthrough capacity for the carbons from North River during 18 months serving as hydrogen sulphide adsorbents. 7383C - caustic impregated (●, ■), 7383F - unmodified (□, ○).

17.6 CONTROL OF ORGANIC ODORANTS (VOCs)

Much of the attention to adsorbent systems for odour control in wastewater treatment plants has been focussed on the removal of hydrogen sulphide, largely because that gas is well known to be highly toxic and odorous, and because it is easy to analyse and therefore a good surrogate for monitoring compliance with regulations. In many or most instances, however, H_2S is not the major odourant affecting the surrounding community, nor does it smell like typical sewer gas. Instead, the deodorization of effluents from wastewater treatment plants requires the removal of a complex mixture of organic gases and vapours that encompass a range of molecular weights, volatilities, and chemical functionalities. They include hydrocarbons, mercaptans and other reduced organic sulphur compounds, amines, and various oxygenates such as saturated and unsaturated carboxylic acids, esters, aldehydes, and ketones. They change from season to season and differ from one plant to another. Acid–base neutralisation or oxidation by air, whether in liquid phase or on an adsorbent bed, can deodorise some of these substances, but no single chemical conversion is applicable to all. It is for this reason that physical adsorption by granular activated carbon, preferably with its capacity undiminished by impregnants, is the method of choice for general control of odours generated by wastewater treatment plants. Of course, a preliminary stage that uses one or more of the chemical methods cited above can be useful in reducing the load on the carbon and thus extending its effective service life.

17.7 REFERENCES

Adib, F., Bagreev, A. and Bandosz T.J. (1999a) Effect of surface characteristics of wood-based activated carbons on adsorption of hydrogen sulphide. *J. Coll. Interface Sci.* **214**, 407-415.

Adib, F., Bagreev, A. and Bandosz, T.J. (1999b) Effect of pH and surface chemistry on the mechanism of H_2S removal by activated carbons. *J. Coll. Interface Sci.* **216**, 360-369.

Adib, F., Bagreev, A. and Bandosz T.J. (2000a) Analysis of the relationship between H_2S removal capacity and surface properties of unimpregnated activated carbons. *Environ. Sci. Technol.* **34**, 686-692.

Adib, F., Bagreev, A and Bandosz, T.J. (2000b) Adsorption/oxidation of hydrogen sulphide on nitrogen containing activated carbons. *Langmuir* **16**, 1980-1986.

Bagreev. A., Rahman, H. and Bandosz. T.J., (2000a) Study of H_2S adsorption and water regeneration of spent coconut-based activated carbon. *Environ. Sci.Technol.***34**, 2439-2446.

Bagreev. A., Rahman, H. and Bandosz. T.J., (2000b) Wood-based activated carbons as adsorbents of hydrogen sulphide: A study of adsorption and water regeneration process. *Ind. Eng. Chem. Res.* **39**, 3849-3855.

Bandosz, T.J and Le, Q. (1998) Evaluation of surface properties of exhausted carbons used as H_2S adsorbents in sewage treatment plants. *Carbon* **36**, 39-44.

Bandosz, T.J. (1999) Effect of pore structure and surface chemistry of virgin activated carbons on removal of hydrogen sulphide. *Carbon* **37**, 483-491.

Bandosz, T.J., Bagreev, A., Adib, F. and Turk, A. (2000) Unmodified versus caustics-impregnated carbons for control of hydrogen sulphide emissions from sewage plants. *Environ. Sci. Technol* **34**, 1069-1074.

Bansal, R.C., Donnet, J.B. and Stoeckli, F. (1988) *Active Carbon.* Marcel Dekker, New York.

Boehm, H.P. (1966) Chemical identification of surface groups. In: *Advances in Catalysis*; Vol. 16, Academic Press, New York, 179-274.

Brinker, C.J. and Scherer, G.W. (1990) *Sol-Gel Science.* Academic Press, New York.

Calgon Carbon Corporation Manual. Carbon Regeneration Using Water: Centaur HSV.

Chang, C.H. (1981) Preparation and characterization of carbon-sulfur surface compounds. *Carbon* **19**, 175-186.

Donnet, J.B., Papirer, E., Wang, W., Stoeckli, H.F. (1994) The observation of active carbons by scanning tunneling microscopy, *Carbon* **32**, 183-184.

Everett, D.H. and Powl, J. C. (1976) Adsorption in slit-like and cylindrical micropores in the Henry's Law region. *J. Chem.Soc. Farad. Trans. I* **72**, 619-636.

Fanning, P.E. and Vannice, M.A. (1993) A DRIFTS study of the formation of surface groups on carbon by oxidation, *Carbon* **31**, 721-730.

Farmerie, J.J. (1985) Regeneration of caustic impregnated carbon. *US patent 4,072,479.*

Ghosh, T.K. and Tollefson, E.L. (1986) Kinetics and reaction mechanism of hydrogen sulphide oxidation over activated carbon in the temperature range of 125-200° C. *Can. J. Chem. Eng* **64**, 969-976.

Gregg, S.J and Sing, K.S.W. (1982) *Adsorption, Surface Area and Porosity.* Academic Press, New York.

Hayden, R.A (1995) Process for regenerating nitrogen-treated carbonaceous chars used for hydrogen sulphide removal WIPO PCT WO9526230A1, 1995.

Hedden, K., Huber, L. and Rao, B. R. (1976) Adsorptive Reinigung von Schwefelwasserstoffhaltigen Abgasen, *VDI Bericht*, **253**, 37-42.

Leon y Leon, C.A. and Radovic, L. R. (1992) Interfacial chemistry and electrochemistry of carbon surfaces. In: *Chemistry and Physics of Carbon* (P.A. Thrower, ed.), Marcel Dekker, New York, Vol. 24, pp. 213-310.

Liu, W.; Vidic, R.D. and Brown, T.D (2000) Optimization of high temperature sulfur impregnation on activated carbon for permanent sequestration of elemental mercury vapors, *Environ. Sci.Technol.* **34**, 483-488.

Matviya, T. M. and Hayden, R. A. (1994) Catalytic carbon U.S. patent 5,356,849

Puri, B.R. (1970) Surface Complexes on Carbon. In: *Chemistry and Physics of Carbons.* (P.L Walker, Jr., ed), Marcel Dekker, New York, Vol. 6, pp. 191-282.

Steijns, M. and Mars, P. (1977) Catalytic oxidation of hydrogen sulphide. Influence of pore structure and chemical composition of various porous substances, *Ind.Eng.Chem.Prod.Res.Dev.* **16**, 35-41.

Turk, A. and Van Doren, A. (1953) Saturation of activated carbon used for air purification. *Agric Food. Chem.* **1**, 145-151.

Turk, A., Sakalis, E., Lessuck, J.,Karamitsos H. and Rago, O. (1989) Ammonia injection enhances capacity of activated carbon for hydrogen sulphide and methyl mercaptan. *Environ. Sci. Technol.* **23**, 1242-1245.

Turk, A., Karamitsos H., Mozaffari, J. and Loewi, R. (1991) Wastes generated from the removal of sulphide odors. In: *Recent Developments and Current Practices in Odor Regulations, Controls, and Technology*. Air and Waste Management Assn. Trans. Series 18.

Turk, A., Sakalis, E., Rago, O. and Karamitsos H. (1992) Activated carbon systems for removal of light gases. *Ann. N.Y. Acad. Sci.* **661**: 221-227.

Turk, A., Mahmood, K. and Mozaffari, J. (1993) Activated carbon for air purification in New York City's sewage treatment plants. *Water Sci. Technol.* **27**(7-8), 121-126.

18

Catalytic oxidation of odorous compounds from waste treatment processes

Piet N.L. Lens, Marc A. Boncz, Jan Sipma, Harry Bruning and Wim H. Rulkens

18.1 INTRODUCTION

18.1.1 Odour removal by oxidation

Oxidation of malodorous compounds such as volatile organic compounds (VOC), hydrogen sulphide (H_2S) and volatile organic sulphur compounds (VOSC) generally leads to a significant decrease or even a complete elimination of their odour nuisance. Unfortunately, the oxidation rates of these compounds are often slow under standard conditions of temperature and atmospheric

 Odours in Wastewater Treatment: Measurement, Modelling and Control edited by Richard Stuetz and Franz-Bernd Frechen. ISBN: 1 900222 46 9

pressure. Even the oxidation rates of easily oxidisable compounds can still be too limited for complete oxidation and elimination of their odour nuisance. The latter is often related to the very low odour threshold values of these compounds, which can easily be in the ppb range. For instance, although H_2S is oxidised quite easily under standard conditions (Kamei and Ohmoto 2000), it is one of the most frequent causes of odour nuisance from waste treatment processes.

Since the oxidation rate under ambient conditions is insufficient to prevent odour nuisance, environmental engineering applies a set of methods, in different technologies, to accelerate the oxidation rates. This can be done by using chemicals (see chapter 16), adsorption (see chapter 17) or microorganisms (see chapter 19). Chemical oxidation processes like hypochlorite oxidation are already applied for a long time. Another group of oxidation techniques, in which the oxidation rates are accelerated using a catalyst, i.e. catalytic oxidation processes, are used more and more (Table 18.1). These processes can be used either directly in the gas phase (e.g. photolysis) or by contact of the gas with a catalyst immobilised on a carrier material (catalytic incineration). Alternatively, Advanced Oxidation Processes (AOPs) like ozonation or Fenton oxidation can be used to treat aqueous gas scrubbing solutions in which the odour compounds are dissolved (Smet and Van Langenhove 1998). In the latter case, the scrubbing solution can be recycled after treatment to absorb newly released odours.

This chapter overviews the principles of catalytic oxidation processes and gives examples of their application in the area of odour abatement. It should be noted that many of these catalytic oxidation processes are studied mainly for their capacity to remove several well-defined compounds. Thus, these studies mainly report on the removal efficiency of these compounds and mostly do not present data on the odour removal efficiency.

Table 18.1. Overview of different categories of treatment methods for catalytic oxidation and/or advanced oxidation of odorous gas streams.

Treatment of gases		Treatment of gas scrubbing solutions and waste waters
Gas/gas interactions	Gas/solid phase interactions	Gas/liquid interactions*
UV Exposure	Catalytic incineration	Wet air oxidation
Ozonation	Catalytic membranes	Ozonation
Electron beam irradiation	Catalysts on support material	Fenton oxidation
	ZnO filtration	Sonolysis
	Fe_2O_3 filtration	Liquid redox processes
	Activated coal	

* Note that treatment of a scrubbing solution implies the application of a gas–liquid transfer method.

18.1.2 Oxidation products

Oxidation processes are mostly suitable to treat amines, phenols, cyanides, H_2S and mercaptans. They can also be used for the removal of halogenated aliphatic compounds and certain pesticides (USEPA 1987). The oxidation products are carbon dioxide, water and an acidic component (HX with X = Cl, Br, F, I) from halogens, SO_x from sulphur compounds, NO_x from amines, nitriles and nitrogen heterocycles and P_2O_5 from phosphorus compounds. Oxidation processes are less suitable for the degradation of highly halogenated organic species like PCBs, as these have a poor reactivity (Table 18.2). As many odorous compounds belong to the groups of compounds that are easily oxidised, oxidation processes can be very suitable deodorization techniques.

Table 18.2. Destructibility of VOCs by catalytic incineration (Nakajima 1991).

VOC	Relative Destructability	Catalytic ignition Temperature (°C)
Formaldehyde	High	< 30
Methanol		< 30
Acetaldehyde		100
Trimethylamine		100
Butanone		100
n-Hexane		120
Phenol		150
Toluene		150–180
Acetic acid		200
Acetone		200
Propane		250–280
Chlorinated hydrocarbons	Low	400

Several catalytic oxidation processes have been described for the catalytic oxidation of H_2S. In the design of systems aimed at removing sulphur-containing compounds, it is important to take into account that conversion to a non-odorous gaseous sulphur species is not possible. Thus, no parallel exists for carbon and nitrogen removing processes, which mainly rely on the conversion of C and N containing compounds into the non-odorous gases CO_2 and N_2, respectively. The end product of the catalytic oxidation of H_2S depends on the applied process conditions (especially pH) and can be S^o, thiosulphate, sulphite or sulphate. S^o is a recoverable solid product that can be reused as fertiliser or in sulphuric acid production (Lens *et al.* 1998). However, chemically produced S^o is hydrophobic and can therefore block piping or filtration materials. Thiosulphate and sulphite are soluble and contribute to the chemical oxygen demand (COD) of the water. Moreover, thiosulphate, sulphite and sulphate are not removed from the water phase and thus remain a potential source of H_2S production when the water is exposed to anaerobic conditions.

18.1.3 Catalytic oxidation and advanced oxidation processes

In catalytic oxidation processes, the chemical oxidation rate is accelerated by a catalyst, which can be both inorganic (e.g. noble metals) or organic (e.g. quinones). In contrast, Advanced Oxidation Processes (AOPs) (Glaze 1987) are oxidation processes where the applied oxidant not only directly oxidises the substrate but also decomposes into radicals like the hydroxyl radical, which are even more reactive oxidants. The presence of these highly reactive radicals (Glaze 1987; Gulyas *et al.* 1995) makes the processes very powerful and capable of oxidising compounds that often cannot be degraded by conventional techniques.

The way the radicals are generated depends on the type of process, and some processes may be more efficient in generating radicals than others, given the conditions under which the process has to be applied. Sources of radicals can be oxygen, ozone or hydrogen peroxide. The processes in which these oxidants can decompose into radicals are wet air oxidation, ozonation, the peroxone process, H_2O_2/UV oxidation and the Fenton process (see 18.3). In some processes, the radicals can be directly generated from the polluting organics, like in the case of electron beam irradiation.

AOPs are mostly employed in the disinfection and treatment of liquid waste streams like scrubbing fluids, although some of these techniques (ozonation, electron beam irradiation, and photolysis by ultraviolet light) are also suitable for destruction of odorous compounds in the gas phase. In case of low concentrations, odorous compounds can be treated by AOPs after a concentration step.

Table 18.3. Methods to concentrate gaseous H_2S prior to its catalytic treatment.

Material	Apparent surface area (g/m^2)	Micropore volume (cm^3/g)	Adsorption capacity ($mol \times 10^5$ /g) Dry	Moist	Reference
Molecular sieve 13X	490	0.25	10.0	0.8	Tanada *et al.* 1982
Activated carbon (untreated)			5.2	6.2	Meeyoo *et al.* 1995
Activated carbon (heat treated at 200 ^{0}C)	1005	0.38	0.2	2.1	Meeyoo *et al.* 1995
Silica gel	407	1.5	1.6	unknown	Chou *et al.* 1986

Table 18.3 overviews a number of concentration methods for H_2S. The AOPs can then be applied to the gas stream evolving during desorption of the adsorbent. When applying off-gas scrubbers, the odours will be dissolved in a liquid phase to which AOPs can be applied (section 18.3). In the latter case, odour removal is a function of its solubility in the solution, the total effective gas-liquid contact area, the concentration and the residence time of the gas in the scrubber.

18.2 CATALYTIC PROCESSES FOR VOC AND H_2S TREATMENT IN THE GAS PHASE

18.2.1 Catalytic incineration

18.2.1.1 Principles of catalytic incineration

Combustion is an effective technique for odour control. Its efficiency depends on the level of complete combustion as incomplete combustion could result in even increased odour nuisance. Despite the fact that oxidation of VOC is exothermic, the reaction heat is insufficient to maintain the temperature required for oxidation, due to the low concentration of VOC present in the gas phase. Therefore, additional heating or a catalyst must be provided to ensure the complete oxidation of the odorous compounds at lower temperatures.

Catalytic incineration (Table 18.4) is the complete chemical conversion of gaseous compounds with oxygen at a certain temperature (below or above 100 °C) and pressure (one or more than one atm.) while the gases are in contact with a solid material (catalyst) to increase the oxidation rate. Gas phase catalytic incineration technology has a lot of advantages over other gas phase treatment techniques. In the concentration range of 20 to 60 ppm, it is sometimes cheaper than granular activated carbon adsorption. It requires less fuel and less expensive construction materials than high temperature treatment processes such as incineration (Snape 1977; Siebert *et al.* 1984).

Most catalytic incineration operations still require elevated temperatures to ensure a complete reaction at a suitable fast rate (Table 18.4). Therefore, either the catalyst bed is heated, or the feed stream is heated. Inlet gas streams are kept at 50 – 150 °C higher than the ignition temperatures (Nakajima 1991). When the preheating uses an internal flame, the process is complex, involving both the products of combustion in the flame and the processes occurring at the catalyst. The operating parameters (temperature and space velocity) for an application depend on the destructibility of the VOC (Table 18.2).

Table 18.4. Overview of some catalytic VOC incineration processes.

Compound	Catalyst	Operation T (°C)	Reference
Benzene, butylacetate, cyclohexane, toluene, methanol, acetylene, butane, chlorobutane, chlorobenzene	U oxide	300–450	Taylor *et al.* 2000
Linear alkanes (C_1-C_4)	U oxide		Taylor and O'Leary 2000
Benzene, n-hexane and oil-refinery gas	Cr_2O_3	240–400	Wang and Chou 2000
Butanone and toluene	Pt, Ni, Cr alloy	120–220	Lou and Chen 1995
Chlorobenzene, Chlorinated methanes, chlorinated aromatics	La, Co, Mn		Sinquin *et al.* 1999
Trichloroethylene	Pt, Pd	250–550	Gonzalez Velasco *et al.* 2000

18.2.1.2 Mechanisms of catalytic incineration

The oxidation of VOC (Figure 18.1) by a catalyst involves species at the surface and/or in the vapour phase, depending on the mechanism of catalysis. The Langmuir-Hinshelwood mechanism (Figure 18.1A) requires the adsorption of each species at nearby sites and subsequent reaction and desorption.

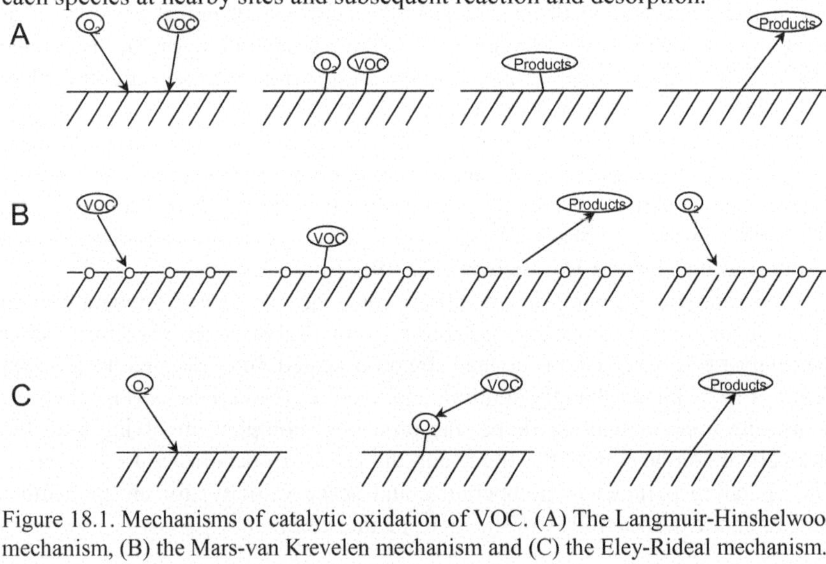

Figure 18.1. Mechanisms of catalytic oxidation of VOC. (A) The Langmuir-Hinshelwood mechanism, (B) the Mars-van Krevelen mechanism and (C) the Eley-Rideal mechanism.

The Mars-van Krevelen mechanism of catalytic oxidations (Figure 18.1B) explicitly requires a redox process in which oxygen is consumed from the catalyst surface by reaction with the compound. The catalyst-surface oxygen is then replenished by oxygen from the vapour phase. The Eley-Rideal mechanism (Figure 18.1C) is similar to the Mars-van Krevelen mechanism, except that the products are formed from adsorbed oxygen and the compound in the gas phase.

For metal or non-reducible oxide catalysts, excess oxygen in the gas phase means that the catalyst surface is well covered with oxygen and that little compound is adsorbed. Thus, the Eley-Rideal mechanism is expected to be important. For metal oxide catalysts containing readily reducible metals, the Mars-van Krevelen mechanism is important.

Metal oxides that are n-type semiconductors are rich in electrons and are generally not highly active as oxidation catalysts. Vanadium pentoxide is a notable exception. In contrast, in p-type semiconductors' conductivity is based on the electron flow into positive holes. The electron-deficient surfaces of such metal oxides readily adsorb oxygen, and if the adsorption is not too strong, they are active catalysts. Insulators which are thermally stable and not friable have value as supports for catalytic active metal oxides or noble metals. Alternatively, ceramic materials or zeolites can be used as support material.

More than one type of surface oxygen species can be involved: adsorbed dioxygen (O_2), ions (O^{2-}, O_2^{2-}) or radical ions ($O^{\bullet-}$, $O_2^{\bullet-}$) on the surface or incorporated into the lattice of the catalyst (Sokolovskii 1990). For all mechanisms, a key factor is the strength of the interaction between the surface and the oxygen (atom, molecule or ion) required for oxidation of the compound. If the oxygen is too tightly bound to a surface, that surface is not highly active as a catalyst. Similarly, if the interaction is too weak, the surface coverage with oxygen is low and the catalytic activity is consequently diminished. Various thermodynamic parameters can be used to describe the strength of adsorption of both oxygen and the compound, i.e. initial heat of adsorption for metal surfaces (Bond 1987) or reaction enthalpy for reoxidation of metal oxides (Satterfield 1991). Since the maximum oxidation rate of a catalysts is a function of these thermodynamic parameters, selection of a catalyst and the operation conditions strongly depend on the compound to be destroyed and its destructibility (Table 18.2).

Complete oxidation of each compound in a multi-component stream is ensured with high temperatures and excess oxygen, and mixed-oxide catalysts are frequently used (Table 18.4). The mixed oxides, especially when promoted with alkali or alkaline earth metal oxides, frequently have activities that are different from the combination of properties of the compounds, and in general activities are higher. This can be attributed to the availability and mobility of the different forms of available oxygen and accessibility of different binding sites with various energy levels.

18.2.1.3 Catalytic incineration of VOC

In a typical application, more than one VOC is degraded. Frequently, oxidation at a catalyst is competitive under stoichiometric or partial oxidation conditions, or one VOC inhibits the catalytic oxidation of another. Using excess oxygen in air overcomes these problems. On the other hand, several odorous compounds, especially sulphurous compounds, can poison the catalysts. For instance, catalytic incineration of a mixture of dimethylsulphide and ethanol, typically emitted by the petrochemical industry, in a Pt/Al_2O_3 fixed-bed catalytic reactor fails because of poisoning of the catalyst (Chu and Lee 1998). The inhibition of the catalyst can be partially overcome by operating the reactor at higher temperatures (> 300 °C).

Many different reactor configurations exist, depending on the location of the heating system, type and number of catalyst bed and operation temperature. Figure 18.2 shows the catalytic solvent abatement (CSA) process, developed to treat waste gases containing both halogenated and non-halogenated VOC. Conversion of each VOC requires a combination of catalysts. Further, scrubbing the effluent from the reactor is necessary to remove the acidic compounds generated.

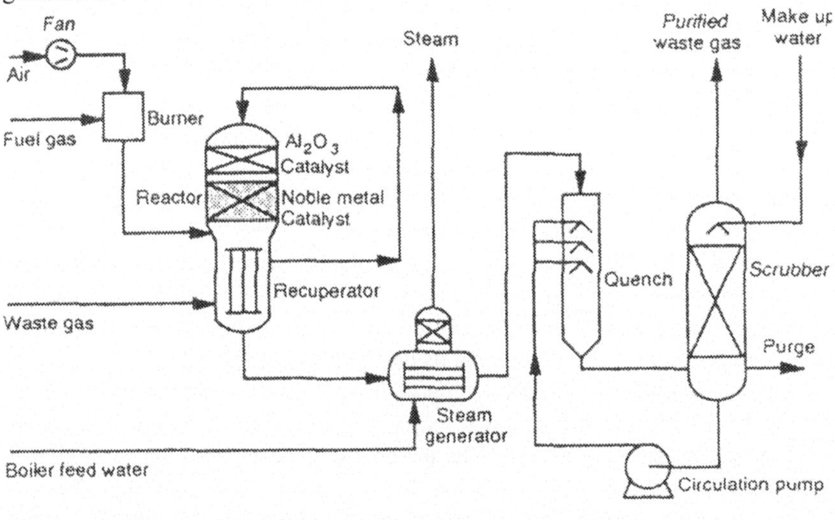

Figure 18.2. Overview of the CSA process for waste gas incineration (after Tebodin, The Hague, the Netherlands; Cha *et al.* 2000).

18.2.1.4 Catalytic incineration of H_2S

Lee *et al.* (1999) report on the catalytic combustion of biogas, containing CH_4 and H_2S (15–40 ppm) using palladium and platinum-based monolithic catalysts at 220 °C (Figure 18.3). Palladium is the best catalyst for methane oxidation, but is partially deactivated by H_2S (Meeyoo *et al.* 1998). In contrast, the platinum catalyst was not deactivated by H_2S.

Chu and Wu (1998) reported on the catalytic incineration of ethylmercaptan in a MnO/Fe_2O_3 fixed bed catalytic reactor. They evaluated the effects of operational parameters such as inlet temperature, space velocity, C_2H_5SH and O_2 concentration. The Langmuir-Hinselwood model was able to describe the observed reaction kinetics.

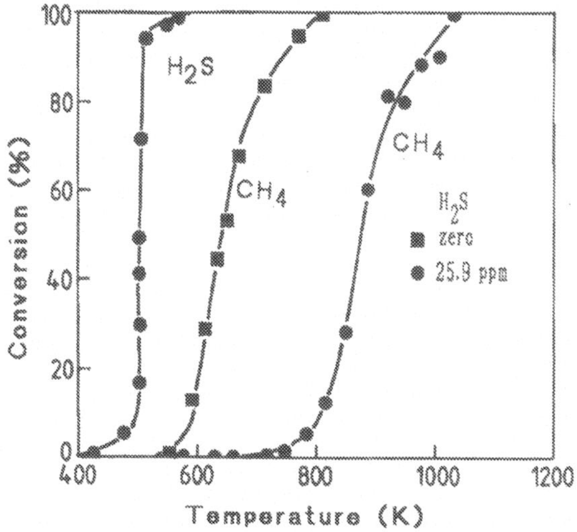

Figure 18.3. Light off curves for hydrogen sulphide, methane and hydrogen sulphide-methane mixtures in the presence of a Pd catalyst (Lee *et al.* 1999).

18.2.2 Dry oxidation processes

VOC, H_2S and VOSC removal by dry oxidation processes mainly relies on filtration of the gas using different types of packing material. The filter material can have catalytic properties (e.g. beds of activated carbon or metals) or be the support of a layer of a catalyst, which is coated onto the filter material.

18.2.2.1 Dry oxidation of VOC

Catalytic membranes combine selective transport of compounds in the gas phase with chemical reactions (Figure 18.4). Moreover, the selective removal of products from the reaction site increases the conversion of product-inhibited or thermodynamically unfavourable reactions (Giorno and Drioli 2000). The catalyst can be flushed along the membrane module, segregated with a membrane module or immobilised in or on the membrane by entrapment, gelification, physical adsorption, ionic binding, covalent binding or cross-linking.

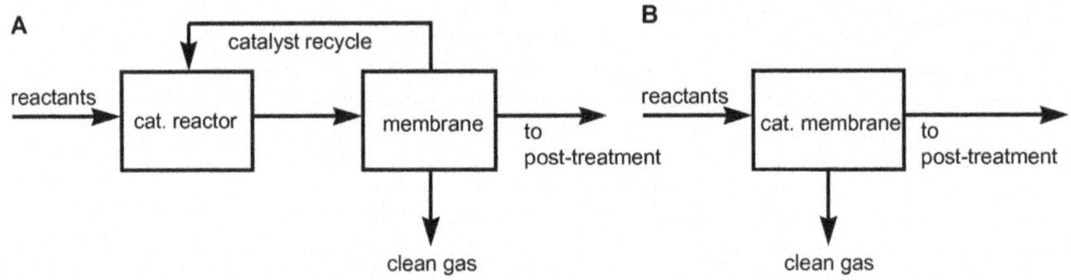

Figure 18.4. Main configuration types of membrane reactors. (A) Reactor combined with a membrane operation unit. (B) Reactor supplied with a membrane active as catalyst and separation unit.

18.2.2.2 Dry oxidation of H_2S

Table 18.5 overviews different catalysts used for H_2S oxidation by dry oxidation. Figure 18.5 shows a simplified scheme of a filtration process. Reactors R1 and R2 are switched alternatively from absorption to regeneration. The process gas passes a blower and enters reactor R1. The gas is heated, the relative humidity is controlled and, if necessary, air or oxygen is supplied. In reactor R2, the catalyst is regenerated which is switched into a recycle loop. A blower recirculates inert gas. Water and desorbed sulphur from the catalyst are condensed and stored in separate sections of the condenser.

Impregnating filter media such as granular activated carbon or activated alumina with a reactive chemical or a catalyst can convert the absorbed contaminants to less odorous compounds. For example, an air filter medium consisting of activated alumina and potassium permanganate can simultaneously remove H_2S, mercaptans, and other VOC.

Much research has been done to optimise the catalytic activity of activated carbon. This mainly involves modifications in the activated carbon production process, e.g. different starting materials (Adib *et al.* 2000b) or inclusion of urea (Adib *et al.* 2000a) or K_2CO_3 (Przepiorski and Oya 1998). K_2CO_3-loaded activated carbon fibres show a high deodorization ability against 30 ppm H_2S gas in air at ambient temperatures (Przepiorski *et al.* 1999). The latter authors found that H_2S did not diffuse to the most interior parts of the fibre, but was oxidised to S^o around the outer regions of the fibre.

Table 18.5. Overview of catalysts used for the removal of H_2S by dry oxidation.

Catalyst	Product	Reference
Transition metals (Mn, Cu, Ni)		Andreev *et al.* 1996b
MoS_2, CdS, $NiPS_3$	$S_2O_3^{2-}$	Iliev *et al.* 2000
Co-phthalocyanine, Metal chalcogenides	$S_2O_3^{2-}$	Iliev *et al.* 2000
Zn-phthalocyanine + hν	SO_4^{2-}	Iliev *et al.* 2000
V_2O_5/SiO_2 and Fe_2O_3/SiO_2	S^o	Park *et al.* 1998
TiO_2/SiO_2	S^o	Chun *et al.* 1998
V/Sb	S^o	Li and Shyu 1997
Ni(OH)/$LiNiO_2$	S^o	Andreev *et al.* 1996a
Au/Fe		Matsumoto *et al.* 1993
Cr(VI) contaminated sediment		Thornton and Amonette 1999

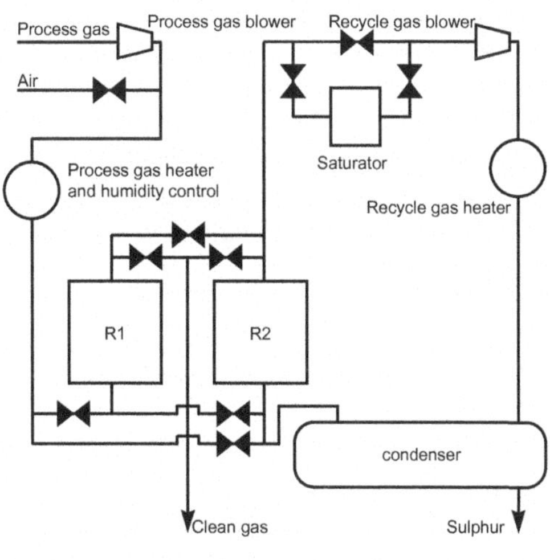

Figure 18.5. Schematic drawing of the Selox Process.

18.2.2.3 Ozonation of odorous compounds

Odorous compounds can be oxidised by ozone (O_3) alone, without the need for a catalyst or UV light (see 18.2.3.1). In the gas phase, ozone has the same preference for reaction with electron-rich parts of the (in)organic compounds present. However, the reaction mechanism may differ as suggested by the different order of reaction in both reactants in the gas phase compared to the situation in aqueous solution (Kuo *et al.* 1997). Amines, which can cause serious odour problems, are known to react fast with O_3, and ozonation has been applied successfully to treat the air from a laboratory-animal breeding facility (Pan *et al.* 1995). Also air contaminated with e.g. tobacco smoke can be purified relatively easy by ozonation.

Ozonation becomes even more effective when combined with UV light. The UV light can dissociate the O_3 molecule, yielding reactive radicals, capable of oxidising (in)organic compounds at rates higher than O_3 itself will do. Thus several volatile organic compounds were shown to be removed effectively by the UV/O_3 combination (Shen and Ku 1997, 1999).

18.2.3 Photolysis processes

18.2.3.1 UV irradiation

Ultraviolet light has played an important role in water and effluent desinfection since the early years of the twentieth century, when the powerful effect of sunlight for oxidation purposes was discovered (Peyton 1990). UV light can also be applied for gas phase catalytic reactions, as UV can penetrate very deep in a gas stream. The ability to produce UV light at or very close to the wavelengths corresponding to the absorption bands causes significant cleaving of the chemical H-S bonds, resulting in powerful oxidation of the odour-forming compounds. Photolysis of sulphurous compounds is initiated by scission of the S-H bond (McClean *et al.* 1999):

$$H_2S + h\nu \rightarrow HS^{\bullet} + H^{\bullet}$$
$$CH_3SH + h\nu \rightarrow CH_3S^{\bullet} + H^{\bullet}$$

The radicals formed in this reaction can effectively be degraded further by reaction with oxygen (air). The efficiency of UV radiation can be considerably improved by combining it with ozonation. When applying photolysis in an oxygen-rich environment, ozonation will already account for some part of the degradation as O_3 is formed from oxygen when UV light is present. This principle is applied in the photolysis of odorous compounds in domestic garbage waste containers (Makaly-Biey and Verstraete 1999).

18.2.3.2 Photocatalytic processes

Photocatalysis is the combination of using a semiconductor, a photo-catalyst, and UV or visible light for the conversion of organic or inorganic compounds. When illuminated a photo-catalyst oxidises and reduces absorbed compounds, which can result in their total mineralisation under ambient and atmospheric conditions. The mechanism is based on excitations of electrons to the conduction band by the photons, which results in electron holes, acting as oxidant, at the illuminated surface of the catalyst particle. At the non-illuminated surface and in the pores, excess electrons act as reductor. Powerful catalysts are TiO_2 and CdS. Typical gas phase compounds that can be degraded are hydrogen sulphide (Borgarello *et al.* 1985; Sabaté Cervera-March *et al.* 1990), organic odour compounds (Suzuki *et al.* 1991), naphthalene (Guillard *et al.* 1993), ammonia (Mozzanega *et al.* 1979; Pichat *et al.* 1982; Cant and Cole 1992) and nitric oxide (Ravindranathan Thampi *et al.* 1990; Cant and Cole 1992).

Figure 18.6 shows a plot of a change in concentration of gaseous acetaldehyde as a function of the UV irradiation time. The initial acetaldehyde concentration was 300 ppm. When 1 mW/cm^2 UV light (365 nm) was irradiated on the TiO_2 powder (0.25 g) laid on a small plate (8.6 cm^2), the acetaldehyde concentration starts to decrease with the concomitant production of CO_2. Photocatalytic processes have also been applied for treatment of off-gases from a soil vapour extraction well at a chlorinated solvent spill site using a TiO_2 photocatalytic reactor (Read *et al.* 1996).

Besides gas phase degradation, also dissolved compounds can be degraded by photocatalysis (Bahnemann *et al.* 1996; Benoit Marquie *et al.* 2000; Klare *et al.* 2000). The efficiency of the process is high enough to allow the use of sunlight as the UV source even in more moderate regions like Western Europe. Figure 18.7 gives an example of the photocatalytic degradation of ethanol dissolved in water.

18.2.3.3 Electron beam irradiation

Electron beam irradiation is a gas phase reaction technique in which a beam of high-energy electrons is led through the gas mixture to be treated. The corona-type discharge that develops generates ozone when oxygen is present (Matzing *et al.* 1996) and this is basically how the Siemens process for ozone generation works. In addition, several radical species like the hydroxyl radical can also be formed. As gases are not very dense the electron beam can protrude well into the gas mixture and larger volumes can be treated. The process can be used effectively to treat gas mixtures including sulphurous and nitrogen compounds, which will be oxidised to form solid ammonium sulphate, which can be sold as fertiliser (Hirota *et al.* 1996; Namba *et al.* 1998; Doi *et al.* 2000). The technique is already in use for the treatment of off-gases of electricity power plants (Matzing *et al.* 1996; Cramariuc *et al.* 2000; Doi *et al.* 2000).

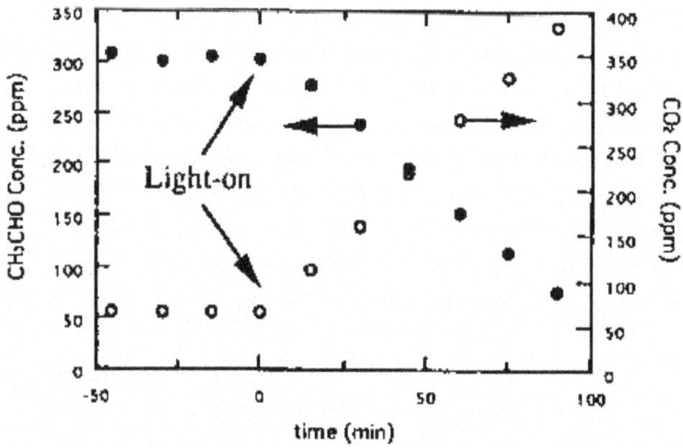

Figure 18.6. Photocatalytic degradation of gaseous acetaldehyde (Noguchi *et al.* 1996).

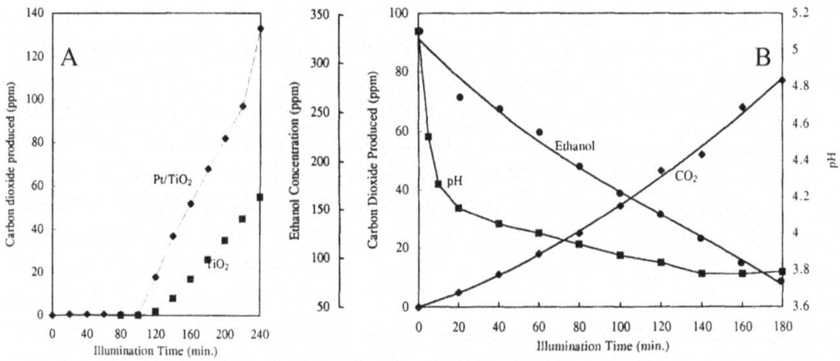

Figure 18.7. Photocatalytic mineralisation of dissolved ethanol. (A) Effect of catalyst type (1.0 g/l catalyst, initial ethanol concentration 670 ppm; initial pH 10.9). (B) Evolution of reaction products as function of time (1% Pt/TiO_2 catalyst, initial ethanol concentration 333 ppm, initial pH 5.1). 15 W black UV light lamp, 600 ml solution, 26 °C (Chen 1997).

18.3 CATALYTIC OXIDATION TECHNOLOGIES FOR SCRUBBING LIQUIDS

The processes described in section 18.2 dealt with the direct catalytic treatment of the gas phase. Catalytic processes are not restricted to this application, but can also be used to treat scrubbing liquors. As in most cases an aqueous solution will be used as the scrubbing liquid, mainly water-soluble compounds will be removed. However, recent advances with extractive membrane reactors also enable the removal of compounds with a low air-water partitioning coefficient by these water based catalytic oxidation technologies (Reij *et al.* 1998). The following section gives an overview of liquid-phase catalytic oxidation processes.

18.3.1 Wet air oxidation / oxidation with oxygen

One of the first AOPs is wet air oxidation (WAO) and this is currently a commercially proven technology. In the WAO process, oxidation is performed with oxygen (air) as the primary oxidant, under conditions of high pressure (3.5–15 MPa) and temperature (150–300 °C). Under these conditions radicals like the hydroperoxide anion radical ($HO_2^{\bullet -}$) and the superoxide anion radical ($O_2^{\bullet -}$) are formed (Mantzavinos *et al.* 1996), which can react (with water) to hydroxyl radicals. This hydroxyl radical, having an oxidation potential of 1.77 V, can subsequently oxidise most organic material present in the solution. Additionally, both radicals ($HO_2^{\bullet -}$ and $O_2^{\bullet -}$) can also directly oxidise organic material by a radical chain reaction, in which the oxygen present plays an important role.

WAO has been successfully used to degrade hydrocarbons (including PAHs), some pesticides, phenolic compounds, cyanides, and other organic compounds (USEPA 1987).

Jagushte and Mahajani (1999) reported on the use of WAO to remove H_2S and $S_2O_3^{2-}$ from spent caustics. They reported a complete conversion of 1×10^{-1} M Na_2S to SO_4^{2-} in 10 min at 0.69 MPa oxygen partial pressure and a temperature of 150 °C in the absence of a catalyst. In the presence of 3.25×10^{-2} M $CuSO_4$ as catalyst, complete conversion of the Na_2S was obtained in 8 min at 120 °C.

The applicability of wet air oxidation for removing odorous compounds from gases is however rather limited as concentrations obtained in the scrubbing liquid are generally too low to make the WAO-process cost effective.

18.3.2 Ozonation

Owing to its high reactivity, ozone has been used as an oxidant in water treatment since the beginning of this century, making ozonation an even older technique than wet air oxidation. Ozonation can be applied both for the treatment of gases and liquids but is most frequently used to treat aqueous solutions.

18.3.2.1 Reaction mechanisms of ozonation

The principles of ozonation only became fully understood after Weiss (1935), Hoigné and Bader (Hoigné and Bader 1976; Hoigné 1998) discovered the role of radical reactions in the ozonation process. Ozone, when dissolved in water, not only acts as an oxidant towards organic solutes, directly oxidising these compounds by addition of oxygen to an (aromatic) double bond (Figure 18.8). It also decomposes in a reaction with the hydroxyl anion or the hydroperoxyl anion, starting a complex sequence of reactions that ultimately yields molecular oxygen, as summarised by:

$$2\,O_3 \rightarrow 3\,O_2$$

The reaction proceeds by a cyclic mechanism as depicted in Figure 18.9.

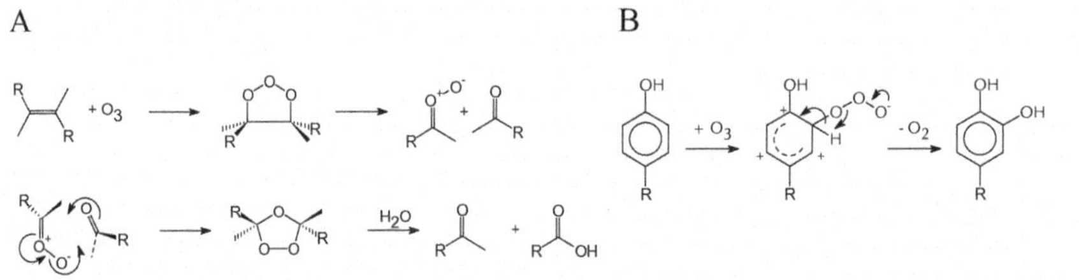

Figure 18.8. Mechanisms for ozonation of (A) alkenes (Criegee 1975) and (B) aromatic compounds (Decoret *et al.* 1984).

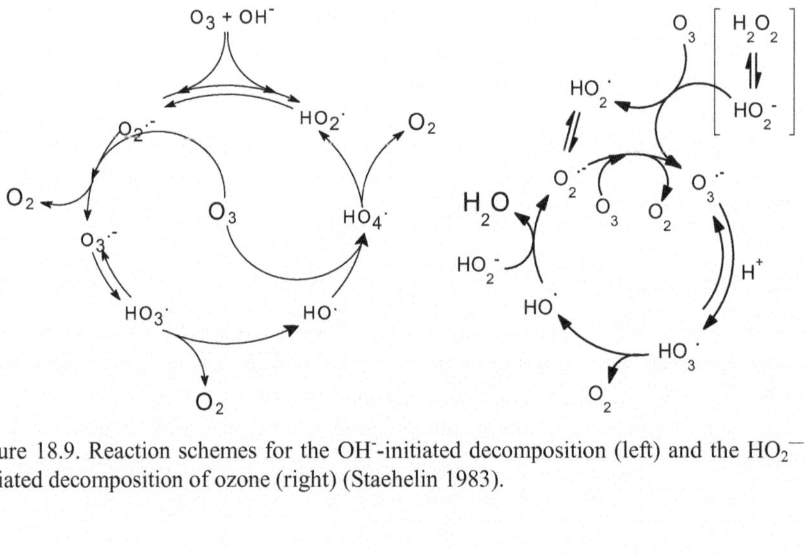

Figure 18.9. Reaction schemes for the OH^--initiated decomposition (left) and the HO_2^- initiated decomposition of ozone (right) (Staehelin 1983).

At steady state, radicals will always be present in the reaction mixture, but the concentration of radicals will depend on (i) pH, (ii) the presence of radical scavengers and (iii) the organic solutes in the solution. As long as no radical scavenging takes place, ozone will decompose at a rate depending on the pH and steady-state radical concentrations will increase at increasing pH. However, radicals can be scavenged from the solution by compounds like the carbonate anion in a reaction sequence given below (Neta *et al.* 1988):

$$HO^{\bullet} + CO_3^{2-} \rightarrow HO^{-} + CO_3^{\bullet-}$$
$$CO_3^{\bullet-} + O_2^{\bullet-} \rightarrow CO_3^{2-} + O_2$$

Hence carbonate catalytically reduces the radical concentration (Hoigné and Bader 1976; Hoigné 1998). The opposite will happen when alkenes are present: oxidation of the double bond by the Criegee ozonation mechanism (Figure 18.8A) yields two aldehyde groups and a hydrogen peroxide molecule. Owing to the low pK_A of H_2O_2 (pK_A= 11.8) and the high reactivity of the hydroperoxyl anion towards ozone the initiation reaction of the right part of Figure 18.8A will become significant and the radical concentration will increase. This effect is especially found when ozonating lignin-containing waters (Gulyas *et al.* 1995; Magara *et al.* 1998). Another reaction sequence that can increase radical concentrations was described by Staehelin and Hoigné (1985) and occurs if compounds are able to react with a hydroxyl radical and oxygen consecutively. The oxygen can than be eliminated from the reacting compound to form a superoxide radical anion ($O_2^{\bullet-}$), which rather selectively transfers an electron to ozone to start another chain of radical reactions (Staehelin and Hoigné 1985).

Whether a direct mechanism (with ozone acting as the primary oxidant) or a radical mechanism (with the hydroxyl radical as the primary oxidant) will prevail, depends on the reaction conditions. In order to have a radical process two conditions have to be fulfilled: the ratio of the initiation over the scavenging reactions needs to allow for radicals to be present, and the rate of initiation needs to be high enough to compete with the rate of the direct oxidation of the organic solutes. When compounds are oxidised that react with ozone at very high rates, initiation reactions may be outcompeted by these oxidation reactions (Boncz *et al.* 1997).

18.3.2.2 Ozonation of organic compounds

Compounds that are known to react fast with ozone are: amines, phenols and alkenes. Figure 18.10 overviews the ozonation rates of 10 groups of frequently occurring compounds and clearly shows that the oxidation with ozone will be far more selective than the oxidation with the hydroxyl radical derived from this oxidant. Where the reaction rate constants of the reaction with ozone differ by 10 orders of magnitude, the reaction rate constants for the radical reaction only differ by 3 orders of magnitude. Which kind of oxidation process (oxidation by molecular ozone or oxidation by hydroxyl radicals) will be preferred, strongly depends on the relative reactivities of the different compounds present in the wastewater.

Figure 18.10 reveals a few trends: aliphatic ketones and alcohols hardly show any reactivity, benzoic acids aren't very reactive as well, but phenols and amines (both aromatic and non-aromatic) react with ozone at a very high rate. The latter compounds will most likely be oxidised by a direct reaction in all situations, even when the reaction conditions are such that a relatively high concentration of radicals is produced. When these fast-reacting compounds are present, the process will most likely be mass-transfer controlled, especially since O_3 is a sparingly soluble gas, and specialised techniques are required to achieve a high degree of dissolution (Paillard and Blondeau 1988). This low solubility however will not immediately significantly affect the selectivity of the process, unless steric factors play a role in the system (Gould 1987). When only radical reactions can occur, much smaller differences in reactivity are to be expected.

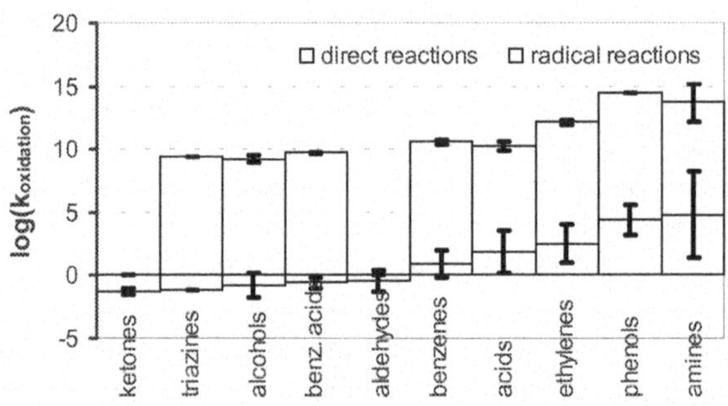

Figure 18.10. Average reaction rate constants for the oxidation of groups of organic compounds with ozone (direct reactions) or the hydroxyl radical (radical reactions) in water (based on reactivity data for 55 organic compounds). No reactivity data were included for the reaction rates of ketones and aldehydes with the hydroxyl radical.

18.3.2.3 Ozonation of sulphur compounds

In scrubbers, chemical oxidation of the VOSC can be obtained by dosing hypochlorite (Van Durme *et al.* 1992), hydrogen peroxide (McNeillie 1984) or ozone to the scrubbing liquid. Hypochlorite is considered to be the most effective oxidant for the chemical scrubbing of VOSC (Van Durme *et al.* 1992). However, scrubbing with hypochlorite is highly undesired when NH_3 and lower amines are present since they are preferentially chlorinated to more odorous or even toxic compounds (chloroamines) rather than being oxidised by hypochlorite (Van Durme *et al.* 1992).

Besides hypochlorite, sulphur containing compounds present in the water phase are efficiently oxidised with ozone (Anderson 1984; Hwang *et al.* 1994). Sulphides can be efficiently removed by ozonation, according to:

$$H_2S + O_3 \rightarrow SO_2 + H_2O \rightarrow H_2O + S + O_2$$
$$CH_3SH + O_3 \rightarrow CH_3\text{-}S\text{-}S\text{-}CH_3 \rightarrow CH_3SO_3H + O_2$$

Sulphonic acids are formed as products, and, depending on the structure of the rest of the molecule, further degraded. However, the sulphonate group is strongly deactivating, and thus the aromatic sulphonate intermediates are resistant to further oxidation. This leads to the accumulation of these well soluble compounds.

Direct gas phase ozonation is usually too slow to be of interest, except for H_2S (Anderson 1984) or in combination with UV radiation (see 18.2.2.3). Nevertheless, Laplanche *et al.* (1984) achieved a complete H_2S and methanethiol removal in emissions of a WWTP at significant lower reagent cost in comparison with hypochlorite oxidation.

18.3.3 Fenton oxidation

In the Fenton oxidation process, hydrogen peroxide (H_2O_2) is added as the oxidant. As is the case with ozonation, in the Fenton process radicals are generated from this added oxidant. Its decomposition into hydroxyl radicals is accomplished in the presence of Fe(II), according to the following equation (Pignatello 1992):

$$Fe^{2+} + H_2O_2 \rightarrow Fe^{3+} + OH^- + OH^\bullet$$

The Fe^{3+} can be recycled to Fe^{2+} by action of UV light or by the following reactions:

$$Fe^{3+} + H_2O_2 \rightarrow Fe\text{-}OOH^{2+} + H^+$$
$$Fe\text{-}OOH^{2+} \rightarrow HO_2^\bullet + Fe^{2+}$$

or by

$$Fe^{3+} + HO_2^\bullet \rightarrow Fe^{2+} + O_2 + H^+$$

The rate of the first set of reactions depends on the Fe^{3+} concentration and pH, as an acid-base equilibrium is involved. The latter reaction is a scavenging reaction, decreasing the number of radicals in solution and is therefore actually an unwanted reaction in this system. It should be noted that iron precipitates at pH values higher than 4, thus restricting Fenton reactions to acid solutions. The maximum dissociation rate is reached at a pH of 3 (Pignatello 1992).

Once radicals have been generated the degradation reactions are more or less comparable to those taking place in ozonation under radical conditions: dechlorination and hydroxylation are the predominant reactions (Aplin and Waite 1999). Thus, the Fenton process can be seen as complementary to ozonation. Its mechanism depends on the presence of soluble compounds and is therefore not hindered by mass transfer limitations, as opposed to the ozonation of reactive compounds. Further, radical scavenging by carbonate can be avoided by lowering the pH (acidification of the reaction medium) to reduce the carbonate solubility.

The Fenton process depends on the presence of (low) catalytic concentrations of iron (Figure 18.11) and can therefore easily be used in situations where iron is already present, like groundwater remediation (Yeh and Novak 1995). When insufficient iron is available, iron salts can be added. Another application for wastewater treatment is the irradiation with UV or visible light during the Fenton reaction (Pignatello 1992). This photo-Fenton process shows a higher efficiency during organic compound oxidation than reactions using only Fenton reagent, because Fe^{3+}-photosensitized reactions occur, which produce Fe^{2+} and hydroxyl radicals.

18.3.4 Electrohydraulic cavitation and sonolysis of wastewater

Electrohydraulic cavitation involves the formation and behaviour of bubbles in liquids (Kotronarou *et al.* 1991). It is induced by applying electrical energy directly in a water phase. The electrical power is provided by a pulse-powered plasma discharge producing pulsed and/or continuous ultrasonic irradiation (i.e. sonolysis) in water. Kinetic and sonoluminescence measurements indicate that an extremely high temperature (> 5000 K) and pressure (> 100 atm) are generated during nearly adiabatic and short-time (< 1 μs) implosions occurring at the cavitation sites (Petrier *et al.* 1992a). When a bubble filled with gas and vapour pulses and collapses, molecules inside the bubble or close to the bubble surface are fragmented, escape into the bulk of the solution and react in various ways outside (or inside) the bubble (Petrier *et al.* 1992b). In this situation, water and H_2S split into radicals according to (Kotronarou *et al.* 1992b):

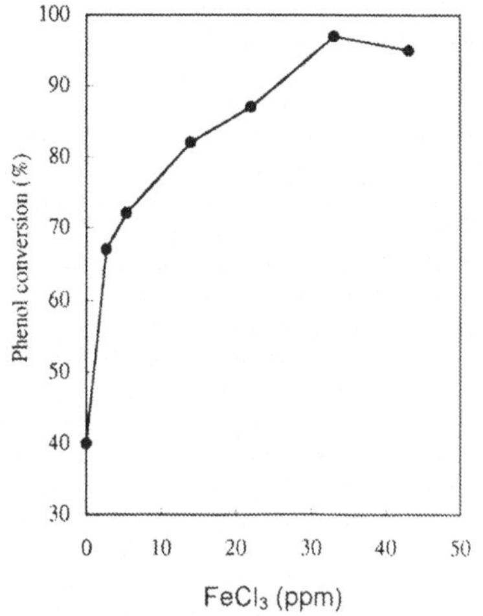

Figure 18.11. The photolytic conversion rate of a phenol solution as a function of the $FeCl_3$ concentration. Initial phenol concentration: 25 ppm, Initial pH: 3.5, Illumination time: 10 minutes (Chen 1997).

$$H_2O \rightarrow H^{\bullet} + {}^{\bullet}OH$$
$$H_2S \rightarrow H^{\bullet} + {}^{\bullet}SH$$
$$CCl_4 \rightarrow Cl_3C^{\bullet} + {}^{\bullet}Cl$$

After the production of these radicals, pollutants such as H_2S and CCl_4 in the water phase can be oxidised to final products such as SO_4^{2-}, Cl^- and CO_2 (Kotronarou *et al.* 1992a, b; Petrier *et al.* 1992b).

18.3.5 Liquid redox processes for H_2S

Several catalytic treatment processes of H_2S scrubbing solutions are proven technologies and commonly practised in the petrochemical industry. Also in this area, several new developments are going on.

18.3.5.1 Vanadium-based catalytic oxidation

In vanadium utilising processes, HS^- is oxidised catalytically to elemental sulphur (S^o) by vanadium (V), which is subsequently reduced to vanadium (IV), according to (Neumann 1984):

$$4\ VO_3^- + 2\ HS^- + H_2O \rightarrow V_4O_9^{2-} + 2S^o + 4\ OH^-$$

In a separate reactor, vanadium (IV) is reoxidised back to vanadium (V) by dissolved oxygen according to:

$$V_4O_9^{2-} + O_2 + OH^- \rightarrow 4\ VO_3^- + H_2O$$

To catalyse the oxygen transfer in the regeneration of the reduced vanadium, different compounds are used. The Stretford process uses anthraquinone disulphonic acid (ADA) and the Unisulf process aromatic sulphonates (Pandey and Malhotra 1999). The fate of non-H_2S sulphur-containing compounds (COS, CS_2) in the Stretford process is not known so far (Pandey and Malhotra 1999).

18.3.5.2 Iron-based catalytic oxidation

A second series of catalytic oxidation processes for H_2S removal relies on iron redox reactions. H_2S is oxidised by the iron ion, according to Hardison (1985):

$$2\ Fe^{3+} + HS^- \rightarrow 2\ Fe^{2+} + S^o$$

In the LOCAT process, a working concentration of 500 ppm total iron is used, which is held into solution by the chelator ethylene diamine tetraacetic acid (EDTA). The Sulferox™ process uses nitrilotriacetic acid (NTA) to keep iron into suspension. In addition, NTA accelerates the reoxidation of Fe^{2+} with O_2:

$$2\ Fe^{3+} + H_2S + NTA \rightarrow 2\ Fe^{2+}.NTA + S^0 + 2\ H^+$$
$$2\ Fe^{2+}.NTA + 2\ H^+ + 1/2\ O_2 \rightarrow 2\ Fe^{3+}.NTA + H_2O$$

18.3.5.3 Other redox processes

Plummer (1987) described the low severity process, in which H_2S reacts with t-butyl anthraquinone to form S^o and the corresponding hydroquinone. Interestingly, the hydroquinone is recycled to anthraquinone by catalytic dehydrogenation and concomitant H_2 production.

18.4 CATALYTIC OXIDATION FOR ODOUR ABATEMENT IN SANITARY ENGINEERING

18.4.1 Catalytic odour removal systems for indoor applications

Matsumoto *et al.* (1993) described the use of a gold/iron oxide catalyst, which removes by catalytic oxidation both H_2S and odorous compounds. The catalyst is coated onto ceramic fibres and the compact units have a wide range of applications. One of them is its incorporation into a toilet, thus removing odour nuisance in the toilet room (Figure 18.12).

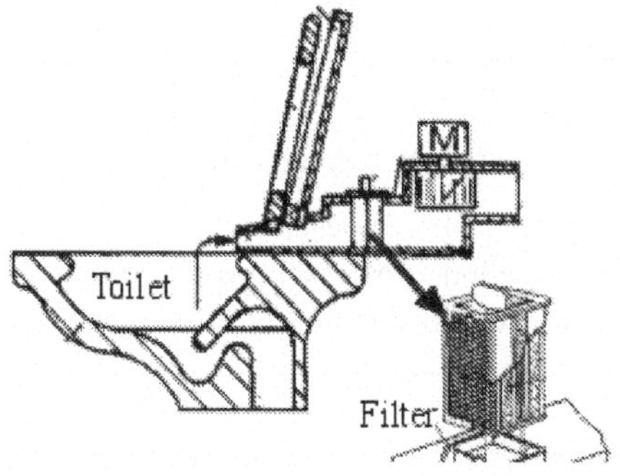

Figure 18.12. Application of a catalyst mounted in a flushing toilet to remove odour nuisance (after: Matsumoto *et al.* 1993).

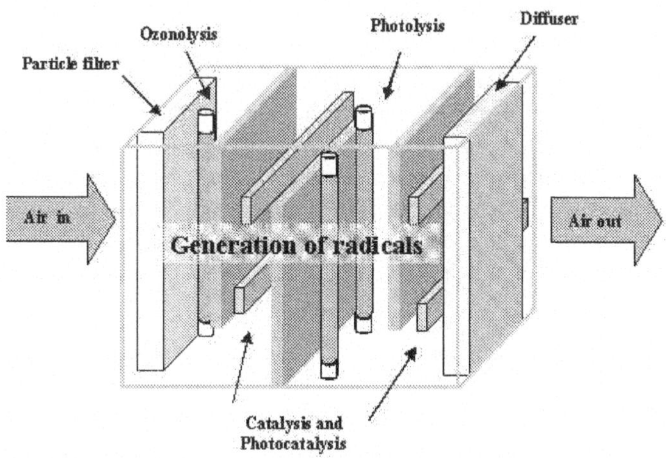

Figure 18.13. Combination of catalytic processes used to deodorise ambient air (BENRAD 2000).

A combination of several catalytic oxidation technologies are applied in a portable air purification unit, marketed by BENRAD (Isaksson *et al.* 2001). Odorous compounds are destroyed by a combination of ozonation, catalysis and photolysis (Figure 18.13). The unit treats 150 m^3/h of contaminated air, and has a net emission of 1.8 mg/min ozone, which is released in the room. Therefore, people are not allowed to stay in the room while the unit is operating and the room has to be ventilated before taking it into use after cleaning. This unit is used to remove the smell of tobacco from rooms or the decontamination of truck and train compartments or boat cabins.

18.4.2 Odour removal from waste containers

Makaly-Biey and Verstraete (1999) describe how UV induced ozone formation can be implemented for odour abatement from vegetable-fruit-garden waste containers. In these containers, UV lamps were mounted, and their effectiveness in reducing odorous liberated from uncontrolled decay in vegetable-fruit-garden waste containers was investigated.

Intermittant operation of a 5W UV-lamp, resulted in a reduction of odour production of 85%, as measured by an electronic nose, within the two weeks of storage in these containers. In all four test runs, the UV-treated samples had no real perceptible odour when using a test panel of five persons, while the control had a very bad odour.

18.4.3 Odour treatment at sewers and sewage works

Lee *et al.* (1999) studied the catalytic incineration of the odorous gases from a sewage pumping station (Table 18.6). These gases are characterised by low concentrations of methane. Detailed kinetic studies were undertaken over a platinum based catalyst and a residence time of 0.31 s was sufficient for complete H_2S removal. There were no signs of deactivation of the catalyst after two years of operation, although the mild steel casing has had to be replaced by stainless steel.

Table 18.6. Catalytic incineration using a Pt-monolith catalyst (operating temperature: 673 K) of odorous gases from Sydney sewage pumping station (Lee *et al.* 1999).

Compound	Average Inlet concentration (ppm)	Average outlet concentration (ppm)
H_2S	< 30	0.56
RSH	< 1	0.09
VOC	< 2	0.20
CH_4	< 1	0.10
Fatty acids	Trace	0.10

McClean *et al.* (1999) report on the use of a combined UV and ozone treatment to reduce the level of nuisance complaints at a sewage treatment plant (Figure 18.14). Application of ozone alone did not succeed to abate the odour, but enhanced UV photolysis was able to reduce the H_2S concentration (influent concentration about 500 ppm) of the air by 98%. The UV lamps are mounted in a plenum, across the airflow, whereas the ozone producing lamp assemblies are located as to introduce the ozone in the receiving section of the plenum.

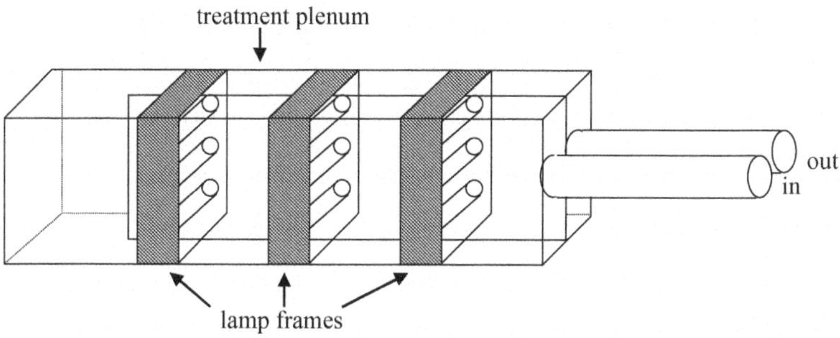

Figure 18.14. View of a ozonation plant for the removal of odours at sewage works (McClean *et al.* 1999).

18.5 REFERENCES

Adib, F., Bagreev, A. and Bandosz, T. J. (2000a) Adsorption/oxidation of hydrogen sulfide on nitrogen-containing activated carbons. *Langmuir* **16** (4), 1980-1986.

Adib, F., Bagreev, A. and Bandosz, T. J. (2000b) Analysis of the relationship between H_2S removal capacity and surface properties of unimpregnated activated carbons. *Environ. Sci. and Technol.* **34** (4), 686-692.

Anderson, R. (1984) Sewage treatment processes: the solution to the odour problem. Characterization and control of odoriferous pollutants in process industries: 471-496. Vigneron, S., Hermia, J. and Chaouki, J. Elsevier Science B.V., Amsterdam, NL.

Andreev, A., Khristov, P. and Losev, A. (1996a) Catalytic oxidation of sulfide ions over nickel hydroxides. *Applied Catalysis B: Environmental* **7**(3-4), 225-235.

Andreev, A., Prahov, L., Gabrovska, M., Eliyas, A. and Ivanova, V. (1996b) Catalytic oxidation of sulphide ions to elementary sulphur in aqueous solutions over transition metal oxides. *Applied Catalysis B: Environmental* **8**(4), 365-373.

Aplin, R. and Waite, T. D. (1999). Comparison of three Advanced Oxidation Processes for Degradation of Textile Dyes. *Proc. IAWQ Int. Specialised Conf. of the Chemical Industry Group - Waste Minimisation and End of Pipe Treatment in Chemical and Petrochemical Industries*, Merida, Mexico, November 14-18, 1999.

Bahnemann, D., Bousselni, L., Freudenhammer, H., Ghrabi, A., Geissen, S.-U., Siemon, U., Saleh, F., Si-Salah, A. and Vogelpohl, A. (1996). Detoxification and recycling of wastewater by solar-catalytic treatment. *Proc. International Conference on*

Oxidation Technologies for Water and Wastewater Treatment, Clausthal, Germany, May 12-15.

Benoit Marquie, F., Wilkenhoner, U., Simon, V., Braun, A. M., Oliveros, E. and Maurette, M. T. (2000) VOC photodegradation at the gas-solid interface of a TiO_2 photocatalyst - Part I: 1-butanol and 1-butylamine. *J. Photochemistry Photobiology A: Chemistry* **132**(3), 225-232.

BENRAD (2000) http://www.benrad.se/.

Boncz, M. A., Rulkens, W. H., Bruning, H. and Sudhölter, E. J. R. (1997). The effect of hydrogen peroxide on the oxidation of chlorophenols with ozone. *Proc. 4th International Conference on Advanced Oxidation Technologies*, Orlando, Florida, USA, Sept. 23-26, 1997.

Bond, G. C. (1987) Heterogenous catalysis: principles and application. Oxford University Press, Oxford.

Borgarello, E., Serpone, N., Liska, P., Erbs, W., Graetzel, M. and Pelizzetti, E. (1985) Photocleavage of hydrogen sulphide in alkaline aqueous media with RuO_2-loaded CdS catalyst supported on a polycarbonate matrix. *Gazetta Chimica Italiana* **115**, 559-602.

Cant, N. W. and Cole, J. R. (1992) Photocatalysis of the reaction between ammonia and nitric acid on TiO_2-surfaces. *J. Catalysis* **134**, 317-330.

Cha, S., Chuang, K. T., Liu, D. H. F., Ramachandran, G., Reist, P. C. and Sanger, A. R. (2000) Pollutants: minimization and control. In: *Air Pollution*, (D. H.F. Liu and B. G. Liptak, eds.), Lewis Publishers, Boca Raton, pp. 107-180.

Chen, J. (1997) Advanced Oxidation Technologies - Photocatalytic Treatment of Wastewater. Environmental Technology. *Wageningen, NL*, Wageningen Agricultural University: 223.

Chou, T. C., Lin, T. Y., Hwang, B. J. and Wang, C. C. (1986) Selective removal of H_2S from biogas by a packed silica gel adsorber tower. *Biotech. Prog.* **2**, 203-209.

Chu, H. and Lee, W. T. (1998) The effect of sulfur poisoning of dimethyl disulfide on the catalytic incineration over a Pt/Al_2O_3 catalyst. *Science of the Total Environment* **209**(2-3), 217-224.

Chu, H. and Wu, L. W. (1998) The catalytic incineration of ethyl mercaptan over a MnO/Fe_2O_3 catalyst. *J. Environ. Sci. Health part A: Toxic and Hazardous Substances and Environmental Engineering* **33**(6), 1119-1148.

Chun, S. W., Jang, J. Y., Park, D. W., Woo, H. C. and Chung, J. S. (1998) Selective oxidation of H_2S to elemental sulfur over TiO_2/SiO_2 catalysts. *Applied Catalysis B: Environmental* **16**(3), 235-243.

Cramariuc, R., Marin, G., Martin, D., Cramariuc, B., Teodorescu, I., Munteanu, V. and Ghiuta, V. (2000) Contribution to electrical discharge electron beam system for flue gas cleaning method. *Radiation Physics and Chemistry* **57**(3-6), 501-505.

Criegee, R. (1975) Mechanismus der Ozonolyse. *Angewandte Chemie* **87**(21), 765-771.

Decoret, C., Royer, J., Legube, B. and Doré, M. (1984) Experimental and theoretical studies of the mechanism of the initial attack of ozone on some aromatics in aqueous medium. *Environ. Technol. Lett.* **5**, 207.

Doi, Y., Nakanishi, I. and Konno, Y. (2000) Operational experience of a commercial scale plant of electron beam purification of flue gas. *Radiation Physics and Chemistry* **57**(3-6), 495-499.

Giorno, L. and Drioli, E. (2000) Biocatalytic membrane reactors: applications and perspectives. *TIBTECH* **18**, 339-349.

Glaze, W. H. (1987) Drinking water treatment with ozone. *Environ. Sci. and Technol.* **21**, 224.

Gonzalez Velasco, J. R., Aranzabal, A., Lopez Fonseca, R., Ferret, R. and Gonzalez Marcos, J. A. (2000) Enhancement of the catalytic oxidation of hydrogen-lean chlorinated VOCs in the presence of hydrogen-supplying compounds. *Applied Catalysis B: Environmental* **24**(1), 33-43.

Gould, J. P. (1987) Correlations between chemical structure and ozonation kinetics: preliminary observations. *Ozone Sci. Engin.* **9**, 207-216.

Guillard, C., Delprat, H., Can, H.-V. and Pichat, P. (1993) Laboratory studies of the rates and products of the phototransformations of naphtalene adsorbed on samples of titanium dioxide, ferric oxide, muscovite and fly ash. *J. Atmos. Chem.* **16**, 47-59.

Gulyas, H., Bismarck, R. v. and Hemmerling, L. (1995) Treatment of industrial wastewaters with ozone/hydrogen peroxide. *Water Sci. Technol.* **32**, 127.

Hardison, L. C. (1985) Go from H_2S to S in one unit. *Hydrocarbon Process* **73**, 70-71.

Hirota, K., Makela, J. and Tokunaga, O. (1996) Reactions of sulfur dioxide with ammonia: Dependence on oxygen and nitric oxide. *Indust. Engin. Chem. Res.* **35**(10), 3362-3368.

Hoigné, J. (1998) Chemistry of Aqueous Ozone and Transformation of Pollutants by Ozonation and Advanced Oxidation Processes. The Handbook of Environmental Chemistry. 5: 83-141. Hrubec, J. Springer Verlag, Berlin, Germany.

Hoigné, J. and Bader, H. (1976) The role of hydroxyl radical reactions in ozonation Processes in Aqueous Solutions. *Water Res.* **10**, 377-386.

Hwang, Y., Matsuo, T., Hanaki, K. and Suzuki, N. (1994) Removal of odorous compounds in wastewater by using activated carbon, ozonation and aerated biofilter. *Water Res.* **28**(11), 2309-2319.

Iliev, V., Prahov, L., Bilyarska, L., Fischer, H., Schulz Ekloff, G., Wohrle, D. and Petrov, L. (2000) Oxidation and photooxidation of sulfide and thiosulfate ions catalyzed by transition metal chalcogenides and phthalocyanine complexes. *J. Molecular Catalysis A: Chemical* **151**(1-2), 161-169.

Isaksson, M., Li, J. and Soremark, R. (2001) An advanced oxidation technology combining UV, ozone and catalysts for decontamination of water. *Water Sci. Technol.* in press.

Jagushte, M. V. and Mahajani, V. V. (1999) Insight into spent caustic treatment: on wet oxidation of thiosulfate to sulfate. *J. Chem. Tech. Biotech.* **74**(5), 437-444.

Kamei, G. and Ohmoto, H. (2000) The kinetics of reactions between pyrite and O_2-bearing water revealed from in situ monitoring of DO, Eh and pH in a closed system. *Geochimica et Cosmochimica Acta* **64**(15), 2585-2601.

Klare, M., Scheen, J., Vogelsang, K., Jacobs, H. and Broekaert, J. A. C. (2000) Degradation of short-chain alkyl- and alkanolamines by TiO2- and Pt/TiO2-assisted photocatalysis. *Chemosphere* **41**(3), 353-362.

Kotronarou, A., Mills, G. and Hoffmann, M. R. (1991) Ultrasonic irradiation of p-nitrophenol in aqueous solution. *J. Physical Chem.* **95**, 3630.

Kotronarou, A., Mills, G. and Hoffmann, M. R. (1992a) Decomposition of parathion in aqeous solution by ultrasonic irradiation. *Environ. Sci. Technol.* **26**, 1469.

Kotronarou, A., Mills, G. and Hoffmann, M. R. (1992b) Ultrasonic irradiation of hydrogen sulfide in aqeous solution. *Environ. Sci. Technol.* **25**, 2940.

Kuo, C. H., Zhong, L., Wang, J. and Zappi, M. E. (1997) Vapor and liquid phase ozonation of benzene. *Ozone Sci. Engin.* **19**(2), 109-117.

Laplanche, A., Bonnin, C. and Darmon, D. (1984) Comparative study of odors removal in a wastewater treatment plant by wet scrubbing and oxidation by chlorine or ozone. In: *Characterization and Control of Odoriferous Pollutants in Process Industries*, (Vigneron, S., Hermia, J. and Chaouki, J. eds.) Elsevier Science B.V., Amsterdam, pp. 277-294.

Lee, J. H., Trimm, D. and Cant, N. W. (1999) The catalytic combustion of methane and hydrogen sulphide. *Catalysis Today* **47**(1-4), 353-357.

Lens, P. N. L., Visser, A., Janssen, A. J. H., Pol, L. W. H. and Lettinga, G. (1998) Biotechnological treatment of sulfate-rich wastewaters. *Critical Reviews in Environ. Sci. Technol.* **28**(1), 41-88.

Li, K. T. and Shyu, N. S. (1997) Catalytic oxidation of hydrogen sulfide to sulfur on vanadium antimonate. *Ind.Engin.Chem.Res.* **36**(5), 1480-1484.

Lou, J. C. and Chen, C. L. (1995) Destruction of butanone and toluene with catalytic incineration. *Hazardous Waste & Hazardous Materials* **12**(1), 37-49.

Magara, K., Ikeda, T., Tomimura, Y. and Hosoya, S. (1998) Accelerated degradation of cellulose in the presence of lignin during ozone bleaching. *J. Pulp Paper Sci.* **24**(8), 264-268.

Makaly-Biey, E. M. and Verstraete, W. (1999) The use of a UV lamp for control of odour decomposition of kitchen and vegetable waste. *Environ. Technol.* **20**(3), 331-335.

Mantzavinos, D., Hellenbrand, R., Livingston, A. G. and Metcalfe, I. S. (1996). Reaction mechanisms and kinetics of chemical pretreatment of bioresistant organic molecules by wet air oxidation. *Proc. International Conference on Oxidation Technologies for Water and Wastewater Treatment*, Clausthal, Germany, May 12-15.

Matsumoto, T., Tabata, K. and Maki, M. (1993) Catalytic composite for deodorizing gases and a method for preparating the same. US Patent no.: 5266543.

Matzing, H., Baumann, W. and Paur, H. R. (1996) Chemistry of the electron beam process and its application to emission control. *Pure and Applied Chem.* **68**(5), 1089-1092.

McClean, J. C., Hamilton, D. J. and Clark, N. (1999). Odour abatement using enhanced UV photolysis: a new application for a proven technology. *Proc. CIWEM and IAWQ conference on Control and prevention of odours in the water industry*, London, UK, September 1999.

McNeillie, A. (1984) The use of hydrogen peroxide for odour control. Characterization and control of odoriferous pollutants in process industries. (Vigneron, S., Hermia, J. and Chaouki, J., eds.) Elsevier Science B.V., Amsterdam, pp. : 471-496.

Meeyoo, V., Adesina, A. A. and Foulds, G. (1995) The H_2S decomposition activity of supported MoS_2 catalysts prepared via PFHS method. *Reaction Kinetics and Catalysis Letters* **56**(2), 231-240.

Meeyoo, V., Lee, J. H., Trimm, D. L. and Cant, N. W. (1998) Hydrogen sulphide emission control by combined adsorption and catalytic combustion. *Catalysis Today* **44**(1-4), 67-72.

Mozzanega, H., Herrmann, J. M. and Pichat, P. (1979) Ammonia oxidation over UV-irradiated TiO_2 at room temperature. *J. Physical Chem.* **83**(17), 2251-2255.

Nakajima, F. (1991) Air pollution control with catalysis - Past, present and future. *Catalysis Today* **10**, 1.

Namba, H., Hashimoto, S., Tokunaga, O. and Suzuki, R. (1998) Electron beam treatment of lignite-burning flue gas with high concentrations of sulfur dioxide and water. *Radiation Physics and Chemistry* **53**(6), 673-681.

Neta, P., Huie, R. E. and Ross, A. B. (1988) Rate Constants for Reactions of Inorganic Radicals in Aqueous Solution. *J. Physical Chemical Reference Data* **17**(3), 1027-1262.

Neumann, D. W. (1984) Oxidative absorption of H_2S and O_2 by iron chelate solutions. *American Institute of Chemical Engineering Journal* **30**, 62-69.

Noguchi, T., Hashimoto, K. and Fujishima, A. (1996). Dependence of product distribution on TiO_2 surface characteristics: photocatalytic decomposition of gaseous acetaldehyde. In: *proceedings of the 2nd International Conference on TiO_2 Photocatalytic Purification and Treatment of Air and Water*, Cincinnati, OH, USA, October 26-29.

Paillard, H. and Blondeau, F. (1988) Les nuisances olfactives en assainissement: causes et remèdes. *T.S.M. L'eau* **2**, 79-88.

Pan, T. M., Shimoda, K., Cai, Y., Kiuchi, Y., Nakama, K., Akimoto, T., Nagashima, Y., Kai, M., Ohira, M., Saegusa, J., Kuhara, T. and Maejima, K. (1995) Deodorization of laboratory animal facilities by ozone. *Experimental Animals* **44**(3), 255-259.

Pandey, R. A. and Malhotra, S. (1999) Desulfurization of gaseous fuels with recovery of elemental sulfur: An overview. *Critical Reviews in Environmental Science and Technology* **29**(3), 229-268.

Park, D. W., Chun, S. W., Jang, J. Y., Kim, H. S., Woo, H. C. and Chung, J. S. (1998) Selective removal of H_2S from coke oven gas. *Catalysis Today* **44**(1-4), 73-79.

Petrier, C., Jeunet, A., Luche, J. and Reverdy, G. (1992a) Unexpected frequency effects on the rate of oxidative processes induced by ultrasound. *Journal of the American Chemical Society* **114**, 3148-3150.

Petrier, C., Micolle, M., Merlin, G., Luche, J. L. and Reverdy, G. (1992b) Characteristics of pentachlorophenate degradation in aqueous solution by means of ultrasound. *Environ. Sci.Technol.* **26**(8), 1639-1642.

Peyton, G. R. (1990) Oxidative treatment methods for removal of organic compounds from drinking water supplies. *Significance and Treatment of Volatile Organic Compounds in Water Supplies* (N.M. Ram, R.F. Christman and K.P. Cantor, eds.) pp. 313-362, Lewis Publishers.

Pichat, P., Herrmann, J. M., Courbon, H., Disdier, J. and Mozzanega, M. N. (1982) Photocatalytic oxidation of various compounds over TiO_2 and other semiconductor oxides: mechanistic considerations. *Canadian Journal of Chemical Engineering* **60**, 27-32.

Pignatello, J. J. (1992) Dark and photoassisted Fe^{3+} catalyzed degradation of chlorophenoxyherbicides by hydrogen peroxide. *Environ.Sci. Technol.* **26**, 944.

Plummer, M. A. (1987) Sulfur and hydrogen from H_2S. *Hydrocarbon Process* **75**, 38-40.

Przepiorski, J. and Oya, A. (1998) K_2CO_3-loaded deodorizing activated carbon fibre against H_2S gas: Factors influencing the deodorizing efficiency and the regeneration method. *J. Materials Science Letters* **17**(8), 679-682.

Przepiorski, J., Yoshida, S. and Oya, A. (1999) Structure of K_2CO_3-loaded activated carbon fiber and its deodorization ability against H_2S gas. *Carbon* **37**(12), 1881-1890.

Ravindranathan Thampi, K., Ruterana, P. and Graetzel, M. (1990) Low temperature thermal and photoactivation of TiO_2-supported Ru, Rh and Cu catalyst for CO-NO reaction. *Journal of Catalysis* **126**, 572-590.

Read, H. W., Fu, X. Z., Clark, L. A., Anderson, M. A. and Jarosch, T. (1996) Field trials of a TiO_2 pellet-based photocatalytic reactor for off-gas treatment at a soil vapor extraction well. *J. Soil Contamination* **5**(2), 187-202.

Reij, M. W., Keurentjes, J. T. F. and Hartmans, S. (1998) Membrane bioreactors for waste gas treatment. *J. Biotechnol.* **59**(3), 155-167.

Sabaté Cervera-March, J. S., Simarro, R., Liska, P., Erbs, W., Grätzel, M. and Pelizzetti, E. (1990) Photocatalytic production of hydrogen from sulfide and sulfite waste streams. *Chem. Engin. Sci.* **45**(10), 3089-3096.

Satterfield, C. N. (1991) *Heterogeneous Catalysis in Practice*. McGraw-Hill, New York, NY, USA.

Shen, Y. S. and Ku, Y. (1997) Treatment of gas-phase trichloroethene in air by the UV/O_3 process. *J. Hazardous Materials* **54**(3), 189-200.

Shen, Y. S. and Ku, Y. (1999) Treatment of gas-phase volatile organic compounds (VOCs) by the UV/O_3 process. *Chemosphere* **38**(8), 1855-1866.

Siebert, P. C., Meardon, K. R. and Serne, J. C. (1984) Emission controls in polymer production. *Chem. Engin. Prog.*, 86-176.

Sinquin, G., Hindermann, J. P., Petit, C. and Kiennemann, A. (1999) Perovskites as polyvalent catalysts for total destruction of C-1, C-2 and aromatic chlorinated volatile organic compounds. *Catalysis Today* **54**(1), 107-118.

Smet, E. and Van Langenhove, H. (1998) Abatement of volatile organic sulfur compounds in odorous emissions from the bio-industry. *Biodegradation* **9**(3-4), 273-284.

Snape, T. H. (1977) Catalytic oxidation of pollutants from ink drying ovens. *Plating Metal Review* **21**(3), 90-91.

Sokolovskii, V. D. (1990) Principles of oxidative catalysis on solid oxides. *Critical Reviews in Science and Engineering* **32**, 1.

Staehelin, J. (1983) Ozonzerfall in Wasser: kinetiek der initiierung durch OH^- ionen und H_2O_2 sowie der folgereactionen der OH und O_2^- radicale. *Zurich, Switzerland*, ETH: 157.

Staehelin, J. and Hoigné, J. (1985) Decomposition of Ozone in water in the presence of organic solutes acting as promoters and inhibitors of radical chain reactions. *Environ. Sci. Technol.* **19**, 1206.

Suzuki, K. I., Shigeyyki, S. and Takashi, Y. (1991) Photocatalytic deodorization (oxidation of organics) on TiO_2 coated, supported on honeycomb ceramics. *Denki Kagau* **59**(6), 521-523.

Tanada, S., Boki, K., Kita, T. and Sakaguchi, K. (1982) Adsorption behavior of hydrogen sulfide inside micropores of Molecular Sieve Carbon 5A and Molecular sieve Zeolite 5A. *Bulletin of Environmental Contamination and Toxicology* **29**(5), 624-629.

Taylor, S. H., Heneghen, C. S., Hutchings, G. J. and Hudson, I. D. (2000) The activity and mechanism of uranium oxide catalysts for the oxidative destruction of volatile organic compounds. *Catalysis Today* **59**(3-4), 249-259.

Taylor, S. H. and O'Leary, S. R. (2000) A study of uranium oxide based catalysts for the oxidative destruction of short chain alkanes. *Applied Catalysis B: Environmental* **25**(2-3), 137-149.

Thornton, E. C. and Amonette, J. E. (1999) Hydrogen sulfide gas treatment of Cr(VI)-contaminated sediment samples from a plating-waste disposal site - Implications for in-situ remediation. *Environ. Sci. Technol.* **33**(22), 4096-4101.

USEPA (1987) A compendium of technologies used in the treatment of hazardous wastes. *Cincinnati, OH, USA*, U.S. Environmental Protection Agency, Center for Environmental Research Information.

Van Durme, G. P., McNamara, B. F. and McGinley, C. M. (1992) Bench-scale removal of odor and volatile organic compounds at a composting facility. *Water Environ. Res.* **64**(1), 19-27.

Wang, J. B. and Chou, M. S. (2000) Kinetics of catalytic oxidation of benzene, n-hexane, and emission gas from a refinery oil/water separator over a chromium oxide catalyst. *J. Air Waste Manag. Assoc.* **50**(2), 227-233.

Weiss, J. (1935) The radical HO_2 in Solution. *Transactions of the Faraday Society* **31**, 668.

Yeh, C. K. and Novak, J. T. (1995) The effect of hydrogen peroxide on the degradation of methyl and ethyl tert-butyl ether in soils. *Water Environ. Res.* **67**, 828.

19

Biotechnological treatment of sewage odours

Herman Van Langenhove and Bart De heyder

19.1 INTRODUCTION

Sewage and sewage treatment has been associated with nauseous odours since ancient times. In fact one of the reasons for building sewage transport systems in ancient cities as Rome (cloaca maxima) was to avoid generation of odours. It was only after the work of Pasteur, who showed the relation between microorganisms, infectious diseases and the presence of biodegradable materials that hygienic considerations became more and more important.

Odorous compounds in sewage mainly originate from two processes: anaerobic decomposition of biodegradable material present in the wastewater or direct emission of specific chemicals with wastewater discharges. The former

 Odours in Wastewater Treatment: Measurement, Modelling and Control edited by Richard Stuetz and Franz-Bernd Frechen. ISBN: 1 900222 46 9

process dominates problems with domestic wastewater, while the latter can be more important in mixed (industrial and domestic) wastewater systems. Because of its specificity odour problems related to industrial effluent have to be investigated case by case. Therefore we will focus on municipal wastewater.

As wastewater becomes more and more anaerobic, different odorous compounds depending on precursor, pH and oxidation–reduction potential are microbiologically formed. Hydrogen sulphide, the product of sulphate reduction, is the odorant most commonly associated with sewage odours. In some cases a good correlation between hydrogen sulphide emission concentrations and odour concentrations was reported (see chapter 6). Nevertheless, next to hydrogen sulphide, mainly dimethyl sulphide and dimethyl oligosulphides significantly contribute to sewage odours. Next to these odorants many more volatile organic carbon compounds (VOC), such as aliphatic and aromatic hydrocarbons, chlorinated hydrocarbons (e.g. tetrachloro-ethylene, dichlorobenzene), aldehydes and ketones have been identified in waste gases from wastewater treatment plants (Zeman and Koch 1983; Van Langenhove *et al.* 1985). Although these compounds can be of importance (e.g. with respect to tropospheric ozone formation) their contribution to sewage odours is limited because of their relatively high odour threshold compared with the threshold of sulphur compounds. So treatment of sewage odours basically means the elimination of hydrogen sulphide and organic sulphur compounds from waste gas containing large numbers of VOC.

Sulphur compounds, relevant to sewage odours, are all part of the bio-geochemical sulphur cycle. In this cycle sulphide is the most reduced and sulphate the most oxidised species. An overview of the biological part of this cycle is given by Brüser *et al.* (2000). From this it is clear that, depending on the environmental conditions, sulphur compounds can be converted from one species into another by biochemical processes. So it is not surprising that as early as 1923 the basic concept of biofiltration for the control of H_2S emissions from sewage treatment plants was discussed (Leson and Winer 1991).

In this chapter basic principles of biological waste gas treatment and the applicability of these techniques for reduction of sewage odours will be discussed.

19.2 TYPES OF REACTOR

To date, three types of reactor are in practice commonly used for biotechnological waste gas treatment: bioscrubbers, biotrickling filters and biofilters. Figure 19.1 overviews these different types of reactor concept. Figure 19.2 gives an example of a biofilter.

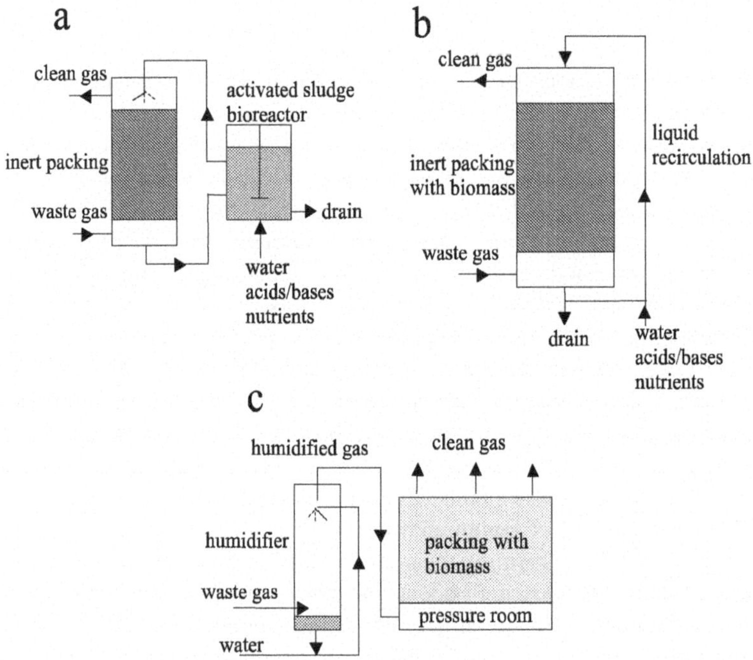

Figure 19.1. Process schemes of (a) a bioscrubber, (b) a biotrickling filter and (c) a biofilter.

In bioscrubbers, the pollutant is absorbed in an aqueous phase in an absorption tower. The aqueous phase containing the dissolved compounds is then treated in a separate activated sludge unit. The effluent of this unit is circulated over the absorption tower in a co- or countercurrent way to the gas stream. The microorganisms in the activated sludge unit (Kennes and Thalasso 1998) degrade the pollutants.

In a biotrickling filter, the waste gas is forced through a packed bed filled with a chemically inert carrier material. The latter is colonised by microorganisms, similar to trickling filters in wastewater treatment. The liquid medium is circulated over the packed bed. The pollutants are first taken up by the biofilm on the carrier material and then degraded by the microorganisms in the biofilm. The liquid medium can be recirculated continuously or discontinuously and in co- or countercurrent to the gas stream.

In a biofilter, the gas to be treated is first humidified and then forced through a packed bed with an organic carrier material (compost, peat, bark or a mixture of these) on which the microorganisms are attached as a biofilm. The pollutants are sorbed by the filter material and degraded by the biofilm.

Figure 19.2. Biofilter for air treatment at the WWTP of Deurne (Belgium).

19.3 BASIC PROCESS MECHANISMS

In all types of the described biotechnological reactors two main processes take place: first, the pollutants are transferred from the gas phase to a liquid medium or to a biofilm, and second the pollutants are degraded by the microorganisms that are present in the liquid medium or biofilm. The combination of the different underlying physicochemical and biological mechanisms results in a complex system. Fundamental parameters of this system remain difficult to quantify (e.g. mixed-order kinetics, sorption in organic media, air flow patterns in reactor, etc.). This implies that in many cases, the design and operation of the treatment system will be based on (semi) empirical knowledge. In the remainder of this section, some basic aspects and the complexity of the mass transfer and biodegradation mechanisms are further illustrated. For further information the reader is referred to the following publications on dynamic modelling of biotechnological air treatment equipment (Deshusses *et al.* 1995a,b; Devinny *et al.* 1999; Hwang and Tang 1997; Okkerse *et al.* 1999; Shareefdeen and Baltzis 1994; Zarook *et al.* 1997a,b).

19.3.1 Mass transfer

Processes such as absorption, adsorption, diffusion and convection will influence the physical transfer of pollutants from the gas phase to the microorganisms. Depending on the type of reactor, these processes can occur in and between several phases : gas, liquid, biofilm, inert solid matrix, inactive organic material, etc.

Because in a bioscrubber the pollutants have to be absorbed in the water phase, this technology is only suitable for pollutants that are highly water-soluble. For a biotrickling filter, the same boundary condition applies, but less stringent than in a bioscrubber. Also the intermittent trickling of a biotrickling filter can enhance the removal of less water-soluble compounds (De Heyder, 1998). In a biofilter, the apolar fraction of the organic carrier material promotes the sorption and subsequent biodegradation of less water-soluble compounds. Table 19.1 characterises the applicability of the type of reactor as a function of the air–water partition coefficient (*m*), which is often used to express equilibrium partitioning of a pollutant between the gas and the liquid phase.

Table 19.1. Applicability of the bioscrubber, biotrickling filter and biofilter as a function of the air-water partition coefficient *m* (Van Groenestijn and Hesselink 1993).

Reactor concept	m [a] $(\text{mole/m}^3)_{air}/(\text{mole m}^3)_{water}$
Bioscrubber	< 0.01
Biotrickling filter	< 1
Biofilter	Maximum 10

[a] m = Cg/Cl at equilibrium
C_g = gas compound concentration
C_l = liquid compound concentration

Air/water partitioning coefficients can be found in the literature (e.g. Staudinger and Roberts 1996). It has to be kept in mind, however, that most of the data apply for pure water systems and that air/water partitioning can be influenced by the presence of salt (Dewulf *et al.* 1995) and dissolved organic matter.

Partitioning coefficients describe equilibrium situations and do not give information on transfer rates. Transport processes in the gas and the water phase and within the biofilm determine transfer rates. Especially in biofilm reactors (biotrickling filter and biofilter), the mass transfer mechanism, on a microscopic scale, can be very complex. Microscopic observation of fully hydrated biofilm sections revealed important heterogeneity, with large channels extending from the gas–liquid interface of the biofilm to the inert solid matrix (Moller *et al.* 1996). Such channels clearly increase pollutant and oxygen availability. In a previous study with submerged biofilms, De Beer *et al.* (1994) evaluated that the supply of oxygen through such voids and channels are roughly 50% of the total oxygen transfer. Although research in this field is going on (e.g. Picioreanu *et al.* 2000) details on mass transfer processes in such heterogeneous systems as biological waste gas treatment remain largely unknown.

Another important mechanism can be the physical sorption of gaseous pollutants in a biofilter medium. This sorption can be due to absorption in the water phase, adsorption onto the solid matrix or absorption in the solid matrix. Adsorption on the surface of carrier material is a fast, reversible process. Pollutants can also permeate into the organic matrix of the biofilter material. Permeation is much slower than surface sorption due to the low diffusion coefficients of pollutants in the organic polymers. The sorbed pollutants can become re-available and support microbial activity to survive periods that the biofilter is not in operation. Smet *et al.* (1996) showed that dimethyl sulphide was sorbed into the organic fraction of bark, peat and compost by a kinetically slow process (equilibration time > 14 days). Carbon dioxide measurements by Deshusses (1997) at the outlet of a biofilter suggested that pulses of acetone, 1-propanol and methyl isobutyl ketone were in first instance sorbed onto the packing material and subsequently degraded within 2 – 5 h.

19.3.2 Biodegradation

Biodegradation is the effect of the microbial metabolism, which is the ensemble of enzymatically mediated reactions by which microbial cells can counterbalance losses of activity caused by wash-out, starvation and decay.

The physico-chemical reactor conditions favourable for the development of microbial activity are listed in Table 19.2. It is important to maintain these favourable conditions during the complete life-span of the air treatment equipment. This can imply that the influent waste gas must be pre-humidified, nutrients must be dosed, pH must be corrected, etc.

Owing to their biogenic origin, compounds contributing to sewage odours can be considered as biodegradable. A Michaëlis-Menten type of equation, mediated to describe enzymatic reactions can in most cases describe the kinetics of biodegradation:

$$R = R_{max}\ C_l /(C_l+K_s) \qquad (19.1)$$

Where:
R = the biodegradation rate,
R_{max} = the maximum biodegradation rate,
C_l = the compound concentration in the liquid
K_s = the compound affinity constant.

In order to maintain a sufficient level of biodegradation, it is important that the compound concentration C_l, is not far below the compound affinity constant K_s. A required level of effluent concentration will therefore probably be difficult to attain if it is lower than the equilibrium gas concentration for K_s (Hartmans 1997).

Most studies concerning biotechnological waste gas treatment reported in literature focus on the treatment of a waste gas containing a single pollutant. In many practical situations, however, the waste gas is likely to be composed of more than one pollutant. The presence of multiple pollutants can result in a decreasing process performance. Smet *et al.* (1997) showed that dimethyl sulphide removal by a biofilter was inhibited when isobuterylaldehyde was added to the waste gas. Deshusses (1997) reported inhibition between 1-propanol, methyl isobutyl ketone and acetone for a compost biofilter. Yet, the presence of multiple pollutants can also result in a positive effect. For example, co-metabolism, whereby a microorganism growing on a particular pollutant gratuitously oxidizes a second pollutant that cannot be used as carbon and energy source, can be the basis for various positive relationships between different types of microorganisms (Beam and Perry 1974).

During the operation of the biotechnological reactor, a natural selection by the pollutant and the physico-chemical reactor conditions gives rise to an adapted microbial community. Yet, little is known about the composition and changes in the microbial community in biotechniques for waste gas treatment and influence on the performance. Investigation by Webster *et al.* (1997) showed that times needed to reach stable microbial conditions in a lab-scale biofilter for removal of low concentrated levels of H_2S and organic compounds were in the order of hundreds of days. The slowly changing pH conditions in the biofilter were put forward as one of the causes of this long stabilisation time. De Castro *et al.* (1996) showed that in a biofilter treating α-pinene a rapid differentiation was observed from the inoculum and also significant differences over the height of the filterbed. A topic related to microbial dynamics is the inoculation of the reactor system with specialised organisms. This can be required if the suitable microbial community does not develop spontaneously from common sources as compost or activated sludge. Smet *et al.* (1996) showed that inoculation of a lab-scale biofilter with a dimethylsulphide-degrading enrichment culture from soil, increased the dimethulsulphide to about 35 g/m³ d in comparison with the initial 10 g/m³ d.

Table 19.2. Overview of physico-chemical reactor conditions favourable for microbial activity.

Parameters		Optimal range in common reactor systems
Temperature	The rate of reaction for microorganisms doubles with about every temperature rise of 10 K, until a limiting temperature is reached.	288 – 303 K
pH	Most microorganisms can not tolerate pH levels above 9.5 or below 4.0. Although there are exceptions e.g. *Thiobacillus* species.	6.5 – 7.5
Water activity	The availability of liquid water is essential for all biochemical processes. This availability is expressed as the water activity of the medium, being the relative humidity of the air in equilibrium with the medium divided by a factor 100.	0.95 – 1
Oxygen	Most pollutants are biodegraded using oxygen as an oxidant. The critical oxygen concentration for aerobic activity of microorganisms lies in the range 0.1 – 1.6 mg/l (Bailey and Ollis, 1986). The solubility of oxygen in an aqueous solution at ambient conditions is in the order of 8 – 10 mg/l. It cannot be excluded however that in biological waste gas treatment conditions can occur favouring biodegradation reaction in which other electron acceptors e.g. nitrate are used.	1 – 2 mg/l
Nutrients	Inorganic elements such as N, P and other trace elements such as K, Ca and Mg are vital to synthesis of microbial cells. The weight ratio in microbial mass equals about C:N:P:K = 50:10:4:1. In aerobic treatment, about one half of the carbon source is assimilated into biomass, while the other half is respired to CO_2.	C:N:P:K 100:10:4:1

19.4 DESIGN AND OPERATIONAL PARAMETERS

The design or operation of biotechnological waste gas treatment equipment can be described using several parameters. Table 19.3 describes some basic parameters used for further definition of the operational and design parameters. The most important operational and design parameters are given in Table 19.4.

When designing the air treatment equipment or describing its operation, it must be realised that many of the parameters defined in Table 19.4 are related and that identical value can be obtained under different operational conditions, e.g.

$$L_{M,V} = C_{g,i} / \tau \tag{19.2}$$

The above equation shows that identical mass loading ($L_{M,V}$) can be obtained at different combinations of pollutant influent concentration ($C_{g,i}$) and empty bed residence time (τ). So, even with identical mass loading rates different removal efficiency (RE) and elimination capacity (EC) can be obtained. Even more, it can be noted that an identical empty residence time (τ) can be obtained for different combinations of reactor section (S_r) and reactor height (H_r):

$$\tau = V_r / Q_g = S_r \times H_r / Q_g \tag{19.3}$$

Different combinations of reactor section and reactor height corresponding to an identical empty bed residence time will correspond to a different superficial gas velocity (v_g). It is possible that different values of v_g result in significant different conditions with respect to mass transfer (laminar-turbulent regime) and thus removal efficiencies or elimination capacities. Therefore, obtained or estimated values of RE or EC should always be specified with the corresponding independent operational parameters allowing to identify the unique operational conditions, e.g.: reactor height (H_r), reactor section (S_r), waste gas flow rate (Q_g), influent pollutant concentration ($C_{g,i}$).

Table 19.3. Basic parameters for definition of design and operational parameters.

Symbol	Units	Description
$C_{g,e}$	mole/m³	Pollutant concentration in effluent air
$C_{g,i}$	mole/m³	Pollutant concentration in influent air
H_r	m	Height of reactor
Q_g	m³/s	Gas flow rate
S_r	m²	Section area of reactor
V_r	m³	Volume of the reactor (i.e. packing volume or volume of scrubber unit)
θ	-	Porosity of the reactor (i.e. volume of void space over volume of reactor)

In practice, the waste gas flow rate (Q_g) and the pollutant influent concentration ($C_{g,i}$) will be determined by the conditions on the wastewater treatment plant. The height of the air contacting equipment will be set to the maximum value not entailing mechanical compaction, high construction costs, less favourable operational control. To obtain the required effluent concentration ($C_{g,i}$) determined by a consent or the plant management, the parameters which have to be defined in the design of the different types of bioreactors are summarised in Table 19.5.

Table 19.4. Definition of the most important operational and design parameters.

Symbol and Equation	Units	Description
$\tau = V_r/Q_g$	s	Empty bed residence time (τ) relates the flow rate to the size of the equipment (e.g. packing volume).
$\tau_\theta = \theta\ \tau$	s	True residence time (τ_θ) is the actual time of residence keeping count of the void space in the reactor .
$v_S = Q_g/S$	m/s	The superficial gas velocity (v_g) is the gas flow rate of gas per unit section area of reactor.
$L_V = Q_g/V_r$	1/s	The volumetric loading rate (L_V) is the flow rate of gas per unit of reactor volume.
$L_{M,S} = Q_g\ C_{g,i}\ /\ S_r$ $L_{M,V} = Q_g\ C_{g,i}\ /V_r$	kg/m^{2}s kg/m^{3}s	The mass loading rate ($L_{M,S}$ or $L_{M,V}$) is the mass of pollutant entering the reactor system per unit section area or volume of the reactor.
$RE = (C_{g,i}\text{-}C_{g,e})/C_{g,i} \times 100$	%	Removal efficiency (RE) is the fraction of the pollutant removed expressed as a percentage.
$EC = (C_{g,i} - C_{g,e}) \times Q_g\ /\ V_r$	kg/m^{3}s	The elimination capacity (EC) is the mass of pollutant removed per reactor volume and per unit time.

Table 19.5 Overview of practical design parameters for each type of reactor.

Type of reactor	Design parameter
Biofilter	Type of packing material Section (S_r) Number of stages
Biotrickling filter	Type of packing material Section (S_r) Liquid recirculation rate
Bioscrubber	Type of packing material Section (S_r) Liquid reciculation rate Activated sludge concentration

As an illustration a typical profile of elimination capacity (EC) and removal efficiency (RE) as a function of S_r (constant Q_g and $C_{g,i}$) for a biofilter is given in Figure 19.3. These profiles are determined on the basis of pilot experiments. For biotrickling filters and for bioscrubbers the pilot experiments should also include the effect of the other design parameters (Table 19.5).

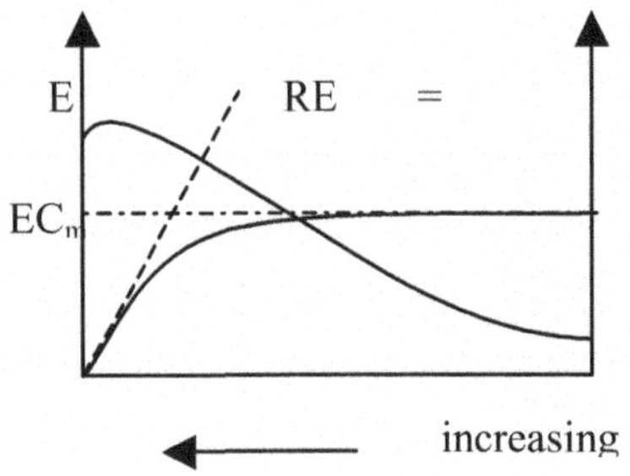

Figure 19.3. Typical profile of elimination capacity (EC) and removal efficiency (RE) as a function of the biofilter section S_r. (Q_g and C_{gi} are) constant.

19.5 PERFORMANCE

Most of the information on the performance of biotechnological waste gas treatment comes from controlled laboratory experiments with biofilters and biotrickling filters. An overview of the data on hydrogen sulphide and organic sulphur compound removal up to 1999 can be found in Herrygers *et al.* (2000). From this overview it is clear that hydrogen sulphide elimination can be performed in biotrickling filters as well as in biofilters. Reported maximum elimination capacities are of the order of 3–3.5 kg $H_2S.m^{-3}.d^{-1}$ for biofilters (Yang and Allen. 1994) as well as for trickling filters (Guey *et al.* 1995). Some systems were not inoculated, some were inoculated with pure strains (e.g. *Thiobacilli* or *Hyphomicrobium* species). Others systems were inoculated with night soil sludge. Inoculation however did not seem to be an important operational parameter. Controls of pH and sulphate concentration were critical factors. Optimal pH values depend on the type of organisms oxidising hydrogen sulphide e.g. some *Thiobacilli* are acidophobic and perform well near neutral pH while other *Thiobacilli* are acidophilic and grow at low pH values (Kasakura and Tatsukawa 1995). Low pH will also affect hydrogen sulphide mass transfer (pK_{a1} = 7). Increasing sulphate concentrations increase the osmotic pressure, which can reduce biological activity. Yang and Allen (1994) experimentally determined that hydrogen sulphide oxidation strongly decreased when sulphate concentrations were higher than 25 g SO_4^{-2}-S.kg^{-1} dry compost. As far as biofilter materials are concerned it seems that similar performances can be obtained with different materials provided that pH and moisture are properly controlled. In general biofilter materials containing an organic fraction perform better than inert carriers (Kim *et al.* 1998). Experimental results also show that hydrogen sulphide elimination is not affected by the presence of organic sulphides (Cho *et al.* 1991).

Microorganisms capable of methyl sulphide oxidation do not seem to been inherently present in currently used biofilter materials. Smet *et al.* (1996) showed that low elimination capacities (< 0.01 kg.m^{-3}.d^{-1}) were obtained with both non-incoulated wood bark and compost biofilters. It has been reported that inoculation with methylotrophic *Hyphomicrobium* spp. and autotrophic *Thiobacilli* spp. significantly increases the maximum elimination capacity of the systems. Compared with hydrogen sulphide, lower maximum elimination capacities (< 1 kg.m^{-3}·d) for organic sulphides. Furthermore the removal of these compounds is more sensitive to environmental factors such as pH and sulphate concentration (Hirai *et al.* 1990). The elimination of organic sulphides is also affected by the presence of other compounds. Hirai *et al.* (1990) showed that the presence of hydrogen sulphide reduced the elimination of dimethyl sulphide. Smet *et al.* (1997) reported that the degradation of dimethyl sulphide by a *Hyphomicrobium* species was inhibited by the presence of isobutyraldehyde, while the presence of toluene had no effect. These data show that most important chemicals contributing to sewage odours can be eliminated from waste gases by biotechnological techniques.

19.6 PROCESS MONITORING

A key to good operational practice is the implementation of a monitoring program for continuous assessment of process performance. The main objective of the monitoring program is to assess the odour reduction by the air treatment equipment. Yet, in most cases budget restrictions will not allow frequent measurement of the odour concentration of the influent and effluent air. Therefore, monitoring process performance of the air treatment equipment is in most cases based on the measurement of parameters which:

- characterise the general operational reactor conditions (e.g. temperature of reactor, relative humidity effluent air, etc.),
- indicate the obtained odour reduction (e.g. personal sensorial observations, simple H_2S measurements, etc.).

Most of this monitoring can be based on analysis techniques already standardised for examination of air, water or soil (e.g. determination of moisture content of biofilter material) (Eaton *et al.* 1995).

A monitoring program must be adapted to the skills and experience of the operational personnel. In many cases, the operational personnel at a wastewater treatment plant will not be well acquainted with air treatment equipment, and dedicated training will be required. Monitoring techniques for which the operational personnel have not the required skills will have to be outsourced to specialised co-operators or external companies.

Keeping count of the aspects discussed above, it can be advisable in practice to set-up a basic monitoring program, circumventing the need for complicated or expensive measurements. The basic monitoring program, however, must allow the collection of data which allows one to (i) estimate basic process parameters such as space time, removal efficiency, etc. and (ii) assess a possible deterioration of the treatment process. In the latter case a more complicated monitoring program can be set up to detect the cause of the deterioration and remediate it. An example of a basic monitoring program is described in Table 19.6. Table 19.7 overviews more complicated monitoring techniques to be performed less frequently as a function of budget and / or operational state of the equipment. It is evident that the use of specialised monitoring techniques to detect the causes of process deterioration, will only be optimal if reference values corresponding to a normal operational state of the reactor equipment are available (e.g. respiration activity biomass, etc.). It can therefore be useful to perform some of these measurements on a yearly basis.

In practice, the implementation of a monitoring program can be hindered by the configuration or the size of the full-scale reactor equipment. For example, a simple and reliable sampling of the effluent waste gas is only possible if the waste gas is guided to a single outlet tube (covered biofilter). On the other hand, if the biofilter is covered, it is impossible to assess the superficial waste gas distribution using a smoke test (Table 19.7). Another example is the determination of the moisture content of the biofilter material. In principle, samples should be taken at different locations and different depths. Regular sampling can therefore result in a disturbance of the packing structure (channel formation) and decrease the biofilter performance.

19.7 PROCESS CONTROL

Depending on the type of reactor, different parameters must be controlled to maintain operational stability. The bioscrubber shows the best control possibilities due to the recirculating water phase that allows the removal of excess biomass, supply nutrients, control pH, etc. The operational state of a biotrickling filter or biofilter, however, can be more critical to control.

Table 19.6. Overview of a basic monitoring program. Parameters should be logged at least once per week in a journal.

Parameter (method)	Evaluation of	Applicable for [a]		
		BS	BTF	BF
Air pressure drop (fixed manometer)	Compaction or clogging	•	•	•
Air velocity influent pipe (handheld air velocity probe)	Air flow	•	•	•
Energy consumption fan (kWh counter)	Air flow	•	•	•
Sensorial quality influent and effluent air (personal sensor observation) [b]	Odour load ; Odour removal	•	•	•
H_2S influent and effluent air (handheld electrochemical probe or colorimetric reaction tubes) [c]	H_2S load ; H_2S removal	(•)	(•)	(•)
Relative humidity influent and effluent air (fixed relative humidity probe)	Prehumidification influent air ; Packing moisture content	-	(•)	•
Temperature packing (fixed temperature probe)	Temperature inhibition	•	•	•
pH leachate (fixed or handheld pH probe)	pH inhibitionr	•	•	(•)
Water consumption (water volume counter)	Consumption make-up or sprinkling water	•	•	(•)

[a] : - = non applicable, • = applicable, (•) = possibly applicable
: BS = bioscrubber, BTF = biotrickling filter, BF = biofilter
[b] : on an arbitrary scale (e.g. 1 = weak odour, 2 = …, 5 = extreme odour)
: not allowable at H_2S concentrations > 10 ppm
[c]: detection limit is typical 1 ppm (electrochemical probe) or 0.25 ppm (colorimetric reaction tubes).

The major operational requirement for the biofilter is maintaining an optimum moisture content in the filter material (Leson and Winer 1991). Moisture content between 40 and 60 percent is considered optimal. Non-optimal water content can result in inactivation of the biomass, compaction of the filter material, breakthrough of incompletely treated waste gas and the formation of anaerobic zones which emit odorous compounds. In practice, the water content of the filter material can be controlled by (i) humidification of the waste gas before entering the biofilter or (ii) irrigation of the filter material. In more than

90% of all large scale biofilter systems built to date, control of the water content relies on manual operations (van Lith *et al.* 1997). Automated control strategies can be based on on-line measurement of the weight of the relative humidity of the incoming and outgoing waste gas or water content of the filter material. The limited use of automated control of water content is mainly related to the limited reliability of the measurement hardware (e.g. measurement of air relative humidity near the saturation point) or limited availability of control strategies that translate the measurement signal into an adequate control action (De heyder 1998).

Table 19.7. Overview of possible monitoring methods in the case of process deterioration.

Monitoring method	Evaluation of
Continuous H_2S monitoring influent and effluent air	Peak loads
Odour concentration influent and effluent air	Odour load and removal
Chemical composition of influent and effluent air (GC-MS)	Organic key compounds
Biodegradation activity (laboratory batch tests)	Microbial activity
Respiration activity (CO_2 measurement laboratory batch tests or on site, gas chromatography, infrared)	Microbial activity
Moisture content packing (gravimetric)	Microbial reaction conditions
State of deeper packing layers (visual inspection)	Scaling of packing
Air flow pattern (tracer injection and detection, e.g. CH_4)	Mean residence time
Smoke test (smoke bomb influent)	Air distribution over packing
Fatty acid analysis (gas chromatography) :	Anaerobic activity
pH packing	Acidification packing

For a biotrickling filter, attention with respect to operational stability is given especially to the prevention of clogging. Clogging can be the result of an excessive biomass formation and also of poor biofilm formation and biofilm detachment from the packing material (Weber and Hartmans 1996). The latter will result in the presence of biomass in lumps between the packing material. The clogging of the packing material will result in the formation of uncontrolled anaerobic zones, a shorter space time of the waste gas and in the end a complete blocking of the flow of the waste gas through the packing. An overview of investigated methods to prevent clogging of the packing of a biotrickling filter is given Table 19.8.

Table 19.8. Methods investigated to prevent clogging of a biotrickling filter.

Method	Reference
Limiting nutrients + with 0.1 M NaOH wash	Weber and Hartmans (1996)
Addition of protozoa	Cox and Deshusses (1997)
Nitrate as sole source of nitrogen + backwashing with packing fluidisation	Smith *et al.* (1996)
Backwashing with packing fluidisation	Sorial *et al.* (1998)

19.8 COSTS

Investment costs are often calculated on the basis of cost per unit of airflow rate treated (EUR/(m³/h)). An overview of reported investment costs for the different types of reactor is given in Table 19.9. These figures illustrate that biofiltration can generally be considered to be the cheapest odour-abatement method. Yet, these figures suggest also that the investment cost can be largely dependent on case-specific boundary conditions.

Operating costs are primarily a function of energy consumption, water consumption and disposal, monitoring requirements, maintenance and media replacement. All of these operating costs vary from case to case. However generalized costs for biofiltration have been reported to range 0.1 tot 3 EUR/(1000 m³ treated) (Devinny *et al.* 1999). For biotrickling filters and bioscrubbers these costs should be increased with the cost of liquid recirculation.

Table 19.9. Overview of reported investment costs for the different types of reactor.

Reactor	Investment costs Euro/(m³/h)	References
Biofilter	2 – 5	Diks (1992)
	5 – 150 (range)	Devinny *et al.* (1999)
	7 – 35 (average)	
	5 – 34 (open reactor)	STOWA (1996)
	10 – 68 (closed reactor)	
Biotrickling filter	5 – 20	Diks (1992)
	23 – 92 (excl. auxiliary equipment)	STOWA (1996)
Bioscrubber	23 – 92 (excl. auxiliary equipment)	STOWA (1996)

19.9 REFERENCES

Bailey, J.E. and Ollis, J.F. (1986) *Biochemical Engineering Fundamentals*, 2nd edition, McGraw-Hill, New York.

Beam, H.W. and Perry, G.G. (1974) Microbial degradation of cyclo parafinic hydracarbons via co-metabolism and commensalism. *J. General Microbiol.* **82**, 163-169.

Brüser, T. Lens, P.N.L. and Trüper, H.G. (2000) The biological sulphur cycle. In: *Environmental Technologies to Treat Sulphur Pollution: Principles and Engineering.* (P. Lens and L. Hulshoff Pol, eds.) pp. 47-85, IWA Publishing. London.

Cho, K.S., Hirai, M. and Shoda, M. (1991) Removal of dimethyl disulphide by the peat seeded with night soil sludge. *J. Ferment. Bioeng.* **71**, 289-291.

Cox, H.J.J. and Deshusses, M. (1997) Increasing the stability of biotrickling filters by using protozoa (233-240) In: *Biological waste gas treatment* (Prins W.L. and van Ham J. eds.), VDI Verlag, Düsseldorf.

De Beer, B., Stoodly, P.R., Roe, F. and Lewandowsky, Z. (1994) Effects of biofilm structure on oxygen distribution and mass transport. *Biotech. Bioeng.* **43**, 1131-1138.

De Castro, A., Allen, D.G. and Fulthorpe, R.R. (1996) Characterisation of the microbial population during biofiltration and the influence of the inoculum source. In: *Proc. 1996 Conference on Biofiltration*, (Reynolds F.E. and Tustin C.A. eds), pp. 164-172, The Reynolds Group.

De heyder, B. (1998) Biotechnological treatment of poorly water soluble waste gases: case study ethene. Ph.D. thesis, Universiteit Gent, Gent, B.

Deshusses, M (1997). Transient behaviour of biofilters : start-up, carbon balances and interactions between pollutants. *J. Environ. Engin.* **123**, 563-568.

Deshusses, M., Hamer, G., Dunn, I.J. (1995a) Behaviour of biofilters for waste air biotreatment. 1. Dynamic model development. *Environ. Sci. Technol.* **29**, 1048 - 1058.

Deshusses, M., Hamer, G., Dunn, I.J. (1995b) Behaviour of biofilters for waste air biotreatment. 2. Experimental evaluation of a dynamic model. *Environ. Sci. Technol.* **29**, 1059-1068.

Devinny, J.S., Deshusses, M.A. and Webster, T.S. (1999) *Biofiltration for Air Pollution Control.* CRC Press, Boca Raton.

Dewulf, J., Van Langenhove, H. and Drijvers, D. (1995) Measurement of Henry's law constant as function of temperature and salinity for the low temperature range. *Atmosph. Environ.* **29**, 323-331.

Diks, R.M.M. (1992). The removal of dichloromethane from waste gases in a biological trickling filter. Ph. D. thesis. Technical University of Eindhoven, The Netherlands.

Eaton, A.D., Clesceri, L.S. and Greenberg, A.E. (1995). Standard methods for the examination of water and wastewater. American Public Health Association, Washington, USA.

Guey, C., Degorce-Dumas, J.R. and Le Cloirec, P. (1995) Hydrogen sulphide removal a biological activated carbon. *Odours VOCs J.* **1**, 136-137.

Hartmans, S. (1997). Biological waste gas treatment: kinetics and modeling. *Med. Fac. Landbouww. Univ. Gent*, **26**(4b), 1501-1504.

Herrygers, V. Van Langenhove, H. and Smet, E. (2000) Biological treatment of gases polluted by volatile sulphur compounds. In: *Environmental Technologies to Treat Sulfur Pollution: Principles and Engineering.* (P. Lens and L. Hulshoff Pol, ed.) pp. 281-304, IWA Publishing, London.

Hirai, M., Ohtake, M. and Shoda, M. (1990) Removal kinetics of hydrogen sulphide, methanethiol and dimethyl sulphide by peat biofilters. *J. Ferment. Biotechnol.* **70**, 334-339.

Hwang, S.J. and Tang, H.M. (1997). Kinetic behaviour of the toluene biofiltration process. *J. Air Waste Manag. Assoc.* **47**, 664 - 673.

Kasakura, T.K and Tatsukawa, K. (1995). On the scent of a good idea for odour removal. *Water Quality International* **2**, 24-27.

Kennes, C. and Thalasso, F. (1998) Waste gas biotreatment technology. *J. Chem. Technol. Biotechn.* **72**(4), 303-319.

Kim, N.J., Hirai, M. and Shoda, M. (1998) Comparison of organic and inorganic carriers in removal of hydogen sulphide in biofilters. *Environ. Technol.* **19**, 1233-1241.

Leson, G. and Winer, A.M. (1991) Biofiltration: an innovative air pollution control technology for VOC emissions. *J. Air Waste Maneg. Assoc.* **41**, 1045-1054.

Moller, S. Pedersen, A.R., Poulsen, L.K., Arvin, E. and Molin, S. (1996) Activity and three dimensional distribution of toluene degrading *Pseudomonas putida* in a multi-species biofilm assessed by quantitative in-situ hybridization and scanning confocal laser microscopy. *Appl. Environ. Biotech.* **12**, 4632-4640

Okkerse, WJH, Ottengraf, SPP, Osinga-Kuipers, B and Okkerse, M (1999) Biomass accumulation and clogging in biotrickling filters for waste gas treatment. Evaluation of a dynamic model using dichloromethane as a model pollutant. *Biotech. Bioengin.* **63**, 418-430.

Picioreanu, C. van Loosdrecht, M.C.M. and Heijen, J.J. (2000) A theoretical study on the effect of surface roughness on mass transport and transformation in biofilms. *Biotech. Bioengin.* **68**(4), 355-369.

Shareefdeen, Z. and Baltzis, BC (1994) Biofiltration of toluene vapor under steady-state and transient conditions : theory and experimental results. *Chem. Engin. Sci.* **49**, 4347 - 4360.

Smet, E, Heireman, B and Van Langenhove, H (1996) The contribution of physical sorption processes to the biofiltration of dimethylsulphide. In: *Biofiltration of organic sulphur compounds*, Ph. D. thesis, Ghent University, Ghent, Belgium.

Smet, E. Van Langenhove, H. and Verstraete, W. (1997) Isobutyraldehyde as a competitor of the dimethyl sulphide degrading activity in biofilters. *Biodegradation* **8**, 53-59

Smith, F.L., Sorial, G.A., Suidan, M.T. Breen, A.W. and Biswas, P. (1996) Development of two biomass control strategies for extended stable operation of highly efficient biofilters with high toluene loadings. *Environ. Sci. Technol.* **30**, 1744-1751

Sorial, G.A., Smith, F.L., Suidan, M.T., Pandit, A., Biswas, P. and Brenner, R. (1998) Evaluation of a trickle-bed air biofilter performance for styrene removal. *Water Res.* **32**, 1593-1603.

Staudinger, J. and Roberts, P.V. (1996) A critical review of Henry's law constants for environmental applications. *Crit. Rev. Environ. Sci. Technol.* **26**(3), 205-297

STOWA (1996). Odour abatement on sewage treatment plants. Report 96-02 (in Dutch). Hageman Verpakkers, Zoetermeer, The Netherlands.

Van Groenestijn J.W. and Hesselink P.G.M. (1993) Biotechniques for air pollution control. *Biodegradation*, **4**, 283-301

Van Langenhove, H., Roelstraete, K., Schamp, N. and Houtmeyers, J. (1985) GC-MS identification of odorous volatiles in wastewater. *Water Res.* **19**(5), 597-603.

Van Lith, C., Leson, G and Michelsen, R. (1997). Evaluating design options for biofilters. *J. Air Waste Manage. Assoc.* **47**, 37-48

Weber, F.J. and Hartmans, S. (1996) Prevention of clogging in a biological trickle-bed reactor removing toluene from contaminated air, *Biotech. Bioengin.* **50**, 91-97.

Webster, T.S., Devinny, J.S., Torres, E.M. and Basrai, S.S. (1997) Microbial eco-systems in compost and granular activated carbon biofilters. *Biotech. Bioengin.* **53**, 296-303.

Yang, Y. and Allen, E.R. (1994). Biofiltration control of hydrogen sulphide. I. Design and operational parameters. . *J. Air Waste Manag. Assoc.* **44**, 863-868.

Zarook, S.M., Shaikh, A.A., Ansar, Z. (1997a) Development, experimental validation and dynamic analysis of a general transient biofilter model. *Chem. Engin. Sci.* **52**, 759-773.

Zarook, S.M., Shaikh, A.A., Ansar, Z., Baltzis, B.C. (1997b) Biofiltration of volatile organic compound (VOC) mixtures under transient conditions. *Chem. Engin. Sci.* **52**, 4135-4142.

Zeman, A. and Koch, K. (1983) Mass spectrometric analysis of malodorous air pollutants from sewage plants. *Int. J. Mass Spec. Ion. Phys.* **48**, 291-294.

20

Activated sludge diffusion as an odour control technique

Robert P.G. Bowker and Joanna E. Burgess

20.1 ACTIVATED SLUDGE ODOUR REMOVAL: DESCRIPTION AND BIODEGRADATION THEORY

A complication of chemical odour treatment arises from the fact that, in most cases, odours emanate from a variable mixture of gases rather than from single compounds, making chemical reactions for specific gases unreliable as control methods. Economic control of odours is best achieved by destroying the gases responsible as opposed to simply moving them from the gaseous phase to the liquid phase. A variety of systems can be employed for this purpose, but they mostly tend to involve the construction of a dedicated processing plant with its associated control systems and high capital costs. Many odorous sites do not justify such expenditure, particularly where the odours result from waste treatment processes. For wastewater treatment sites, activated sludge

 Odours in Wastewater Treatment: Measurement, Modelling and Control edited by Richard Stuetz and Franz-Bernd Frechen. ISBN: 1 900222 46 9

diffusion offers a low cost alternative. By collection of the odorous gas and its diversion into an activated sludge aeration basin, odours can be eliminated using relatively low cost technology.

The inherent properties of aeration basins make them particularly effective in removing gases from an odourant / air mixture. Activated sludge treatment of wastewater is an aerated oxidation process. It is one of the best established and widespread biological wastewater treatment processes in the developed world for both domestic and industrial wastewaters (Clark and Stephenson 1998), and as such its adaptability to accommodate new demands in effluent quality is of great importance. The process relies on the suspension of a microbial population mixed with wastewater under aerobic conditions. Microbial growth brings about the removal of organic matter from the liquid as the compounds present are oxidised by the micro-organisms in the sludge. The end results are microbial biomass and products of oxidation such as CO_2, NO_3^-, SO_4^{2-} and PO_4^{3-}. Activated sludge plants have been used to treat a wide range of industrial wastes by effectively accelerating natural processes involving chemical, biological and physical agents, as the biomass is able to acclimatise to, and oxidise, a large number of contaminants provided they are present in soluble form.

An activated sludge plant for simultaneous treatment of wastewater and odour can be schematically represented (Figure 20.1). In the aeration tank the wastewater is added to the microbial biomass and air supplied via diffusers. This aerates and mixes the suspension, allowing maximum contact between the flocs and the wastewater. Complete mixing ensures an adequate food supply for the microbial cells and maximises the oxygen gradient to optimise mass transfer and disperse the products of metabolism from inside the flocs. Wastewater entry displaces mixed liquor into a clarifier, where the flocculated biomass separates into sludge and clarified effluent. The floc nature of the biomass is very important as it controls the efficient absorption and adsorption of organics from the waste and the separation of the sludge from the water in the settling tank.

The aeration in an activated sludge plant speeds up the growth of the bacteria present at the outset and increases the number of collisions between flocs and hence their chance of aggregation into larger flocs containing non-living particles. This process occurs within a set range of environmental conditions, which limit the activity of the organisms responsible for the treatment process. For this' reason, biological wastewater treatment requires control of certain environmental parameters, such as dissolved oxygen (DO) levels, mixing regime, provision of nutrition, trace element supply and physical conditions such as temperature and pH.

There are fewer examples of liquid-based odour control systems than media-based systems (WEF/ASCE 1995), although the advantages and disadvantages of such systems differ and so their suitability to the conditions in different wastewater treatment plants (WWTPs) also differ. Activated sludge diffusion is used as an alternative to more established bioreactors for waste gas treatment, such as biofilters, bioscrubbers and biotrickling filters. Contaminant removal mechanisms in activated sludge diffusion of waste gas include absorption (the solution of gases into the mixed liquor; limited by bubble size and gas residence time), adsorption (high molecular mass compounds with low solubility adsorb onto flocs) or condensation (volatile organic compounds in warm air condense on contact with the cooler mixed liquor), followed by biodegradation. Foul air is collected from its source and transferred via blowers through a delivery pipework system to submerged nozzles in the activated sludge aeration tank (Figure 20.1). The odorous air bubbles diffuse into the mixed liquor where the contaminants dissolve and are subsequently adsorbed or absorbed and biodegraded.

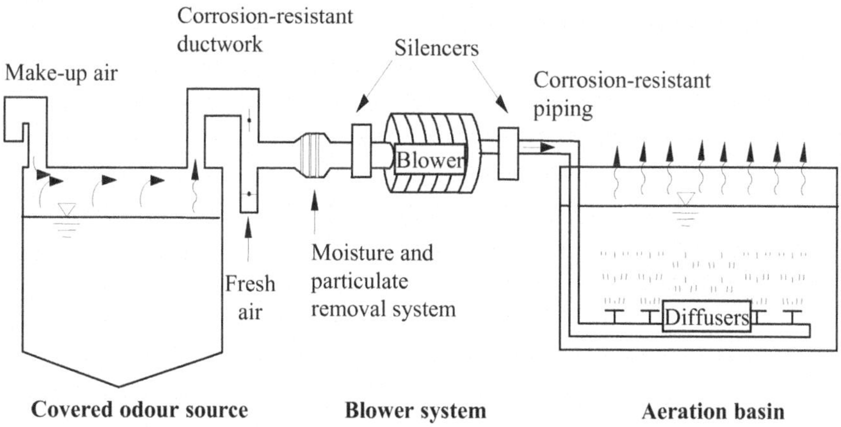

Figure 20.1. Schematic representation of a typical activated sludge plant.

20.2 DESIGN / OPERATION CONSIDERATIONS

20.2.1 Odorous air pre-treatment

An air pre-treatment system should be included in any foul air diffusion system. The system should be designed to remove free moisture and condensate that could be acidic in the presence of hydrogen sulphide (H_2S), as well as particles consisting of dust and grease aerosols. Normally, a mesh pad or chevron demister of the type used in packed tower scrubbers will be adequate for moisture removal. Such devices normally provide 99% removal of droplets greater than 50 microns.

For fine bubble diffusion systems, filter systems normally specified for activated sludge aeration applications will be adequate to remove particles and grease aerosols. In Los Angeles County, USA, a two-stage filter system has been found successful in protecting the blower. This consists of a 2.5 cm deep, pleated glass-fibre pre-filter followed by a 30 cm deep, pleated fibreglass filter. The system is designed to remove 95% of particles 0.3 microns and greater in size. Typical face velocities are 0.6 to 2.5 m/sec. The modular filter panels are easily replaced in a corrosion resistant housing. Figure 20.2 is a sketch of a foul air pre-treatment system.

It is important that all components including the demister, filter frames, and filter housing be constructed of materials that are resistant to attack by H_2S or dilute sulphuric acid. Such materials include fibreglass, stainless steel and plastics such as PVC, polypropylene, and polyethylene.

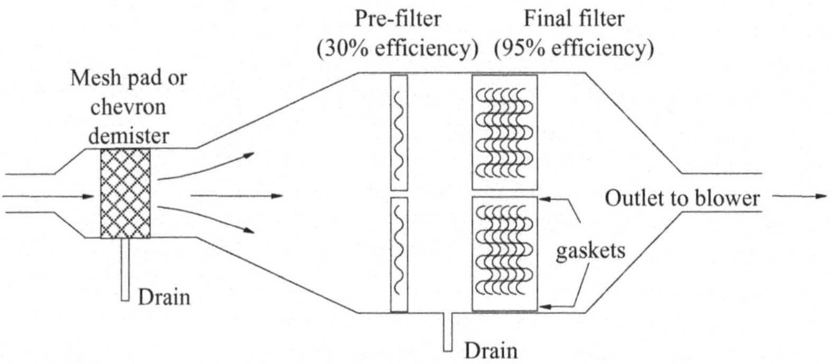

Figure 20.2. Schematic of foul air pre-treatment system.

20.2.2 Blowers

Both centrifugal and rotary-lobe positive displacement blowers have been used for the diffusion of odourous air into activated sludge basins. Some practitioners have recommended using centrifugal blowers because the positive displacement blowers have close tolerances between the lobes and the casing which may be more susceptible to clogging with the organic "tarry" material that has been reported. This problem occurred at the Valley Forge WWTP, USA (section 20.7), causing the positive displacement blowers to shut down after several weeks of operation. This was rectified by preventing grease from entering the ductwork and by improving the filtration system.

Blower corrosion is perhaps the biggest concern with handling foul air, although incidences of blower corrosion directly linked to H_2S are very limited. For new installations, blowers can be specified with a protective coating or metal plating to protect against corrosion. Different manufacturers offer different corrosion protection systems, including a phenolic coating and nickel plating. Manufacturers should be contacted for recommendations once the characteristics of the odourous air stream are defined. Normally where blowers already exist, no special precautions are taken other than removal of moisture and particulates. However, at Annapolis, Maryland, USA, existing aeration blowers were shipped back to the manufacturer to be coated before handling odourous air from the sludge thickeners and primary clarifiers. At another location, steam injection ports were specified for a new centrifugal blower to allow periodic removal of any contaminants that built up on the blower volute.

20.2.3 Diffusers

A variety of diffusers have been successfully used in odourous air diffusion including coarse bubble diffusers and both flexible membrane and ceramic dome fine bubble diffusers. For particularly strong or difficult-to-treat odours such as from sludge storage tanks, fine bubble diffusers provide superior performance. At Concord, New Hampshire, USA, experimentation with both coarse and fine bubble diffusers showed approximately 96% odour removal and 92% H_2S removal with coarse bubble diffusers, and 99.9% odour removal and 99.7% H_2S removal with fine bubble diffusers.

There have been no reports of diffuser clogging or corrosion associated with handling foul air. In the United States, some engineering firms have specified the use of flexible membrane diffusers because they are resistant to attack by H_2S or sulphuric acid.

The greater the depth of the diffuser, the greater the driving force available to drive the odourous gas into solution, and the longer the residence time of the gas bubble. Normally, designing a diffuser system based on supplying process air for biological oxidation will be adequate for odour treatment with regard to diffuser depth and spacing. Diffuser depth should be a minimum of 3 m unless pilot testing indicates that a shallower depth will provide adequate odour treatment.

Successful odour treatment by diffusion requires an active biological population and a healthy mixed liquor. Since most applications of odourous air diffusion use an existing aeration basin already equipped with diffusers, this is seldom an issue. For dedicated odour diffusion systems that are not designed to supply process oxygen, odours should not be diffused into channels or basins where an active mixed liquor population does not exist.

20.2.4 Corrosion protection

Corrosion can be a problem with odourous air diffusion unless materials are carefully selected. Concrete and carbon-steel items suffer from exposure to H_2S and sulphuric acid (Ryckman-Siegwarth and Pincince 1992; WEF/ASCE 1995), however, fibreglass, stainless steel, polyvinyl chloride, or high-density polyethylene are all suitable for the foul air delivery system. Condensate drains should be provided at low points to remove acidic condensate.

Concerns regarding potential corrosion damage to blowers are historically the largest impediment to utilising existing blower/diffuser systems for foul air treatment. However, based on the experience at some 30 facilities in the USA, such concerns are not well founded. Isolated reported cases of blower corrosion may have been due to the failure to remove acidic condensate from the ductwork leading to the blower. As discussed in Section 20.2.2, several coating systems, such as phenolic coatings or nickel plating, can be used to provide additional protection against corrosion. Inlet filters and filter housings must be constructed from corrosion resistant materials or they may deteriorate rapidly. Mild steel or galvanised steel should be avoided in favour of 316 stainless steel, fibreglass, or plastic. Blower discharge piping should be stainless steel above the water surface.

Some sites reported corrosion of the concrete aeration tank, ameliorated by the provision of a protective coating at the waterline. Corrosion of diffusers (both coarse and fine bubble) was not found to be a significant problem in a survey of WWTPs employing activated sludge odour diffusion (Ryckman-Siegwarth and Pincince 1992). Diffusion minimises the amount of equipment involved in introducing the air into the sludge, but could increase the amount of system maintenance required. Activated sludge diffusion of odourous air works well in situations where the activated sludge plant is not heavily loaded and DO levels are maintained, and has been in use at several sites in North America for several years.

20.2.5 Increased odour emission

Activated sludge diffusion of odourous air reduces the presence of liquid phase odourants via biological oxidation, but can produce odours via gas stripping, if systems are overloaded (Ryckman-Siegwarth and Pincince 1992; Vincent and Hobson 1998). However, this is not a significant operating problem for two reasons. First, in most cases there is no detectable difference between the odour off an activated sludge plant treating odourous offgas and the odour from an activated sludge plant operating "normally", provided sufficient DO is maintained in the mixed liquor. Second, even in cases where aeration basin odours do increase, odours are significantly reduced at the wastewater treatment site, as the odour monitored at site boundaries is the product of the entire site as opposed to the activated sludge tanks alone.

Full-scale sites using activated sludge diffusion found that aerating with off-gas from grit chambers and primary clarifiers and fine bubble nozzles could affect the tank air emissions, effluent concentrations and the quantity of volatile organic compounds biodegraded. In the cases where odour emission from the aeration tank increased, the emissions from the site as a whole decreased owing to the odours from the grit chambers and primary clarifiers being eliminated. The concentrations of volatile organic compounds emitted to the environment via the reactor effluent increased, but the total emissions from the site decreased as a substantially higher proportion of the total volatile organic compounds received by the site were biodegraded. Use of foul air for aeration carries many advantages for sites at which all emissions to the atmosphere must be treated before discharge.

It has been stated that aeration tanks cannot always accept the total volume of foul air generated at a wastewater treatment works (WEF/ASCE 1995). However, a survey of several North American full-scale treatment plants using activated sludge foul air diffusion was carried out to assess the extent of the problems experienced with odour treatment. Foul air accounted for 20–100% of aeration air supply, but in no case was an excess of foul air reported (Ryckman-Siegwarth and Pincince 1992).

20.3 FACTORS AFFECTING PERFORMANCE

As with the treatment of wastewater, treatment of waste gas is influenced by a number of factors, including the characteristics of the aeration tank, the nature of the contaminants to be degraded and the operating regime of the individual site.

20.3.1 Depth of the aeration basin

One reported disadvantage of activated sludge diffusion is the need for a deep aeration tank to provide a long gas residence time, when a shallow reactor represents a reduction in energy requirements. However, shallow activated sludge basins have been shown to effectively degrade a mixture of benzene, toluene, ethylbenzene and xylene (BTEX). The BTEX was treated in a bench-scale activated sludge reactor with a working volume of 2 l and liquid depth of 40 cm. The reactor was run with sludge ages of 1.7, 2.7 and 9.2 days (hydraulic retention time was equal to sludge retention time (SRT)) with 15–17 mg/l BTEX in the air entering the reactor. The BTEX in the off gas was below the limit of detection (0.01 mg/l), indicating >99% removal in all cases and showing that shallow activated sludge tanks are able to biodegrade BTEX in contaminated air (Bielefeldt *et al.* 1997).

In further studies on the effects of mixed liquor depth on odour treatment, a pilot activated sludge plant was run to treat foul air from the headspace of a dissolved air flotation sludge thickener. The 35 l working volume reactor held 127 cm depth of activated sludge, with 250 mg/l mixed liquor volatile suspended solids (MLVSS). The contaminated air contained low levels of H_2S, amines, ammonia and mercaptans, all of which were removed to <0.1 ppm in the tank off-gas. Reducing the height of the liquid to 60 cm had no effect on the levels of contaminants present in the effluent gas. As in wastewater treatment, acclimation of the activated sludge was crucial to effective gas treatment, as unacclimated activated sludge gas treatment was biodegradation limited, removing only ~45% of some contaminants. Longer SRTs allowed degradation of indole and skatole.

20.3.2 Bubble size

Bubble size is also an important factor in gas treatment efficiency: Concord, New Hampshire WWTP used activated sludge diffusion to treat air with high levels of H_2S. Coarse bubble diffusers positioned at 3 m depth provided 95% odour reduction (measured by olfactometry) and 92% H_2S reduction, but the odour from the activated sludge plant was detectably higher than before activated sludge diffusion of foul air began. Changing the system to fine bubble diffusers resulted in >99.5% reduction in both odour and H_2S, the activated sludge plant odour being indistinguishable from its odour with no waste gas treatment.

Experiments at lab-scale have shown the activated sludge tanks can effectively degrade sulphurous compounds, aliphatic amines, toluene and low relative molecular mass compounds (Fukuyama *et al.* 1986). Lab-scale experiments using odourous air rather than sample odour gases achieved ~99% removal efficiencies; the pilot plant approximately 90%. Later work at pilot-scale (Fukuyama *et al.* 1986) reported that the pilot plant coped well with variable loads which had not been an issue in the lab experiments. The WWTP in the study received >25% of its load from industrial sources, so has a high proportion of odourous components.

During one study, continuous activated sludge tank deodorisation of exhaust gas from wastewater treatment and night-soil treatment plants was carried out for several months. Efficiency was measured in terms of the concentrations of the main odourants prior to and after treatment; influent concentrations varied greatly, but outlet concentrations were more consistent. Mean removal efficiencies were 90% for aromatic hydrocarbons and dimethyl sulphide, 96% for H_2S (mean influent concentration of 7 mg/g mixed liquor suspended solids (MLSS)/d) and 100% for ammonia (Fukuyama *et al.* 1986).

20.3.3 Aeration intensity

Investigation into the effect of aeration intensity employed an aeration tank (depth 1.0 m, working volume 150 l, MLSS 11.20 g/l, SRT of ∞) receiving two levels of aeration intensity. Aeration intensity was found to affect the degree of removal of VOCs by gas stripping (Table 20.1) and Fukuyama *et al.* (1986) concluded that the decrease in measured components at an aeration intensity of 12 m^3 air/ m^3 /h (except carbon disulphide, which did not contribute to odour as the outlet concentration was below the limit of human detection) and simultaneous increase in odour units (OU) leads to the conclusion that increased aeration strips out other, unmeasured odourants present in the wastewater.

Table 20.1. Effect of aeration intensity on activated sludge odour treatment.

Odourant	Aeration intensity	
	6 m^3 air/ m^3 tank volume/h	12 m^3 air/ m^3 tank volume/h
Total aromatic hydrocarbons	87.50–91.40%	85.34–93.33%
Dimethyl sulphide	80.00–93.10%	80.83–92.56%
Carbon disulphide	31-.58–45.45%	15.05–47.73%
Odour unit	74.29–90.65%	62.86–76.92%

Night-soil treatment plant foul air was also treated using an activated sludge aeration tank and odourant removal was compared to a control tank filled with clean water. The first run was carried out using an aeration intensity of 4.7 m^3 air/ m^3 tank volume/day, MLSS of 16.28 g/l and SRT of ∞. The second run was carried out using a lower loading rate, finer bubbles and an aeration intensity of 2.0 m^3 air/ m^3 tank volume/day, MLSS of 15.55 g/l and SRT of ∞, and resulted in greater contaminant removal than the first run (Table 20.2). The control tank attained comparable removal efficiencies when loading rates were consistent and normal, but did not remove peak loads which were removed by the activated sludge tank (up to 0.58 ml H_2S /ml air), resulting in a consistent outlet

H_2S concentration from the test aeration tank. This indicates that a proportion of the odourants are dissolved in the liquid and go no further, but that much of the odour removal reported is dependent on biodegradation to avoid saturation of the liquid with the odour compounds.

Table 20.2. Removal of night-soil treatment plant odours by activated sludge.

Odourant	Mean removal during run 1 (%)	Mean removal during run 2 (%)
Ammonia	99.12	99.92
H_2S	87.24	95.3
Methyl mercaptan	78.43	93.93
Dimethyl sulphide	29.41	74.03
Dimethyl disulphide	-0.80	-0.11–23.30

Fukuyama *et al.* (1986) established the relationship between H_2S loading and removal rates as:

$$y = -0.981x + 99.26 \tag{20.1}$$

Where:
y = removal efficiency (%),
x = H_2S load (mg/g MLSS/d).

There must be a threshold top loading rate at which equation (20.1) ceases to be true, as H_2S exerts a toxic effect on biomass when present to excess. This threshold value clearly exceeds 7 mg H_2S/g MLSS/d, the mean loading rate applied, but has yet to be established.

20.3.4 Operating parameters

The same authors also studied the effectiveness of a two-stage diffusion process. Two identical aeration tanks (depth 1.0 m, working volume 150 l, MLSS 8.82 g/l, SRT of ∞, aeration intensity of 30 m^3 air/ m^3 tank volume/day) were used in series. The results (Table 20.3) showed large standard deviations in the extra removal obtained in the second stage and low mean values for extra contaminant removal, which mean that the cost of duplication in adding a second aeration tank is not justified in most cases. The use of two-stage treatment is most useful for very variable loads of airborne contaminants, instead of recycling the outlet air.

Table 20.3. The effect of a second diffusion treatment stage.

Odourant	Removal after 1st stage	Extra removal after 2nd stage	Average extra removal gained
Total aromatic hydrocarbons	81.91–88.53%	0.00–11.24%	5.76%
Dimethyl sulphide	80.95–94.38%	3.37–7.26%	4.89%
Carbon disulphide	0.00–11.11%	0.00–40.00%	4.17%

Investigation into the effects of sludge reaeration was carried out, using one aeration tank (depth 1.0 m, working volume 150 l, MLSS 4.65 g/l., two different SRTs, 1 h and 4 h) and an aeration intensity of 30 m^3 air/ m^3 tank volume/day (Fukuyama *et al.* 1986). The removal efficiencies obtained (Table 20.4) indicated that increasing SRT in this experiment led to decreasing removal efficiencies, but the removal efficiencies at both SRTs were very low in comparison to the authors' other experiments, in which SRT = ∞. This leads to the conclusion that the SRTs were both too low to compare to the typical SRTs used in wastewater treatment (6–10 days for domestic wastewater, longer for industrial effluent (Eckenfelder and Grau 1992)) and the data are not representative of the effects of SRT on odour treatment operated in a 'real' system.

Table 18.4. The effects of SRT variation on activated sludge odour treatment.

Odourant	Removal with 1h SRT	Removal with 4h SRT
Total aromatic hydrocarbons	9.26–22.91%	0.11–16.25%
Dimethyl sulphide	21.69–35.00%	10.00–21.05%
Carbon disulphide	13.04–33.90%	0.00–26.09%

Loading rates of 15 mg H_2S/g MLSS/d were degraded very well (~95% removal efficiency) in the laboratory. The pilot plant was subjected to variable loading rates and the effects of other odourants not present in the laboratory experiments, but still achieved ~90% removal efficiency up to 7 mg H_2S/g MLSS/d (<90% over 7 mg H_2S/g MLSS/d). The order of sulphur removal by the sludge was: H_2S > methyl mercaptan > dimethyl sulphide > dimethyl disulphide. Removal efficiencies may suffer when peak concentrations occur, as foul air compounds have to be acclimated to just as wastewater components do, but this rarely affects aeration tank outlet concentrations of odourant (Fukuyama *et al.* 1986), and particularly high concentrations can be degraded by recycling the air during peak loads (Frechen 1994).

Performance data relating to volatile organic compound removal are reported by Oppelt *et al.* (1999) operating an activated sludge plant (liquid depth 6.6 m,

MLVSS 2242 mg/l, DO 2.0 mg/l) through which headspace air from a lift station was diffused, mixed with fresh air. The wastewater passing through the aeration tank contained concentrations of the volatile organic compounds present at the wastewater treatment site far in excess of the concentrations measured in the headspace air, so it was not possible to calculate volatile organic compound removal efficiencies during normal operation of the plant. Instead, the system's effectiveness was measured during the construction phase, with no wastewater flowing through the activated sludge plant. The aeration tanks were filled with mixed liquor from another activated sludge tank and allowed a two-week period for acclimation to the foul air contaminants, after which the aeration tank headspace was sampled. Eleven volatile organic compounds entered the aeration tank (Table 20.5) and were biodegraded, but the variation in the aeration tank emission data is so great that longer term results, generated with a working system are still required to build on these very promising data.

Table 18.5. Lift station and aeration tank volatile organic compound emission data.

Compound	Lift station headspace air concentration (μg/m³)	Aeration tank headspace air		
		Mean concentrati on (μg/m³)	Standard deviation	Relative standard deviation (%)
Benzene	315	20	7.9	39.5
Chloroethane	912	81	40.4	49.9
Chloroform	596	84	19.4	23.0
Ethyl benzene	1589	14	12.5	87.0
Hexane	20,059	3815	2917.9	76.5
Toluene	11,106	121	167.5	138.3
1,2,4-Trimethylbenzene	4097	30	32.5	110.7
Vinyl acetate	82,308	45	40.7	89.4
Xylene	4325	17	12.7	74.7
o-Xylene	9708	21	20.0	95.5

20.4 EFFECTS ON WASTEWATER TREATMENT

Introduction of odourous air into heavily loaded activated sludge plants can cause loss of process performance, although foul air drawn from waste treatment processes such activated sludge compost systems can be high in oxygen, thus providing an advantage (WEF/ASCE 1995).

Activated sludge plant operation is affected by the amount of sulphide entering the reactor. All sulphurous compounds are inhibitory to nitrification

(Henze *et al.* 1995); if the prevailing pH drops below 7, then nitrification declines (Æsøy *et al.* 1998). H_2S input into activated sludge either via air or wastewater inputs has been seen to result in nitrification inhibition and bulking sludge (Bentzen *et al.* 1995, Æsøy *et al.* 1997). Sulphide inhibition depends on the composition and acclimation of the biomass, the concentration of H_2S and other components in the wastewater, and temperature (as it affects solubility and bacterial growth rates). Laboratory-scale activated sludge reactors showed higher concentrations of filamentous bacteria (responsible for sludge handling problems) when high aqueous loadings of H_2S were applied (Johnson *et al.* 1995). Increases in the levels of filamentous bacteria present in the mixed liquor of full-scale activated sludge plants accepting foul air have been reported, but no cause and effect has been identified by the WWTPs allegedly experiencing this problem. This may be due to the fact that sulphide concentrations in wastewater are usually higher than H_2S concentrations in foul air, and the impact of the gaseous sulphide load is therefore less that that of the aqueous load.

General effects of aeration with foul air from wastewater treatment on the activated sludge were noted throughout a pilot-scale trial of odour diffusion (Fukuyama *et al.* 1986). The mixed liquor pH remained constant (around neutral), the MLSS decreased, effluent suspended solids increased and changes were seen in the community structure of the biomass (decreased numbers of Protozoa and increased Euglypha). The authors also observed the effects of night-soil treatment odourous air on the activated sludge. They found that pH was reduced from 7.6 to 3.15 over one 33 day experiment, and from 6.3 to 5.5 over 22 days, with consequential effects on nitrification. Autolysis of the biomass at pH 3.15 led to ammonia in the reactor effluent in excess of the influent ammonia concentrations. Mixed liquor volatile suspended solids and MLSS decreased by 130-160 mg/l and the sulphur and nitrogen content of the biomass increased during the experiments. 22–39% of the sulphur present was metabolised to SO_4^{2-} and no residual sulphides were measured in the wastewater.

Metal salts commonly present in activated sludge systems where they are employed as coagulants will form an insoluble precipitate with sulphides. Iron salts are most commonly used because of their low cost and minimal toxicity to the activated sludge biomass. Ferrous chloride is one option for co-precipitation (Clark *et al.* 2000), but any iron salt will react with dissolved sulphides. In this case, pH can be significantly reduced and should be monitored to ensure that the level of wastewater and waste gas treatment obtained remain satisfactory. Adjustment to mixed liquor pH can increase the solubility of sulphides, making them bioavailable and reducing the odour emitted by the activated sludge tank. Although H_2S gas is only slightly soluble in water, the ionised species HS^- and

S^{2-} are highly soluble in water. The high pH values at which this level of solubility is attained can be maintained only by compromising wastewater treatment, so pH adjustment alone can not be used to optimise activated sludge treatment of odourous air in activated sludge plants employing co-precipitants.

In treating foul air from waste treatment plants, some control of pH and MLSS would be required to maintain ongoing performance. The constant input of new wastewater and routine sludge surplussing and recycling will provide this control and avoid the accumulation of toxic metabolites. It has been found that sulphurous compounds are converted to sulphate and partly taken up by the sludge; nitrogenous compounds are converted to nitrate and nitrite. High loading rates, plus a pH of >5.0, facilitate nitrification of ammoniacal-N, and as no nitrate or nitrite was found in the effluent or MLSS supernatant, denitrification was also occurring.

20.5 ADVANTAGES OVER MEDIA-BASED SYSTEMS

Odour control for off-gas from sludge composting have been studied and the methods of wet scrubbing, biofiltration and activated sludge diffusion compared (Ostojic *et al.* 1992). Wet scrubbing is one of the most popular methods of odour control in the USA, but suffers from recurring problems and averages 70–75% removal efficiencies. Biofiltration using compost or wood chips averages around 90–95% removal efficiency, and has replaced wet scrubbing at some sites, but suffers badly when media humidification fails (45% removal efficiency). Activated sludge averages ~100% removal efficiency at full scale, where 2.0–2.5 m depth of MLSS is maintained (Springfield, Massachusetts and Orlando, Florida), and can reduce the level of background odour in cases where surface mechanical aerators are replaced by submerged nozzles when the foul air diffusion system is fitted. Activated sludge outperforms wet scrubbing for treatment of air from sludge composting, as the air contains a number of odourants in addition to sulphurous compounds (alcohols, ketones, aldehydes, acids) which are biodegradable but which persist after wet scrubber treatment.

Activated sludge diffusion avoids the problems with biofilters, biotrickling filters and membrane bioreactors i.e. media plugging, excess biomass accumulation, gas short-circuiting, moisture control and maintaining a correct biofilm thickness (Bielefeldt *et al.* 1997). The advantages and disadvantages of activated sludge diffusion are summarised in Table 20.6. Any filter consisting of a bed of media has to be supplied with pre-humidified waste gas to prevent dehydration of the filter micro-organisms. The biofilter and trickling biofilter both consist of a packed-bed of media onto which water is sprayed. The odourants then diffuse into the thin water layer within the filter, from which they are taken up by micro-organisms. Pollutants with low water solubility may not diffuse into the thin layer, as the water surface area is small by comparison to the area available in activated sludge diffusion using small bubbles.

Filter biotreatment of gases containing chlorinated pollutants, sulphur compounds or ammonia results in accumulation of chloride, sulphate or nitrate ions and subsequent acidification of the biofilter; acidification can be buffered by chemical additions such as lime, but the mineral end products can neither be neutralised in nor removed from the filter. The use of the micro-organisms in suspension in the liquid means that toxic end-products are removed from the liquid phase as components of the reactor effluent or as solids incorporated into the biomass removed for disposal. Humidity does not require control, the volume of mixed liquor stabilises the reactor temperature and nutrients are supplied in the wastewater.

Table 20.6. Summary of activated sludge diffusion.

Advantages	Disadvantages
Simple and effective.	Increased blower maintenance.
Low O&M, low capital cost.	Gas dissolution is rate limiting step.
Easily controlled via the wastewater.	Ability to treat odorants other than H_2S limited.
Removal of the degradation products by washout (avoids biomass inhibition).	Process can be difficult to control, as composition of wastewater is not controlled.
Biomass acclimation capacity provides efficient pollutant degradation.	Some question consistency of performance.
Excess biomass removed routinely.	Useful only where the sludge is aerobic, nitrifying and the concentration of H_2S is low.
Use of existing facilities (no footprint) and equipment (operator familiarity).	Overloaded systems can produce odours via gas stripping.
Economical treatment of large volumes.	Odourants inhibit nitrification.
No chemical requirements.	H_2S input may result in bulking sludge.
Can treat up to ~100 ppm H_2S long term.	
Avoids media plugging, gas short-circuiting and moisture control .	
>99.5% reduction in odour and H_2S.	

20.6 ECONOMICS

20.6.1 Using existing blowers and diffusers

The vast majority of cases in which odourous air is treated in activated sludge basins involve use of existing blowers and diffusers designed to provide process oxygen for biological oxidation. The additional operation and maintenance costs associated with handling foul air are minimal, and the odour diffusion process eliminates the storage and handling of hazardous chemicals such as sodium hypochlorite and sodium hydroxide.

Capital expenditures for using an existing aeration system for odour treatment are limited to the ductwork for odourous air conveyance, and if not already present, an air pre-treatment system for removal of moisture and particulates. The capital cost of the ductwork is largely a function of the distance between the odour source and the aeration blowers, and the complexity of the run (i.e., obstructions such as process equipment, roads, buildings, etc.). Ductwork is costly, particularly in the larger diameters, and significant ductwork runs can potentially be more expensive than a conventional wet scrubber located close to the odour source. However, the huge savings in chemical costs as well as operation and maintenance and labour costs can easily off-set the additional capital expenditures for ductwork over the lifetime of the system. A cost-effectiveness analysis should be conducted in order to properly weigh these factors.

20.6.2 Using dedicated blowers and diffusers

The economic advantage of odour treatment by activated sludge diffusion may be lost with a blower and diffuser system dedicated to odour treatment. The reason for this is the high energy costs of diffusing air three or more metres below the water surface. This additional cost is not present with an existing blower/diffuser system that already supplies process air required for biological treatment of the wastewater. The power required for a blower to diffuse 100 m^3/min of air to a depth 3 m below the surface is approximately 60 kW.

The cost to supply this energy must be factored into the cost-effectiveness analysis, as well as the capital cost of the blowers and diffusers. Although capital costs may be less than other technologies such as wet scrubbers, the annual energy costs of the diffusion system is likely to be significantly greater than the annual chemical costs of the wet scrubber.

20.7 CASE HISTORIES

20.7.1 Valley Forge Sewer Authority

The Valley Forge Sewer Authority operates a 30 m^3/d WWTP in Phoenixville, Pennsylvania, USA. The plant had experienced high odour emissions from the influent structures and primary clarifiers due to septic conditions in the collection system that promoted the generation of hydrogen sulphide gas. The Authority made the decision to cover the influent structure, the feed wells and effluent launders of the primary clarifiers and the primary effluent splitter box and diffuse the odourous air into the aeration basins. As the aeration tanks were aerated using mechanical surface aerators, this required the installation of blowers and diffusers dedicated to odour treatment.

Table 20.7 summarises the design criteria of the odour control system at the Valley Forge WWTP. Performance testing showed that the system provided over 99.9% removal of odours and H_2S. Inlet odour concentrations of 19,000 odour units (ou) were reduced to 5 to 7 ou at the surface of the basin above the diffusers, equivalent to the background levels measured at other locations in the mechanically aerated basins. Inlet H_2S levels of 77 ppm were reduced to approximately 0.1 ppm, the detection limit of the electrochemical H_2S analyser.

Initially, there were problems experienced in the build-up of a tarry material on the blower lobes that caused the blowers to shut down after several weeks of operation. This was found to be caused by grease being pulled into the ductwork from the primary clarifier feed wells and an inefficient filter mechanism to remove aerosols. After modifications were made to correct these deficiencies, maintenance requirements have been minimal.

Table 20.7. Key design criteria for odour control system at Valley Forge WWTP.

Parameter	Value	
Odour sources	Influent chamber	
	Primary clarifier feed wells	
	Primary clarifier effluent launders	
	Primary effluent splitter box	
Air flow	62 m^3/min (2,200 cfm)	
Air exchange rate below covers	12 AC/hr	
Inlet H_2S (estimated)	120 ppm (summer)	
Blowers	2 - 45 kW (60 hp) positive displacement	
Diffusers	394 tubular, flexible membrane type at 4.3 m (14 ft) depth	
Materials of construction	Covers	- FRP
	Ductwork	- PVC, 316 SS
	Blower filters, silencers	- 316 SS
	Blowers	- steel
	Discharge piping	- 316 SS
Concrete protection	Min. 1 mm vinyl ester coating above water line	

20.7.2 Concord WWTP

The Hall Street WWTP in Concord, New Hampshire, USA is designed to treat 39 m^3/d of wastewater using primary clarification and the activated biofilter

(ABF) process. The ABF process involves redwood-media trickling filters followed by activated sludge basins. The facility had experienced objectionable odours from screening and grit removal processes, primary clarifiers, and sludge storage tanks. The City had experimented with diffusion of odourous air from the sludge holding tanks into the activated sludge basins. This was found to be very successful. With fine bubble diffusers, odour concentration in the air from the sludge holding tanks was reduced from 39,000 ou to 18 ou (equivalent to background levels), for a removal efficiency of greater than 99.9%. Hydrogen sulphide was reduced from greater than 100 ppm to approximately 0.3 ppm. Although a two-stage wet scrubber was ultimately constructed for this air stream, the City investigated using activated sludge diffusion for other air streams. In 1998, the decision was made to replace the ageing mechanical surface aerators with blowers and fine bubble diffusers, and to use the new aeration system to treat odourous air from the influent channels, aerated grit chambers, and primary clarifier effluent launders.

Table 20.8 summarises the design criteria for the system installed at Concord. As opposed to Valley Forge, the blower and diffuser system is designed to provide process oxygen for biological oxidation as well as treatment of the odourous air.

Table 20.8. Key design criteria for odour control system at Concord, NH.

Parameter	Value	
Odour sources	Influent channels	
	Aerated grit chambers	
	Primary clarifier effluent launders	
Odourous air flow	70 m^3/min	
Air exchange rate below covers	6 AC/hr	
Inlet H_2S (estimated)	200 ppm (summer)	
Blowers	2–93 kW, 1–149 kW centrifugal	
Diffusers	2,928 tubular, flexible membrane type, 4.2 m depth	
Materials of construction	Covers	- Aluminum and FRP
	Ductwork	- FRP and HDPE
	Blower filters, silencers	- 316 SS
	Blowers	- steel
	Discharge piping	- 316 SS
Concrete protection	None	

20.7.3 Los Angeles County

At least eight WWTPs operated by the County Sanitation Districts of Los Angeles County utilise activated sludge diffusion as a means of treating odourous air. These WWTPs, all located in Los Angeles County, California, range in size from 49 to 240 m^3/d. Table 20.9 provides a summary of information on these plants. Overall, the use of activated sludge diffusion is considered by the County to be an effective and economical means of odour control. No major problems have been reported, and the use of a simple two-stage air filtration system has minimised problems associated with the accumulation of tarry material on the internal components of the blowers.

Table 20.9. Summary of Los Angeles county WWTPs practising odourous air diffusion.

Treatment plant and location	Plant capacity m^3/d	Odour source	Inst-alled	Foul air flow m^3/min	Comments
Los Coyote WWTP Cerritos, CA	140	Primary clarifiers Influent wet well	1970	280	Filters cleaned every 6 months. Blowers rebalanced and cleaned every year. No corrosion reported. Coarse bubble diffusers. Steel blowers w/coal tar epoxy. SS ducting.
Long Beach WWTP Long Beach, CA	95	Primary clarifier	1973	170	No filters on compressor suction. Have to steam clean suction and compressor once a year. Coarse bubble diffusers; no clogging. Concrete corrosion. 100% removal of H_2S.
Pomona WWTP Pomona, CA	49	Primary clarifiers	1965	170	Fine bubble diffusers. Change filters quarterly.
Whittier Narrows WWTP So. El Monte, CA	57	Primary clarifiers	1962	140	Clean or replace filters on blower suction quarterly to annually. No corrosion reported. Fine bubble diffusers, no clogging.
San Jose Creek WWTP Whittier, CA	240	Primary clarifiers	1971	570	Recently switched from coarse to fine bubble diffusers. New filter system on blower suction. Steel blower.

20.8 REFERENCES

Æsøy, A., Ødegaard, H. and Bentzen, G. (1998) The effect of sulphide and organic matter on the nitrification activity in a biofilm process. *Wat. Sci. Tech.* **37**(1), 115-122.

Æsøy, A., Storfjell, M., Mellgren, L., Helness, H., Thorvaldsen, G., Ødegaard, H. and Bentzen, G. (1997) A comparison of biofilm growth and water quality changes in sewers with anoxic and anaerobic (septic) conditions. *Wat. Sci. Tech.* **36**(1), 303-310.

Bentzen, G., Smith, A.T., Bennet, D., Webster, N.J., Reinholt, F., Sletholt, E. and Hobson, J. (1995) Controlled dosing of nitrate for prevention of H_2S in a sewer network and the effects on the subsequent treatment process. *Wat. Sci. Tech.* **31**(7), 293-302.

Bielefeldt, A.R., Stensel, H.D. and Romain, M. (1997) VOC treatment and odour control using a sparged shallow activated sludge reactor. In *Proceedings of WEFTEC '97, Vol. I. Research: Municipal Wastewater Treatment* p 93-101. WEF, Alexandria.

Clark, T. and Stephenson, T. (1998). Effects of chemical addition on aerobic biological treatment of municipal wastewater. *Env. Tech.* **19**, 579-590.

Clark, T., Burgess, J.E., Stephenson , T. and Arnold-Smith, A.K. (2000). The influence of iron-based co-precipitants on activated sludge biomass. *Trans. IChemE*, **78**(B) 405 - 410.

Eckenfelder, W.W. and Grau, P. (1992). *Activated sludge process design and control: theory and practice.* vol. **1**. Technomic Publishing, Inc., Lancaster.

Frechen, F-B. (1994) Odour emissions of wastewater treatment plants - recent German experiences. *Wat. Sci. Tech.* **30**(4), 35-46

Fukuyama J, Inoue Z and Ose Y (1986) Deodorization of exhaust gas from wastewater and night-soil treatment plant by activated sludge. *Toxicol. Env. Chem.* **12**, 87-109.

Henze, M., Harremoes, P., la Cour Jansen, J. and Arvin, E. (1995) *Wastewater Treatment.* Springer Verlag, Berlin.

Johnson, L.K., Waskow, C.E.G., Krizan, P.A. and Polta, R.C. (1995) Suspended growth bioscrubber for hydrogen sulphide control. *Proc. Specialty Conference on Odor / VOC Control*, p. 181-190. Air Waste Management Association, Pittsburgh.

Oppelt, M.K., Tischler, L., Levine, L. and Kowalik, J. (1999) Clearing the Air. Water *Env. Tech.*, **11**(11), 43-47.

Ostojic, N., Les, A.P. and Forbes, R. (1992) Activated sludge treatment for odor control. *BioCycle* April, 74-78.

Ryckman-Siegwarth, J. and Pincince, A.B. (1992) Use of aeration tanks to control emissions from wastewater treatment plants. *Proc. WEF 65th Annual Conference, New Orleans, Louisiana, USA. Sept. 20 - 24. Session #22: VOC and Odor Control II: Emissions Evaluation and Control.* pp. 83 - 94. WEF, Alexandria.

Vincent, A. and Hobson, J. (1998) *Odour Control.* CIWEM Monographs on Best Practice No. 2, Terence Dalton, London.

WEF/ASCE (1995) *Odor control in wastewater treatment plants.* Water Environment Federation (WEF) Manual of Practice No. 22, American Society of Civil Engineers (ASCE) Manuals and Reports on Engineering Practice No. 82.

Index